Concurrent Simultaneous Engineering Systems

The Way to Successful Product Development

Springer
London
Berlin
Heidelberg
New York
Barcelona
Budapest
Hong Kong
Milan
Paris
Santa Clara
Singapore
Tokyo

H.-J. Bullinger and J. Warschat (Eds)

Concurrent Simultaneous Engineering Systems

The Way to Successful Product Development

With 115 Figures

Springer

Hans-Jörg Bullinger, Dr-Ing, Dr-Ing habil
Joachim Warschat, Dipl-Ing, Dr-Ing

Fraunhofer-Institut für Arbeitswirtschaft und Organisation,
Nobelstrasse 12, 70569 Stuttgart (Vaihingen), Germany

ISBN 3-540-76003-2 Springer-Verlag Berlin Heidelberg New York

British Library Cataloguing in Publication Data
Concurrent Simultaneous Engineering
Systems:Way to Successful Product
Development
I. Bullinger, Hans-Jörg II. Warschat,
Joachim
620.0042
ISBN 3-540-76003-2

Library of Congress Cataloging-in-Publication Data
Concurrent simultaneous engineering systems : the way to successful
product development / Hans-Jörg Bullinger and Joachim Warschat. eds.
p. cm
Includes bibliographical references.
ISBN 3-540-76003-2 (pbk. : alk. paper)
1. Concurrent engineering. 2. Computer integrated manufacturing
systems. I. Bullinger, H.-J. (Hans Jörg) II. Warschat, Joachim.
TS176.C6537 1995 95-41704
658.5-dc20 CIP

Printed in Great Britain

Typesetting by the authors
Imageset by T&A Typesetting Services, Rochdale
Printed by Henry Ling Ltd., The Dorset Press, Dorchester.
Bound by Hartnolls Ltd., Bodmin, Cornwall
69/3830-543210 Printed on acid-free paper

Preface

Concurrent Simultaneous Engineering Systems – The way to successful development

The White Paper on growth, competitiveness and employment, introduced by the European Commission, expresses its ambition in the substitute: "Challenges and paths for entering the 21st century". One important path will be Europe's way to the information society on prominent challenge the appropriate innovation policy.

The future competitiveness of nearly all European companies will be closely linked to their capability to create innovative products. Customer needs are becoming increasingly sophisticated, individualised and diverse. New products have to be found for a world which is characterised by:

- often saturated markets, where cost reduction and partly improvements are not sufficient but the generation of new concepts in technologies, products and business as innovation is necessary,
- more heterogeneous markets than ever before, where the composition of target groups will have to be redefined,
- the age of information and communication where new products and applications will be possible and new market will be open.

The constantly growing demands are generating a corresponding necessity for the management of the product development processes in European companies. Product development processes must be more efficient and effective in the sense of reduced complexity, concentrating on more competence. They have to be more flexible towards changing requirements.

One of the key concepts meeting those challenges is Concurrent Simultaneous Engineering. In recent years companies have been discussing and implementing solutions of re-engineering and improvement of their product development processes. The European-wide importance of Concurrent Simultaneous Engineering induced us to initiate the ESPRIT Project CONSENS (Concurrent Simultaneous Engineering System) in 1992. The objective of the project was to develop an organisational and information-technological concept in order to realise Concurrent Simultaneous Engineering in European Companies.

Due to the information society the nature of the firm has to change. Organisation has to be more flexible and to react quicker to changes.

- By now it is clear that companies will find that competitiveness will be more and more defined by their ability to adapt to a new market framework. New global and inter-connected thinking is required.
- This will extend to the traditional factors of competitiveness, such as entrepreneurial spirit, team work, and co-operation inside the firms' organisations. Their positive impact can only be fully brought to bear by a successful use of information technology (IT).
- Firms using IT will be on the competitive edge, pushing others to rethink their organisation and strategies in a similar way: there is no longer a true choice as to wether or not to adapt to the developments in the field of information technology.

This book presents the result of leading edge research in data and tool integration as well as team management. Furthermore, first experiences of companies involved in the project are described. Now, other European Companies should have the chance to profit from CONSENS. The book helps industrial management to reflect their IT-strategies with respect to new organisational approaches. It also contains information for scientists on the demands of industry for IT-systems, their typical problems and application scenarios, and therefore encourages further research.

Brussels, October 1995

Dr. Martin Bangemann

Contents

1. Introduction 1

2. A Concept for Revitalisation of Product Development
by *S. Berndes, A. Stanke* 7
2.1 The Worldwide Tendencies in Economy 7
2.2 The New Orientation for Product Development – Interactions between Time, Costs, and Quality 11
2.3 Concurrent Simultaneous Engineering (CSE) – Strategy and Methodology of New Oriented Product Development 15
2.4 Topics of Concurrent Simultaneous Engineering 18
2.4.1 Strategies for Concurrent Simultaneous Engineering 18
2.4.2 CSE Areas: Potentials for Improvements in Product Development 24
2.5 Implementation of Concurrent Simultaneous Engineering – A Business Process Re-Engineering Approach 41
2.5.1 Culture – The Factor of Success in CSE-Implementation 41
2.5.2 A guideline for CSE-Implementation 45

3. The Organisational Environment of Concurrent Simultaneous Engineering (CSE) by *S. Berndes, A. Stanke* 57
3.1 Required Characteristics of a CSE Organisation 59
3.2 Project-management in a CSE Organisation 63
3.2.1 The Project-Team and its Characteristics 63
3.2.2 The Project-Manager and his Role 66
3.2.3 Rules of Teamwork 67
3.2.4 Possible Organisation Arrangements for R&D 70
3.2.5 Core Team – Heavyweight Team Structure 73
3.2.6 Portfolio of Approaches 75
3.3 Team Organisation 76
3.3.1 The Concept of the Team Organisation 76
3.3.2 Open Questions Arising with the Team Organisation 79
3.3.3 Advantages of a Team Organisation 81
3.4 Incentive & Motivation in a Team Organisation 85
3.5 Organisational Learning in a Team Organisation 91

3.5.1 Characteristics of a Learning Organisation 91
3.5.2 A Concept for Organisational Learning in a Team Organisation 93
3.6 Conclusion 99

4. Enabling Technologies I: The CONSENS Platform 103
4.1 Support of CSE by the CONSENS Platform by *S. Kessler* 103
4.1.1 Support of CSE by the Framework SIFRAME 105
4.1.2 Support of CSE by the Information Management System 107
4.1.3 Support of CSE by the Product Information Archive 109
4.2 The main Components of the CONSENS Platform by *S. Kessler* 111
4.2.1 The Framework SIFRAME 112
4.2.2 Information Management System (IMS) 114
4.2.3 Product Information Archive (PIA) 115
4.3 Modules of the CONSENS Platform 117
4.3.1 SIFRAME – The CONSENS Framework by *S. Kessler* 117
4.3.2 IMS – Information Management System by *L. Miotti* 138
4.3.3 Product Information Archive (PIA) by *D. Koch* 146
4.4 Summary by *S. Kessler* 155

5. Enabling Technologies II: CONSENS Tools 157
5.1 Introduction by *E. Sleeckx* 157
5.2 Support of CSE by Product Oriented Tools by *E. Sleeckx* 159
5.3 Product Oriented Tools 164
5.3.1 Intelligent Computer Aded Design: KnobieCAD by *P. Raiteri* 164
5.3.2 Design for Manufacturing by *J. L. T. Santos, S. Tilly* 175
5.3.3 Design for Assembly by *R. Menges, U. Eigenmann* 190
5.3.4 Design to Cost by *E.Engelborghs, J. Frech* 204
5.3.5 Design of Production Facilities by *R. Menges, U. Eigenmann* 215
5.3.6 Design for Quality by *K. Jeschke* 227
5.4 Support of CSE by Project Oriented Tools by *E. Sleeckx* 247
5.5 Project Oriented Tools 249
5.5.1 EPM – Engineering Process Manager by *S. Berndes, A. Stanke* 249
5.5.2 MDS: A System for Decision Support in the Economic Efficiency Analysis and Controlling of the Product Developing Process by *G. Korn* 267
5.6 Conclusion by *E. Sleeckx* 281
5.6.1 The Tools in Relation with the PSI-Strategies 281

5.6.2 The CONSENS-Tools and CSE ... 283

6. Case Studies of Successful CONSENS Implementation ... 285
6.1 TEMIC TELEFUNKEN microelectronic GmbH by *J. Bergner* ... 285
6.1.1 Description of Products and Projects ... 285
6.1.2 Description of the CONSENS Environment ... 286
6.1.3 Description of Enabling Information Technology ... 297
6.1.4 Improvements of the Product Development Process with Respect to CSE ... 300
6.2 Dasa Military Aircraft Division by *J. Vilsmeier* ... 302
6.2.1 Description of Products and Projects ... 302
6.2.2 Description of CONSENS Environment ... 303
6.2.3 Description of Enabling Information Technology ... 314
6.2.4 Improvements of the Product Development Process with Respect to CSE ... 321
6.2.5 Future Improvements ... 323
6.3 HIDROSOREFAME by *J. M. Camacho* ... 326
6.3.1 Description of Produts and Projects ... 326
6.3.2 Description of the CONSENS Environment ... 328
6.3.3 Description of the Test Environment ... 331
6.3.4 Improvements of the Product Development Process with Respect to CSE ... 333
6.3.5 Future Improvements ... 340
6.4 ALCATEL BELL by *E. Engelborghs* ... 341
6.4.1 Description of Products and Projects ... 341
6.4.2 Description of CONSENS Enviroment ... 344
6.4.3 Description of Enabling Information Technology and Improvements of the Product Development ... 347
6.5 AEG Schienenfahrzeuge GmbH by *L. Bertling, C. Köhler* ... 351
6.5.1 Description of Products and Projects ... 351
6.5.2 Description of the CONSENS Environment ... 353
6.5.3 Description of Enabling Information Technology ... 365
6.5.4 Improvements of the Product Development Process with Respect to CSE ... 369
6.5.5 Future Improvements ... 370

7. Future Trends in Concurrent Simultaneous Engineering
by *D. Koch, J. Vilsmeier, J. Warschat* ... 373
7.1 Conclusion of the CONSENS Project ... 373
7.2 Co-operation in Engineering Based on Core Competences ... 374
7.3 Co-operation via a Global Engineering Marketplace ... 379
7.4 Multi-company Development Co-operations ... 382

7.5 Conclusion 384

Editors and Authors 387

1 Introduction

In the age of more heterogeneous markets, more specific customer demands, complex technologies and reductions in product life cycle, Concurrent Simultaneous Engineering (CSE) has been recognised as a strategy for product development since the 1990's. The idea of CSE is to develop high quality products and offer them at a lower price and in significantly shorter time to the competitive global market. However, practising CSE requires a wide range of approaches, methods and tools.

This book addresses central issues associated with Concurrent Simultaneous Engineering (CSE). The Concurrent Simultaneous Engineering System (CONSENS) approach is presented as an integrated solution for Concurrent Simultaneous Engineering within a company.

CONSENS, was developed in an European project, which is to be seen as the flagship project in the fields of concurrent engineering of the European Union and of European Strategic Program for Research and Development of Information Technology (ESPRIT). It was intended to link innovative concepts with practical experience. On June in 1992, the CONSENS project started as ESPRIT III Project EP 6896. The consortium consists out of three different kinds of partners: Research Institutes, partners belonging to the Information Technology (IT) Industry and European multinational companies as users and demonstrators. They are:

- AEG Daimler Benz Industrie – Produktionstechnik und Logistik Consult (PLC); Frankfurt, Germany
- AEG Schienenfahrzeuge GmbH (ASF); Hennigsdorf, Germany
- ALCATEL BELL; Hoboken, Belgien
- Daimler-Benz Aerospace AG (Dasa); München, Germany
- DELOS S.p.A; Torino, Italy

- DELTA Industrie Informatik; Fellbach-Schmieden, Germany
- Fraunhofer Gesellschaft – Institute for Industrial Engineering (IAO); Stuttgart, Germany
- Heinz Nixdorf-Institut; Paderborn, Deutschland
- HIDROSOREFAME; Lisboa, Portugal,
- Instituto Superior Tecnico (IST); Lisboa, Portugal
- Siemens Nixdorf Informationssysteme AG; München, Germany
- Syntax Factory Automation S.p.A.(SyFA); Torino, Italy
- Technische Universität Braunschweig – Institut für Werkzeugmaschinen und Fertigungstechnik; Braunschweig, Germany
- TEMIC TELEFUNKEN microelectronic GmbH; Ingolstadt, Germany
- Wetenschappelijk en Technisch Centrum van de Metaalverwerkende nijverheid (WTCM); Leuven, Belgien

The objective of the CONSENS project was to develop a software platform on which several application programs for CSE can be easily integrated. This integration of different tools in one platform ensures an increase of efficiency in research & development, construction, planning of production and in activities concerning the suppliers and customers. The integration platform enables the optimisation of the organisational structure and the planning & steering of the product development process regarding time, cost and quality.

CONSENS includes therefore three main elements – the organisational environment, the CONSENS platform and the CONSENS tools – which will be described in detail. The basic concept of CONSENS is given by the PSI strategies which stand for parallelisation, standardisation and integration of the product development process and by the CSE areas. The overall objective of CSE in terms of shortening lead time, improving in quality, and reducing costs can be reached by applying PSI strategies. The CSE areas which are the process, product, organisation and resources have an influence on the product development process and, therefore, are expected to contain potentials that can improve this process. Both issues are explained in Chapter 2 before CONSENS is described in its entirety. The management of changing processes and structures towards a CSE oriented product development has a enormous importance for practice. In this context, the cultural aspects in particular must be reflected in a guideline for implementation, which is given in section 2.5.

Chapter 3 covers the organisational environment as one specific element of CONSENS. Beginning with the characteristics of a CSE organisation and a general description about project-management in a CSE organisation, a Team Organisation Concept is derived in which CSE can be successfully applied. New incentive & motivation systems are necessary in this form of organisation since hierarchy is of little importance. An approach for this is introduced in section 3.4. In order to achieve a permanent and systematical development and optimisation of the Team Organisation, a concept for a Learning Team Organisation is presented at the end of the chapter.

The CONSENS platform will be embedded in such an organisation. Chapter 4 explains the architecture of the platform and the integration of the modules. From an information technology point of view the focus is to implement a standard platform where various CONSENS tools can be integrated depending on the specific company demands. This can be realised by using a framework (SIFRAME), an information management system (IMS) and a product information archive (PIA).

A large number of highly specialised design methods have been developed to meet the technological needs of complex design tasks, such as design analysis for manufacturability, assembly, cost estimation or others (Chapter 5). Moreover, these design tools should be integrated into the platform to provide a variety of functions and services for the engineer. Furthermore project-oriented tools dealing with information about management processes will be described.

Chapter 6 presents case studies of successful CONSENS implementations. Companies such as Daimler-Benz Aerospace AG, TEMIC Telefunken microelectronic GmbH , Hydrosorefame, Alcatel Bell and AEG Schienenfahrzeuge ASF report their specific CONSENS approach and their experiences.

The book closes with an outlook of trends and a view on the future developments in the field of CSE. Three topics are discussed in chapter 7:

- Cooperation via a global electronic marketplace
- Multi-company development cooperations
- Cooperation in engineering based on core competences.

The book summarises the results of the CONSENS project, which ran over three years.

In this context, grateful acknowledgements are due to all the involved people and institutions throughout Europe. Especially to the project partners named here as authors, who enabled this book to come into being, as well to those, who are not mentioned in this book, but do practice a great form of European co-operation throughout Europe. All have made valuable contributions during the entire length of the project. We want to thank Dr. Schemmer (AEG) for the fruitful discussions about Concurrent Simultaneous Engineering in 1991, which lead to the idea CONSENS and its initiation. Hans Günter Thonemann (SNI) together with Bärbel Peither (IAO) were our driving force in the process of writing the proposal. We would like to thank them for their engagement.

We wish to give our special thanks to the project manager, Siegfried Buck (SNI), who coordinated the project very successfully and efficiently. His way of working was the basis for the successful co-operation.

We are also grateful to the project reviewers Prof. Reiner Anderl, Keith Shaw and Prof. Guy Doumeingts for their thorough project reviews and for numerous pertinent ideas and suggestions.

Grateful acknowledgements are likewise due to the EU project officers Ronald MacKay and Erastos Filos of the European Community, who advised us on the project with numerous useful hints and comments.

Special thanks go to Alexander Stanke (IAO), assisted by Sabine Kessler (SNI) and Eric Engelborghs (WTCM), who gave this book its structure and support to the authors. Their work enabled the realisation of this book.

Furthermore, grateful acknowledgements are due to Ulrike Fempel, who organised the coordination between the authors and to Andreas Thaler, who managed the difficult task of integrating the illustrations into unified design for this book. The desktop publishing was executed by Klaus-Peter Hellweg (IAO) in a perfect and excellent way.

In particular, we wish to thank Imke Mowbray, Erdmuthe Raufelder and Nicholas Pinfield of Springer-Verlag for their excellent support.

Hans-Jörg Bullinger
Joachim Warschat

2 A Concept for Revitalisation of Product Development

S. Berndes, A. Stanke
supported by J. Piller, U. Wagner

The general objective of chapter two is to provide a theoretical background to the method known as Concurrent Simultaneous Engineering (CSE). The first section describes the current common tendencies which provide the framework for the development of CSE. The following deals with the strategies and areas in a company using CSE. The last section is dedicated to a practical guideline for implementing CSE.

2.1 The Worldwide Tendencies in Economy

Decreasing profit-window

In the last 10 years, the life-cycle time of products from different branches of industry decreased while the time spent on product development greatly increased [1]. Compare figure 2.1: Time-trap. Due to these changes the pay-off period between market entry and amortisation extended as well [2].

Three main tendencies

In order to meet the challenges of successfully competing in innovative markets the development and design of new products has become one of the most significant factors. Today's competitive situation can be characterised by three main tendencies:

Shift from a seller´s to a buyer´s market

Companies have to cope with a tremendous shift from a seller's to a buyer's market, i.e. the supply of products is higher than the demand. This implies that the customer's expectations have become more integrated and demanding. Standard products are rejected in favour of individual solutions to problems which meet the specific wishes of the

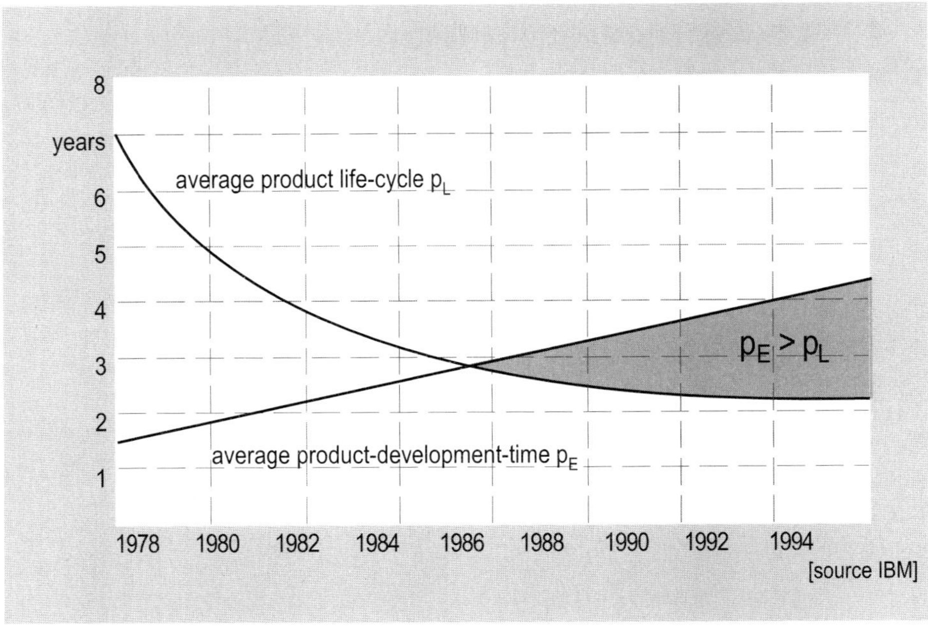

Fig. 2.1: Time-trap

customer. Due to the minuscule nuances in the customer's requests the differentiation of the products and the innovation have to become finer as well [3]. Therefore, originally stable, homogeneous markets will separate into numerous smaller and rapidly changing markets.

Increasing globalisation

The increasing global spread of markets and branches is leading to a larger number of powerful competitors in regional markets. On the one hand, the regional differences in markets still exist, on the other hand, international competitors raise their competence and enlarge their experience. This allows them to enter a regional market in another way. With an increasing number of competing products and more flexible and powerful competitors, it will be vital for a regional producer to quickly react with differentiated products.

Change in the importance of technology

Another change occurred with respect to the significance of technology. Nowadays, the control of technology is a basic requirement for participating in innovative markets, but it is not sufficient for the success of a new product. Due to the worldwide build-up of scientific and technological skills and the international cross-linking between research-

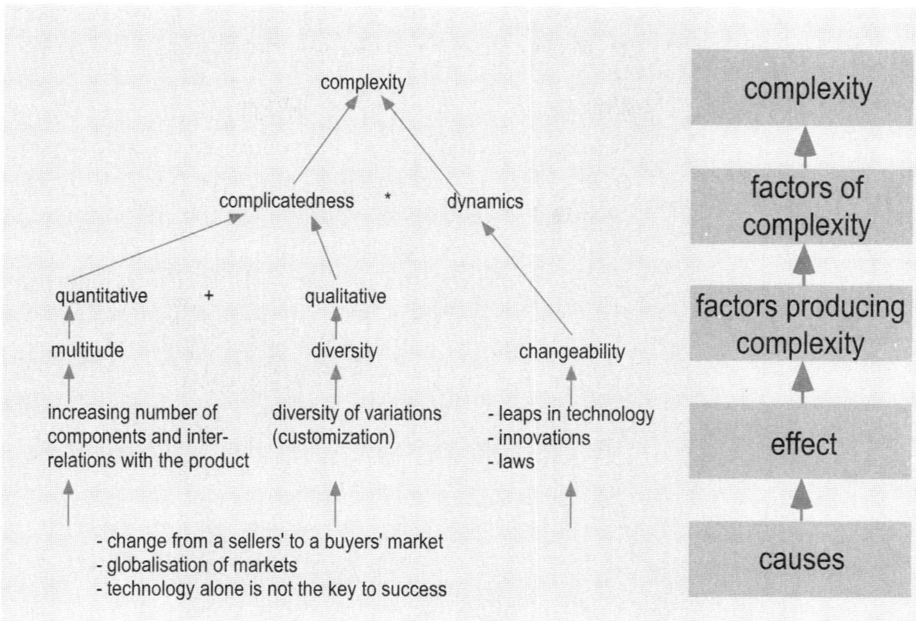

Fig. 2.2: How complexity consists of

institutes and companies, the access to this knowledge is easier and more consistent than before. As a result of this development, it will be more and more difficult for a company to attain competitive advantages which are only based on new technologies [1].

Companies in innovative markets which are not able to react in an appropriate and effective way to the above tendencies will lose competitiveness. These tendencies result in an increasing complexity for a company and has to be handled with appropriate structures and processes. This complexity is shown in figure 2.2.

Key-targets for the product development process

A variety of products in the market and short reaction-time at a low cost level must be ensured. This has lead to the following key targets for the product development process:

- reduction of the "Time to Market" to raise the market share,
- improvement of the overall quality, which meets the needs and wishes of the customers,
- reduction of product and process development costs to decrease the pay-off period.

Generally, there are two possible approaches for a corporation to overcome the challenges of the markets:
Either,

Effectiveness

- the effectiveness can be improved (do the right things) which is a strategic issue and is directed towards the correct choice of technologies, the definition of the best product programme, and the proper investment of resources on the right projects,

Efficiency

- or the operative issue efficiency (do the things right) can be improved. The improvement of efficiency is concerned with the optimisation of processes or structures within the corporation. The method of Concurrent Simultaneous Engineering is a basic approach to improve the efficiency in the product development process.

2.2 The New Orientation for Product Development – Interactions between Time, Costs, and Quality

Interactions between time, costs and quality

Undoubtedly, the optimisation of the "Magic Triangle", which is shown in figure 2.3, consisting out of costs, quality and time is necessary in order to face competition and complexity in the changed environment described above.

Industry benchmarks with "best-in-class" companies have shown that these companies have performed the above key targets in recent years [5].

Parallelisation, integration, standardisation

The targets can be reached by

- parallelisation of sub processes and tasks to eliminate delays,
- integration of departments and persons via matrices or better dedicated cross-functional team work to overcome today's department and to improve communication,
- standardisation of the product development process to improve better understanding and to minimise confusion, and misunderstanding.

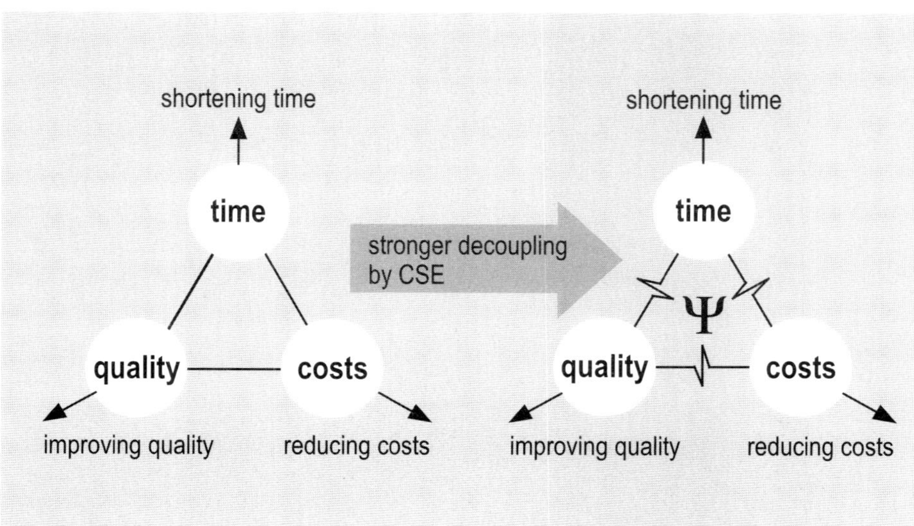

Fig. 2.3: The "Magic Triangle" [4]: decoupling of time, cost and quality

The Key Targets in Detail:

"Time to Market":
As mentioned earlier, one aspect of the new orientation is to cut the "Time to Market". The fact that resources are committed within the company to research and development (R&D) projects is directly affected by the duration of the time put into the development. As a result the "Time to Market" is defined as the period of time from the idea of a product until its appearance on the market. This time is used as a measure [6] for the capital tie-up in a company, the ability to reach the customer, and the ability to react to changes in the company's environment with respect to new products. Consequently, the timing is a crucial criterium for rentability and profitability. "Time to Market" defines the moment of the product introduction. This moment is significant for the initiation of technical developments and for the introduction of technical innovations [7].

Two strategies for launching products

In general, there are two strategies possible for launching new products:
- either a company aspires to be first on the market with a new product generation ("First to market"-strategy) or
- it chooses a "follower"-strategy.

"First to market"-strategy

"First to market"-strategy is suitable for companies which compete in a highly competitive and dynamic market. They can achieve a temporary monopoly with an innovation due to its uniqueness. Thus, a platform for sustainable competitive advantages can be build up which ensure success potentials even after the market entry of competitors. Important advantages [8] of pioneers are:
- a much better cost situation due to the exploitation of "experience-curve" effects
- the image of the pioneer in the events that the product becomes the prototype in its category
- mature relations to the market partner, e.g. suppliers or customers
- time for establishing industry standards with the objective of dominant design.

Follower-strategy

Companies which enter the market at a later moment can find a significantly smaller proportion of potential customers. On the other hand, these followers can benefit from gained experience of the pioneers for free and are able to improve their product in a way to beat the leading product. A study carried out by Mansfield [8] investigated several branches with the result that the costs of imitating followers are on average only 65% of the pioneers' costs. The followers needed only 70% of the time the pioneers needed for the imitation.

Comparison of several studies investigating the market-entry-strategies

Taking several other studies about introduction strategies into account, one cannot generally identify one strategy as superior. The studies of Robinson/Fornell [9], Buzzel/Gale [10] and Lambkin [11] are based on PIMS-data. All of them produce a similar result in that the pioneers are more successful than the followers. Other studies came to the result that there is no correlation between the moment of market entry (Cooper [12]) or that follower strategies are superior (Vanhonacker/Day [13]).

Neither strategy is superior

In relation to correct timing, besides the two generic strategies, situative conditions are relevant which depend on the branches. These situative conditions include for example, market development & competition situation or characteristics of the innovations, e.g. complexity and degree of innovation [7]. For successful companies it is important to be able to manage and to dominate the time aspect and not only to try to be the first one.

Quality:
Another important task is the improvement of the customer orientated product-quality.

Quality, that has to be fulfilled by the company, is not only related to production-oriented product-quality any more [14]. Nowadays, the term "quality" has a broader definition: Quality refers to the degree of fulfilling the customers' requirements. This definition involves requirements for punctuality and reliability combined with regularity of the company's performances (e.g. the proper product offered in response to the customer's requests on time and price).

Broader definition of quality

While in the traditional concept the main aim of quality-control is the minimisation of costs for quality assurance, the target has now to be the reduction of the follow-up costs caused by errors. The sooner quality failure is recognized or prevented in the product development process, the lower the costs for amendment will be.

Costs:
The last task to be solved is the decrease of the product and product development costs.

The market determinies the price today

According to the traditional principle applied to calculation [15], the price of a product is derived from the development costs plus a certain margin. This equation is no longer valid. The price is now determined by the market and thus the price of a specific product is defined by what the customer is ready to pay and not by the total cost which goes into a product. At the early stage of defining and developing a new product almost 70% of the costs and through-out-time of the production [16] are fixed. Taking these facts into account, a consequent cost optimisation can be worked out by two approaches [17]:

- reduction of expenses caused by amendments during the stage of product development
- orientation of obtainable prices while designing the product.

If the company is able to reduce expenditures caused by amendments during the product development, the development time will be cut due to the deliberate usage of resources.

2.3 Concurrent Simultaneous Engineering (CSE) – Strategy and Methodology of New Oriented Product Development

How CSE became into being

In the 1970's and 1980's many companies invested heavily in production automation and computer integrated manufacturing (CIM), but in most cases each department within these firms remained responsible for only its own costs. This practice encouraged everyone to reduce the own costs and to continue with the "over-the-fence" relationship to all other departments, e.g. between marketing and development. The high investments have been made with barely any changes within the organisation. Investigating these phenomena, the American Institute for Defense Analyses (IDA) defined the term "Concurrent Engineering" in 1986. Following this, there was an initiative for military equipment called Computer-Aided Acquisition and Logistics Support (CALS) [18] by the American Department for Defense. CALS represents a union between the department of defense and several companies, mainly belonging to the US arms industry. They defined and worked out the method of "Concurrent Engineering" (CE). The objective of the program was the integration of several systems, i.e. integration of all the subsystems in the final product during the developing stage. To meet these requirements, the integration of technical and organisational interfaces is necessary. Thus, CE is molded by the time aspect as a competitive advantage and describes the character of integration.

Simultaneous Engineering focuses on the simultaneity of process

Simultaneous Engineering (SE) is based on reactions of European industry to the time advantage of Japanese companies launching new products or product generations. This approach describes rather the simultaneity than the aspect of integration.

Simultaneous versus Concurrent Engineering

The difference between Concurrent and Simultaneous Engineering is more or less insignificant. Because of the similarity and relationships between the methods CE and SE, they will be known under one term called Concurrent Simultaneous Engineering (CSE).

Several Japanese companies have successfully been using the method of Concurrent or Simultaneous Engineering for more than 30 years. Some of them have developed their own techniques with different names which are basically Concurrent or Simultaneous engineering. Others do not even have a name for this method, but are using it unconsciously [19] such as Nissan.

CSE is more than a strategy

In literature Concurrent or Simultaneous Engineering is quite often described as a strategy for product development, but practical experience shows that CSE is much more than a strategy. "A strategy defines where the organisation wants to go to fulfil its purpose and achieve its mission. It provides the framework for guiding choices which determine the organisation's nature and direction. A strategy is therefore a declaration of intent." [20]

In the CONSENS Project [21], CSE was defined as a concept of a product development strategy under the consideration of consumer focus.

The definition of CSE in this book is more widespread than in the CONSENS Project. Here, CSE is rather a methodology and an approach to an integrated technique of solving problems involving all resources. Applied in the product development process, CSE focuses not only on the development phase, but on the whole product development process.

The product development process

The product development process refers to the entire product life cycle time. It covers the process from the idea for a product and its production through to its market entry and disposal. The process is a defined order of specific tasks which is required to generate the necessary information for every stage of the process.

Human beings in the center

The most important elements when applying CSE are human beings and the design of product development processes. Co-operation and communication are regarded as the most important success factors by companies which are successfully practicing CSE. This involves [22]:

- cutting back barriers among departments and hierarchies
- promoting interdepartmental co-operation
- building up close links between suppliers and customers
- supporting CSE by the top management.

New orientation of conscieness and behaviour

To achieve the required success, it is not enough to decide on and implement CSE from the top down, as more than the basic rules are demanded. A new orientation of conscieness and behaviour of the complete staff is necessary. CSE cannot work properly without a deep understanding of the problems and performance of each individual team member. The intensive training and preparation with tools as well as the methods of CSE are of great importance for its successful implementation. Each member the staff has to be committed and must "live" the principles of CSE which are teamwork, quality, and customer orientation, incremental improvement of product and production process and an integrated view of design, product, and production. Therefore CSE can be regarded as a methodology, in that it creates greater awareness for day to day business.

Reasons for the success of companies applying CSE

What are the reasons for the success of companies all over the world who are applying CSE in the product development process? The answer can be seen in the differences between the Traditional Apporach and CSE. These differences are shown in figure 2.4.

traditional approach:	concurrent/simultaneous engineering:
• functional task sharing	•multi-functional task fullfilment
• sequential product-/ process development	•parallel product-/process development
•supplier as opponent	•supplier as cooperation partner
•product modification	•out-put-adapted product development
•costs as considerable control instrument	•time as considerable control instrument
•guideline for achievement of objectives	•profit as main-objective
•hesitant reaction to market shifts	•scenarios for market trends
•hidden organisational weaknesses	•disclosure of organisational weaknesses
•parallel working on several projects	•concentration on one project

Fig.: 2.4: Comparison between the traditional approach for product developement and CSE [17]

2.4 Topics of Concurrent Simultaneous Engineering

How and where to realise CSE in a company

As mentioned above, the benefits of CSE can be seen in the optimisation of the product development process regarding the "Magic Triangle" consisting of time, costs, and quality. Thus, it is obvious what can be achieved by adapting CSE. Now, the question is how and where CSE should be realised in the company. To answer these questions, several strategies (how) and areas (where) will be discussed in the following. These strategies and areas represent a new approach for CSE which was worked out in the CONSENS Project.

2.4.1 Strategies for Concurrent Simultaneous Engineering

Three strategies for CSE: PSI

Generally, three possible strategies can be identified as CSE guiding principles (see figure 2.5):

- Parallelisation,
- Standardisation,
- Integration.

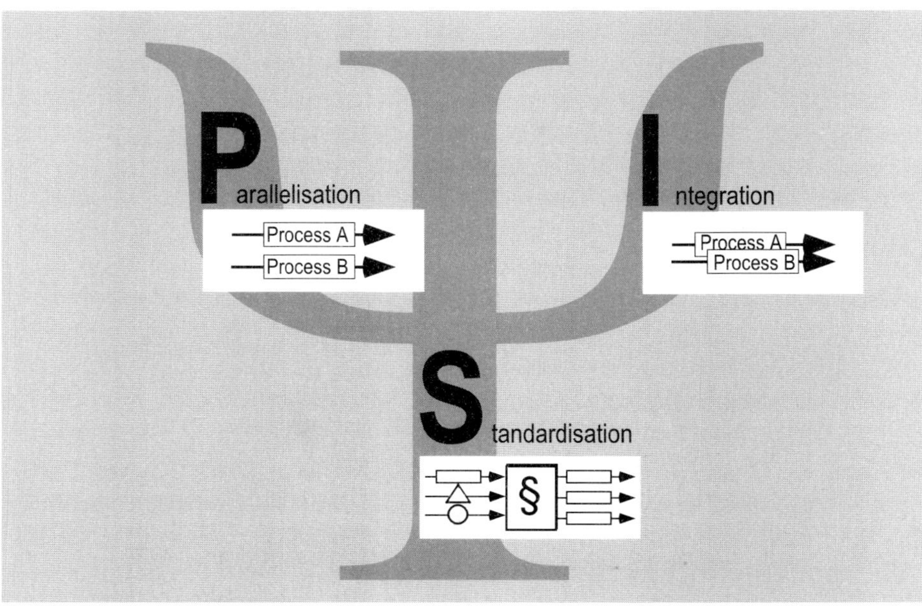

Fig. 2.5: PSI-Strategy

These PSI-strategies determine the direction of designing the product development process and enable a selection of suitable necessary measures to be taken.

2.4.1.1 CSE Means Parallelisation in the Product Development Process

Parallelisation

Parallelisation in the product development process implies the cutting and optimisation of time (compare figure 2.6: Parallelisation). The first step is to remove existing float time in the development process. This means that processes which do not have any dependencies on other processes are carried out simultaneously. In practice, most processes depend on others. In this case the dependent process has to be started before the preceding process is finished. An earlier start of the succeeding process is possible in most cases, because it can be carried out without having completed the preceding process, i.e. not all information is required to start a new process.

Accelerated execution of linked processes

The result of this approach is the advantage of an accelerated execution of linked processes, but also the disadvantage of a higher decision complexity. This higher decision

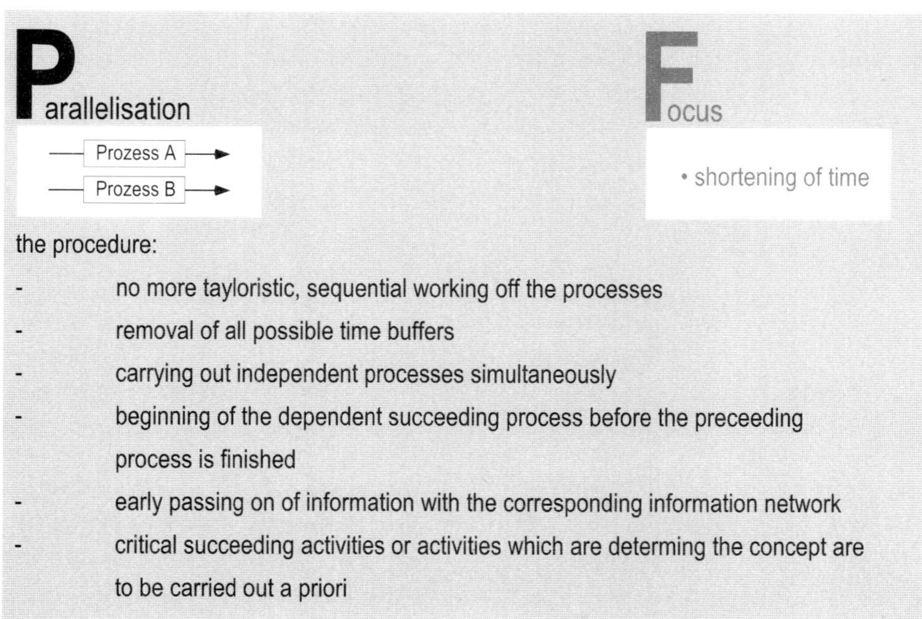

Fig.2.6: Parallelisation

Increased amount of information and a higher decision complexity

complexity is caused by an increased amount of information transfer between departments or teams. Additionally, the proportion of uncertain and incomplete information is also higher due to the fact that not all parallel processes are finished which give inputs to other processes when they are started. This fact is demonstrated with the following example. Necessary tools in the specification and development phase are the duties record book and the design specification book. In order to enable the wanted parallelisation both have to be parallelised. The way this can be carried out is shown in figure 2.7.

The base is the design specification book with all customer and external requirements. The technological functions and solutions realising these requirements are described in the duties record book. After the requirements have been worked out in the first version of the design specification

	design specification book:	duties record book:
function:	Description of: • the system parameters • performance factors • function • characteristics of the system which will be developed	detailed description of the system and combined effects of the system components
responsibility:	product planing	project manager
staff envolved:	product planing, project manager, system team, modul teams, quality assurance	product planing, project manager, system team, modul teams, quality assurance
content:	general description of the system: functionality, technical key data, quality targets, environmental conditions, concepts for service, for installation and for tune up description of the system components combined effects of the components approach for solving realisation problems risk analysis	general detailed description based on the design specification book detailed description of interfaces, information flows, control ways for the system and the system environment audit standards and procedures system test plan risk analysis

Fig. 2.7: Parallelisation of duties record book and design specification record book

book, a first version duties record concerning the system is made describing the interfaces, solutions, and functions between the modules within the system. In the duties record book concerning the modules all concerns of the modules are described. The whole process of improving these two tools is permanent and parallel.

Differences to the Tayloristic principle

Compared to the Tayloristic principle, not only time can be cut due to parallelisation, but also amendment costs because of a lower number of mistakes. One of the most important examples of parallelisation is the synchronisation of the development of product and means of production. The approach of parallelisation should be carried out under the principle: "Parallelisation does not mean to work side by side but to work with one another."

2.4.1.2 CSE Means Standardisation in the Product Development Process

Standardisation

Standardisation is defined as the unity of aspects in the product development process which show a high degree of similarity or the possibility of repetition [23]. Behind the strategy of standardisation are two different ways of proceeding.

Structuring of processes

Firstly, the definition of particular orders of activities, i.e. the structuring of processes. Thus, processes which are often repeated are specified and generalised [23], i.e. they can be handled independently on people and events over a long time. Secondly, the standardisation of products with all its systems, elements, and construction kit.

Structuring of product

Standardisation is related to:

- technical/structural aspects such as the usage of modules or components in the final product, e.g. standard parts,
- procedural aspects as phases or structuring of operations, i.e. the definition of sequences of activities,
- software-Standards, e.g. STEP®,
- aspects relating to the organisation of the structure such as interfaces between projects and departments. A clear definition of the organisation, i.e. standardised structures, is demanded to reduce and control the increased outlay of providing required information due to the implementation of CSE.

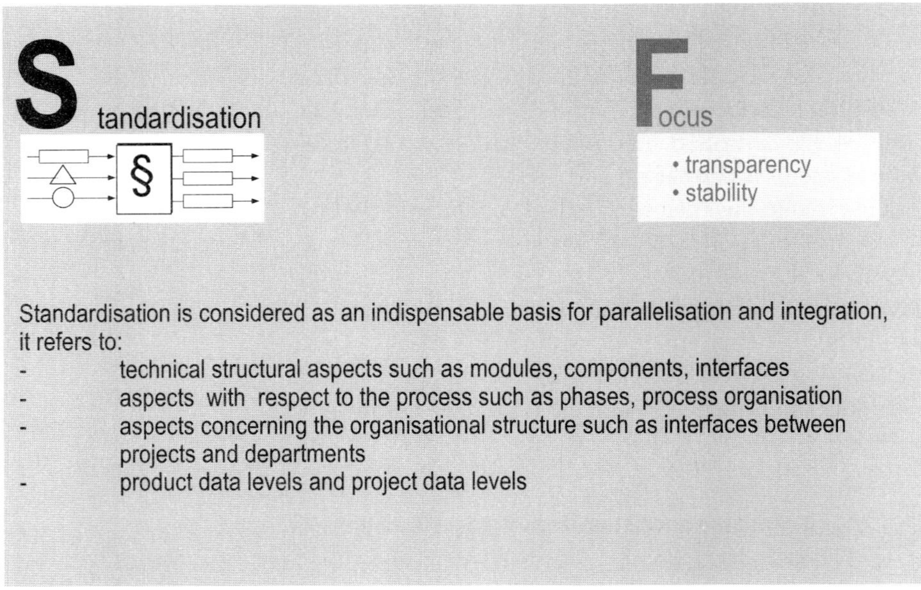

Fig. 2.8: Standardisation

Objectives of standardisation

The objectives of standardisation are to avoid repetition and needless work as well as to learn from existing experience of the company. Thus, repetitive and similar decisions will be taken off the staff on the one hand and on the other hand a better coordination will be achieved. If routine tasks are reduced, more time will be available for innovative and creative work and for the management of unpredictable events. Standardisation can be seen as an indispensable necessity for a CSE-optimised product development process. Certain standards are resulting in a Tayloristic execution of R&D-tasks and thus they are contrary to the CSE-methodology. Nevertheless, these well considered standards can be regarded as a guarantor for a successful parallelisation and integration within the product development process (see figure 2.8). A standardisation should only be carried out if it is really necessary for a parallelisation and integration. Too much standardisation can lead to the above mentioned problems of increasing bureaucracy.

Avoid repetition and needless work

Avoid to much standardisation

However, the form of standardisation can vary from ordinary guidelines to compulsory arrangements and rules, or to fixed detailed operations [24]. It depends on the situation which form of standardisation is suitable.

2.4.1.3 CSE Means Integration within the Product Development Process

Integration

Problems of interfaces

Regarding the product development process as a uniform value-added chain one can realize that beside the R&D-department several functional departments such as sales, marketing, production and service of the company are involved in the development of the product. This allocation of development tasks in different functional areas is combined with increasing problems at the interfaces resulting loss of information [25]. The reasons for the information loss are non harmonized time scales, different interpretation of tasks, and ignorance of the requirements at the other side of the interface. Integration is the measure to overcome interface problems.

Process-oriented working in interdisciplinary teams

Integration demands working in interdisciplinary teams, thinking and behaving in a process oriented way, realising a common objective instead of several objectives that are department specific, and finally being entrepreneurial instead of putting through interests of departments. In order to define a common objective from several ones

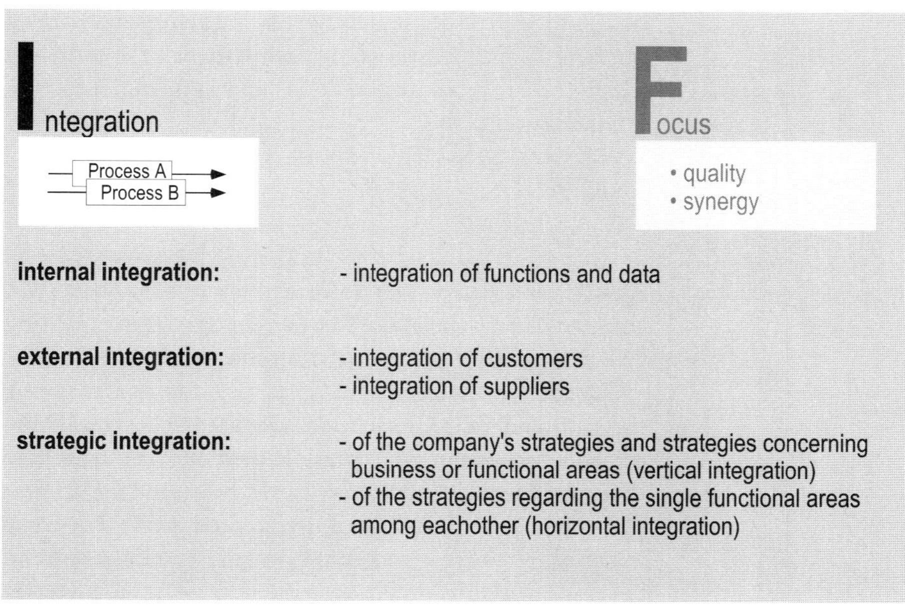

Fig. 2.9: Integration

Shift from functional thinking to a process-oriented thinking

which are competing with each other and to ensure its achievement it is important to work in interdisciplinary teams and to include other departments. Thus, the staff involved have to reduce their thinking which is functional and departmental oriented and to establish a view about the whole process. Nevertheless, with the increasing number of departments involved during the development process of the product – especially during the definition and conception phase – the probability for success is increasing as well.

Dataintegration

Another important aspect of integration is dataintegration within a company. See herefore in detail the report of CALS, e.g. [18].

2.4.2 CSE Areas: Potentials for Improvements in Product Development

Changes must be made in order to improve

If you want to optimise the product development process in a company, i.e. to improve the effectiveness and efficiency, it is inevitable that changes must be made in the organisation's structure, changing processes, investments in new facilities and further training of staff, as well as new solutions to the product structure. These changes are carried out in susceptible and controllable areas of a company which can be summarized as it is done: ([26] and [18])

- Organisation,
- Processes,
- Resources,
- Product.

The importance of Culture

Another area can be seen in the culture of the company. A company's culture is an essential factor of implementing CSE but it cannot be directly controlled or changed [27].

In order to write a useful guideline for practical use, methods and tools for each area are discussed in this chapter. Due to the immense variety of methods and tools, only those are dealt with which are successfully used in corporations that have already implemented CSE. They are assigned to the areas and strategies and therefore used as a basis for building-up a plan for a successful CSE-implementation. See for this chapter 2.5.

2.4.2.1 CSE Area: Organisation

The organisational structure of a company determines its internal structures, i.e. formation of organisational units as well as their coordination. Duties of report, decision com-

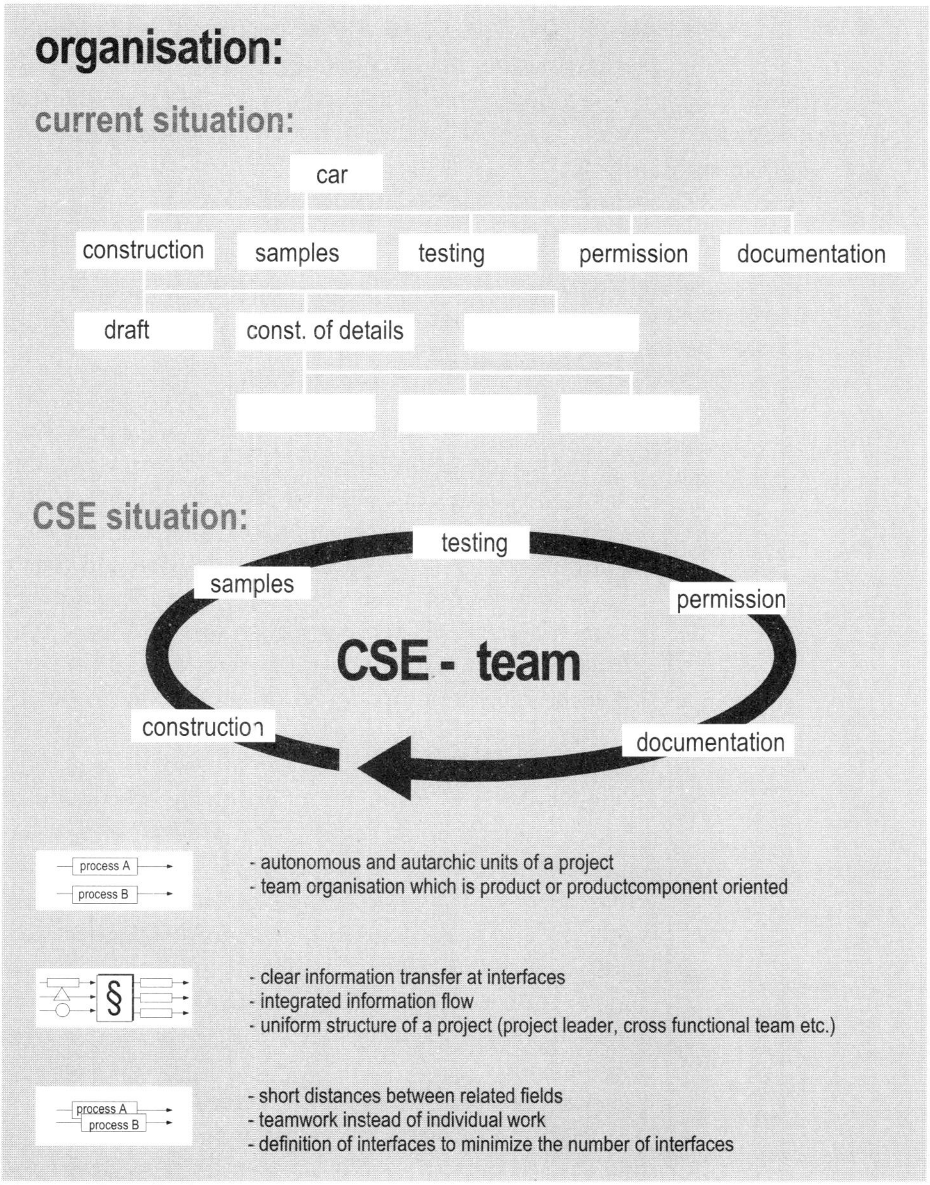

Fig. 2.10: Organisation

petencies, and procedures of coordination are stipulated by formation of hierarchies and positions. For a smooth product development process there is a relevance of informal structures besides the organisational structures [28]. The aspects of an appropriate organisation are described in chapter 3. The figure 2.10 illustrates the differences between the CSE-condition and today's conventional condition concerning the area organisation.

Tools and methods for designing the product development process

The following tools and methods are the most necessary for designing the product development process within the organisational structure.

- project-management as a method,
- Teamwork instead of individual work and career aspirations,
- Committees for coordination of several projects.

Project-Management

A significant characteristic of CSE is the effort to structure a scheme into a unit. Project-management is the most suitable approach for this structuring [29]. Project-management is the guiding of a unique scheme by a project leader based on a fixed method, with interdisciplinary teamwork in flexible teams. This enables parallelisation and thus a time-optimised process within the product development process, because all decisions in an organisational unit result in project requirements. Of importance is the standardisation of sub-processes in the product development process, especially the transfer of information. Standardisation, which is the frequent use of the same organisational structure or form, is supporting the orientation of staff in several product development process. Integration can be achieved by putting together all functions and processes involved in a project. Therefore, all relevant aspects of the product development process can be discussed at an early stage and decisions can be made. In relation to their big proportion of development tasks, suppliers and producers of production means are integrated as well.

Teamwork Instead of Individual Work and Career Aspirations

The success of project work depends on the participation and co-operation of staff. It is necessary to moderate the advantages of the new work approach and to demonstrate how it leads to a new method of success. This means above all to have the relationship between teamwork and career prospects to discretion. Teamwork is a kind of co-operation and characterised by a group of people who are responsible for the execution of their tasks, for the allocation of the tasks among the team members, for the solution of occurring problems, and for communication with other organisational units. The team is represented by a team leader or speaker. Advantages of teamwork can be seen in the shortest possible closed loops, which is leading to a higher degree of process parallelisation, of integration of knowledge among the involved people, and of motivation and performance. Another advantage of teamwork is a standardisation of the team member's qualification at a higher level if the team is assessed as whole. Motivation of staff members is one of the most important factors of performance; better payment alone is not enough for higher motivation [30]. Much more can be achieved with changed prevailing conditions. Creating things, self-realisation, and gaining a high personal status is almost automatically motivating. A career in projects instead of in the traditional structure is an attractive alternative, which enables the workforce to work on changing and demanding projects. The aspects of teamworking are described in chapter 3.2.3 in detail.

Advantages of teamwork

Committees for Coordination of Product Development Processes

The task of the coordination committees is to coordinate all the product related activities of the functional areas for the complete product life cycle. Therefore, they contribute substantially to the improvement of the efficiency of the product development process. The first step is to ensure that the requirements of the market/customers can be realised with the smallest amount of developments. This is within the scope of duties of the product management which is thereby supporting the integration of several development processes. Otherwise, processes would be concurrent. Furthermore, standardisation in the product development process is supplied by the product management with the definition of a few components of the product which are reusable. Another aspect of the benefits of those committees is the possibility to use experience from existing products for defining new products, e.g. with systems based on the unit

Higher efficiency of the product development process

Project steering committee

construction principle. Since the different projects rely on the same resources, it is necessary to implement a management which is coordinating the projects. This management is placed above the level of the development processes and can be realised with a so-called project steering committee. To assist in this, a strategic project controlling is established [31] and the information system is adapted.

2.4.2.2 CSE-Area: Processes

Co-operation and communication as a significant performance factor

While overcoming the functional barriers by reducing boundaries between departments, the necessity for a changed way of thinking among employees is becoming obvious. Thinking in terms of processes along the value-added chain requires co-operation and communication as a significant performance factor. The figure 2.11 shows the current condition and the condition of the CSE in the area processes.

Performance of an organisation subjekt to the implemented processes

Procedures implemented in the company determine the product development process and influence effectiveness and efficiency. Because of this, analysis of existing procedures and processes are a basic precondition for implementing CSE in a company. It has been useful for designing operations that are process oriented to nominate a process owner who is a staff member in charge of a process. It is the task of the process owner to implement the method-based optimisation of cross-functional processes. They ensure the uniform execution of repeating standard operations. The operations are divided into the following two categories which aim to simplify the immense amount of methods and operations:

- Processes for creating product data,
- Management supporting processes.

Processes for Creating Product Data – Operative Processes

First of all, operations will be mentioned which are describing the product and the production process. The effects of those operations are above all in the area of standardisation because of the possible usage of already existing components or parts and the possibility of recourse to the results of other engineers. Within this context one should not forget operations which enable a simulation of product and

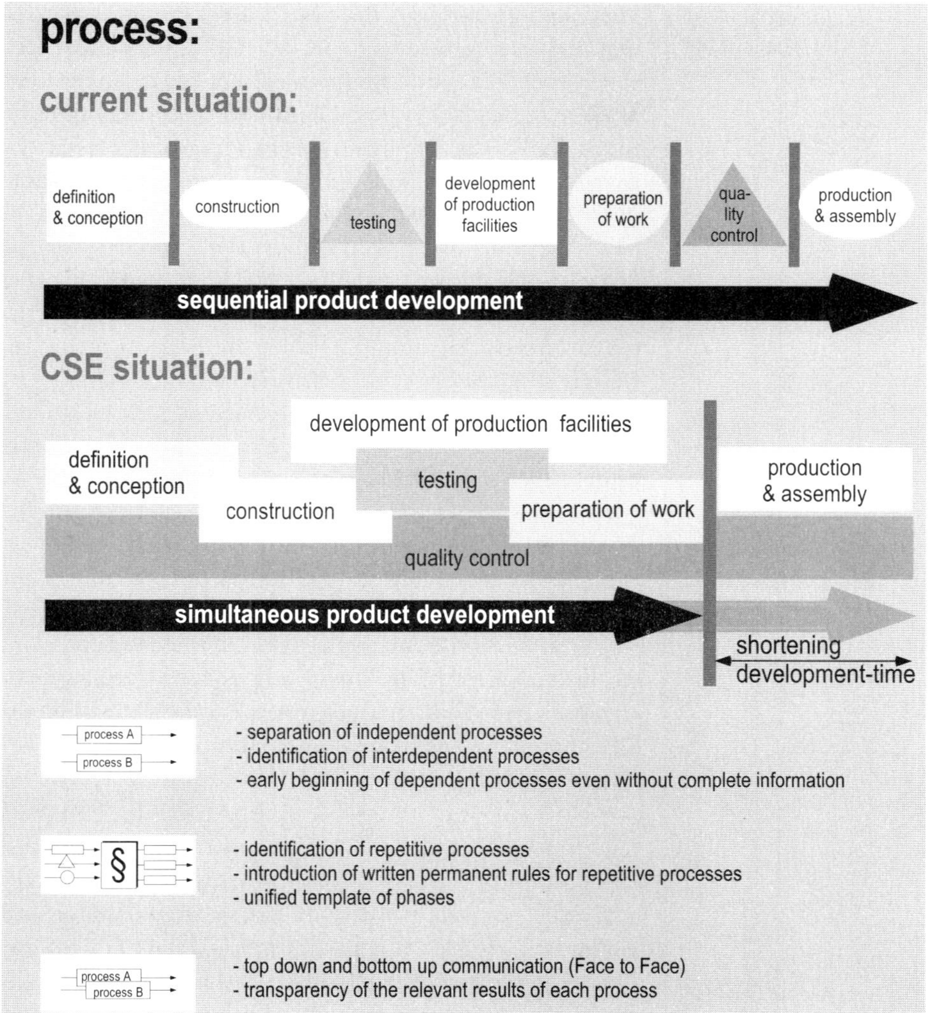

Fig. 2.11: Processes

process characteristics as well as the consideration which can be expressed with algorithms of later aspects in the product development process or product life cycle. In the context of CSE standard operations and methods for optimisation are represented as

Methods for optimisation of the product development process

- Design for Manufacture,
- Design for Assembly,
- Design for Quality,
- Design to Cost.

Optimisation on the basis of DfX

They support engineers in evaluating the product costs, the required time to assemble, the capability to assemble, etc. already being in the design-phase of a product. On the basis of these methods it is possible to find an optimal solution within the "Magic Triangle". Design for Quality enables the development of products with precisely those characteristics which are relevant to the customers. With Design for Manufacturing, it is possible to develop products in a fast and cost-effective way combined with high quality. In order to offer products with the highest possible number of customers' demands (variations) at lowest costs, the method of Design of Assembly is important. This can be achieved by means of modulation with the objective of little additional time spent in the development phase. These methods will be described in detail in chapter 5.

Integration, parallelisation and standardisation supported by these methods

Those methods are supporting integration in the product development process through interdisciplinary work. Furthermore, they enable parallelisation of tasks due to the early evaluation of the capability to assemble and other characteristics with uncertain data. On the one hand standardisation is simplified by these methods using systematic and structured approaches (procedural standard) on the other hand, it is supporting the creation of modules.
Referring to the methods of design for manufacture, for assembly, for quality and design to cost, the following requirement is suggested. As mentioned above, it is necessary that the problems and specific sections of them can be dealt with separately and with distinctive methods. Such methods cannot describe all concerns in a standardised way since the product development process is a complex whole, in which different views and interests are to integrate in one result.

Possible conflicts applying these methods

Due to interdisciplinary teams and the usage of results from several methods each of which is optimising a different problem conflicts can occur generating one optimal solution. It can happen that these methods will be applied in several development units and therefore trigger arguments among the units. The automotive development is an example, therefore, or already existing procedures for optimisation of the construction from the point of view of aerodynamics as well as from the capability to assemble. These procedures have to be linked with design-aspects. Concerns of aerodynamics and the capability to assemble can be assessed by engineers, and decisions can be taken on a rational base. Additional consideration of the aesthetics and qualities of the product can only be ensured by the discussion and decision process among the development units. The objective should be to design aspects in the product which are causing conflicts in an open, i.e. quite flexible way on the one hand and in a standardised way on the other hand, which is supporting the process with known tools. Progress, quality and expenditures of meetings are influenced by the style of discussion and the time sequence of the meeting.

Management Supporting Processes

The coordination of posts concerned is essential for carrying out parallel processes. Therefore, standardised descriptions of the carried out sub-processes are necessary. Since managerial processes are laregly designed by the organisation, the planning and coordination procedures will only be mentioned very briefly. They should be adjusted in view of the complexity of the product development. Tools for a comprehensive project scheduling, -management and -controlling, project information systems, cost recording systems and decision support systems are supporting members of staff. They describe their activities and refer to past experience by way of old plans, checklists or others. The description is followed by the coordination and scheduling of activities which are dealt with by different organisational units.

Project-management supporting tools

Most modern tools – see chapter 5.5.1 Engineering Project Manager for a detailed description – of this category are supplying procedures for planning and controlling as required by complex R&D-projects. The inherent uncertainty in the product development process is not eliminated by planning. A consistent description of activities or of the planned submittal of results refers only to the near future of the product development process. This submittal of results is located at the organisational interfaces, giving a precise definition of the results which takes the product structure into account. Thus, expenditures for planning and maintenance of information can be reduced to a minimum. These tools serve to achieve comprehensive information and consequently the integration of the project team.

Project-planning does not eliminate uncertainty in the product development process

All members are informed about managerial processes, which are important for their work.

Receiving information is combined with the obligation to report amendments, results and the level of the own work, etc. In order to restrict project-meetings to decision-making, these tools assist in demanding and processing information in an event-controlled way.

2.4.2.3 CSE Area: Resources

Production factors are determining the efficiency and effectiveness

The efficiency and effectiveness of the production of a company's goods and services are dependent on its traditional production factors: labour, capital, land, and the new factor information. These factors are mainly determined by the resources available within the company. Changes in the product development process are made possible by developing and designing these resources, thus achieving improvements.

Human Resources

Human beings have professional qualifications, certain behavioural patterns, and the ability to socialize. It is of outstanding importance for CSE to employ the right people and to encourage to develop the required skills.

CSE depends on interdisciplinarity and understanding

The development of human resources greatly supports these methods helping to prepare employees for the changes to their work induced by CSE. Interdisciplinarity and the understanding of other views are required for CSE. Job rotation, e.g., is an integrative means for project leaders or generalists in the project team to achieve the necessary understanding. Common projects and intensive co-operation are the best training places for specialists in the operating departments to develop the requested abilities. Figure 2.12 shows the differences between the two approaches.

"Learning-by-Doing" instead of theoretical training

Team-oriented organisations demand staff capable to work as part of a team. This skill can rarely be taught by theoretical training. The approach "Learning-by-Doing" is much more effective. This involves the coaching of new project team members and an additional training for employees. It makes sense to introduce new working techniques and to train the staff parallel to the projects. Above all, techniques for moderation and presentation are important.

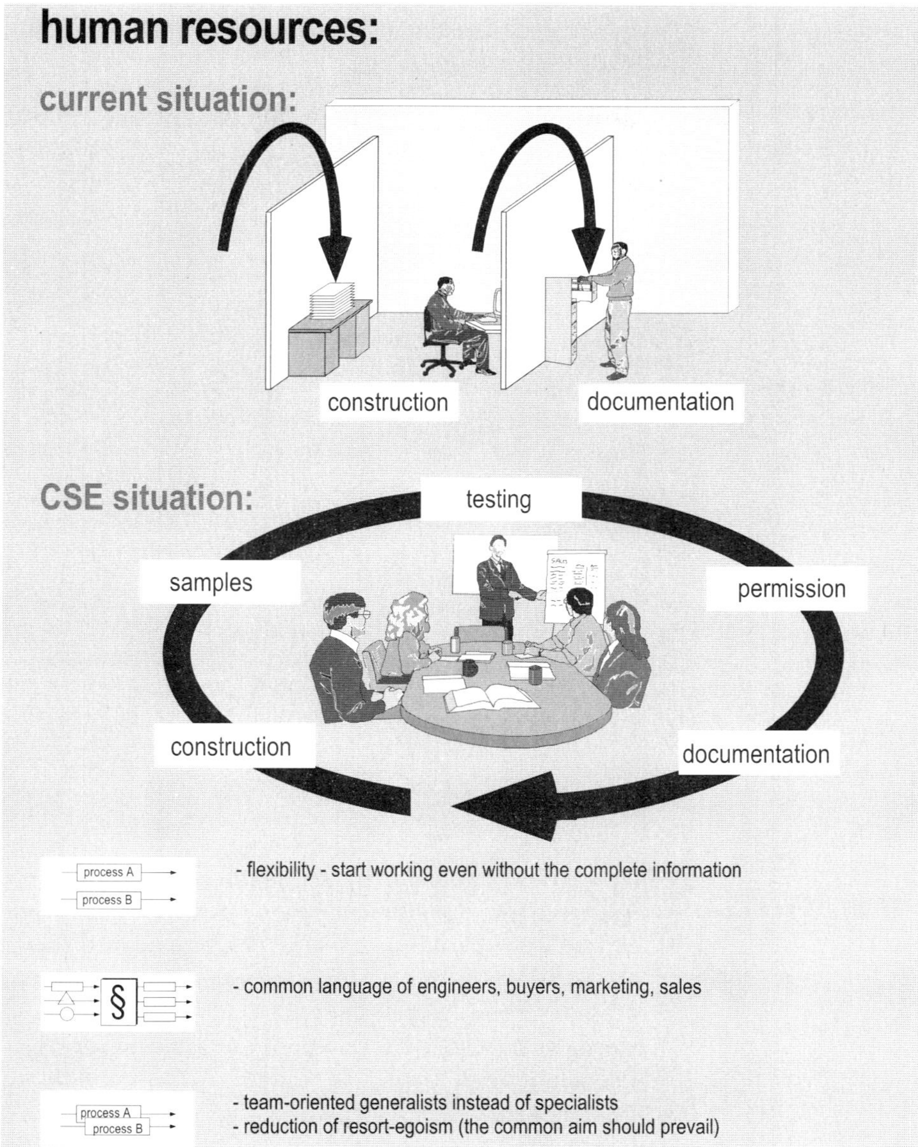

Fig. 2.12: Human Resources

Projects which are carried out in the sense of CSE require technical, methodical, and social competence from staff members and managerial employees (see figure 2.13). The knowledge of every necessary discipline must be represented in the team. A marketing problem should be dealt with

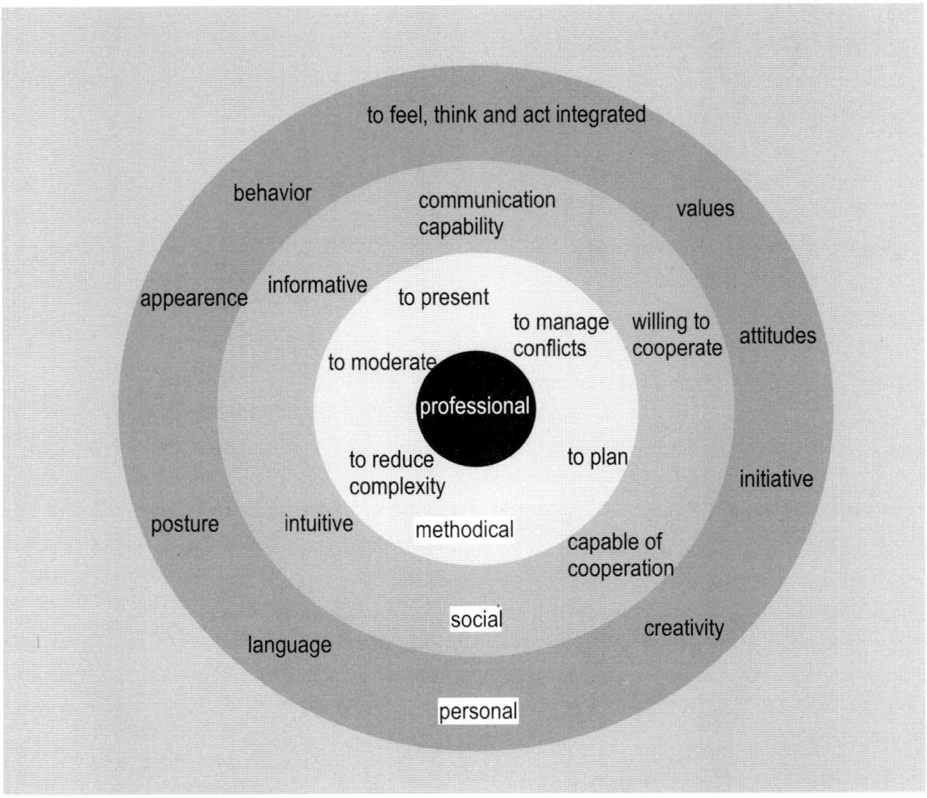

Fig. 2.13: Circle of competencies

by a market and not by an engineer and vice versa. Approaches to the different design for X-methods (manufacture, assembly or quality) must be mastered by the team. The same applies to discussion processes, decision-making procedures and generally procedures of project and process management. Other preconditions for team work are the

- capability and
- willingness

to interact.

Team capability

Team capability involves, among other things, the handing-over of information and the ability frankly to admit weaknesses. Furthermore, team capability means to take over the responsibility for own processes and decisions, to take decisions in a disciplined way, to carry them out punctually, thus achieving maximal parallelisation.

The principle of participation instead of delegation is applied to managing processes. The shift to a CSE-capable product development is a permanent challenge to the management of human resources. The personnel management has to deal with the appropriate training of the and the recruitment of suitable candidates. Furthermore, it has to keep the specialists in the company, since they have the know-how.

Physical Resources

Physical resources are to be considered for the implementation of CSE besides the human resources.

Investments in data processing

Investments in the data processing will be one of the highest quantities of the necessary capital expenditure. The data processing aims to ensure that existing data will be memorised only once and that it is consistently available at every location in the right format for the required tool. Today, several approaches are traced back in order to transfer these requirements into products. The ESPRIT-Project CONSENS makes an important contribution to this. It develops a platform for the integration of CSE-tools (SIFRAME). This platform contains a database with all the product data saved in a uniform way. This platform is described in detail in chapter 4.

The principle of picking-up information should be supported more to be able to always treat processes with the right input and control information. This means that the data must be accessible everywhere in a flexible way. This implies the acceleration of the transformation process of the most important data. This data can be handed over in a logical way (push-principle) or fetched (pull-principle) on the basis of a scheme, which has already been determined.

Necessary equipment for communication and co-operation within the team

Therefore, the strategies of parallelisation, integration, and standardisation are supported. For the frank co-operation of team members working on a product development process and progress in the project work it is necessary to provide the appropriate equipment. This includes conference rooms equipped with adequate facilities for presentations such as white-boards, meta-plan boards, and overhead projectors. This equipment should be permanently available. Additionally, team members should work in the same location, i.e. the same office or floor to increase the proximity and thus to improve the information flow.

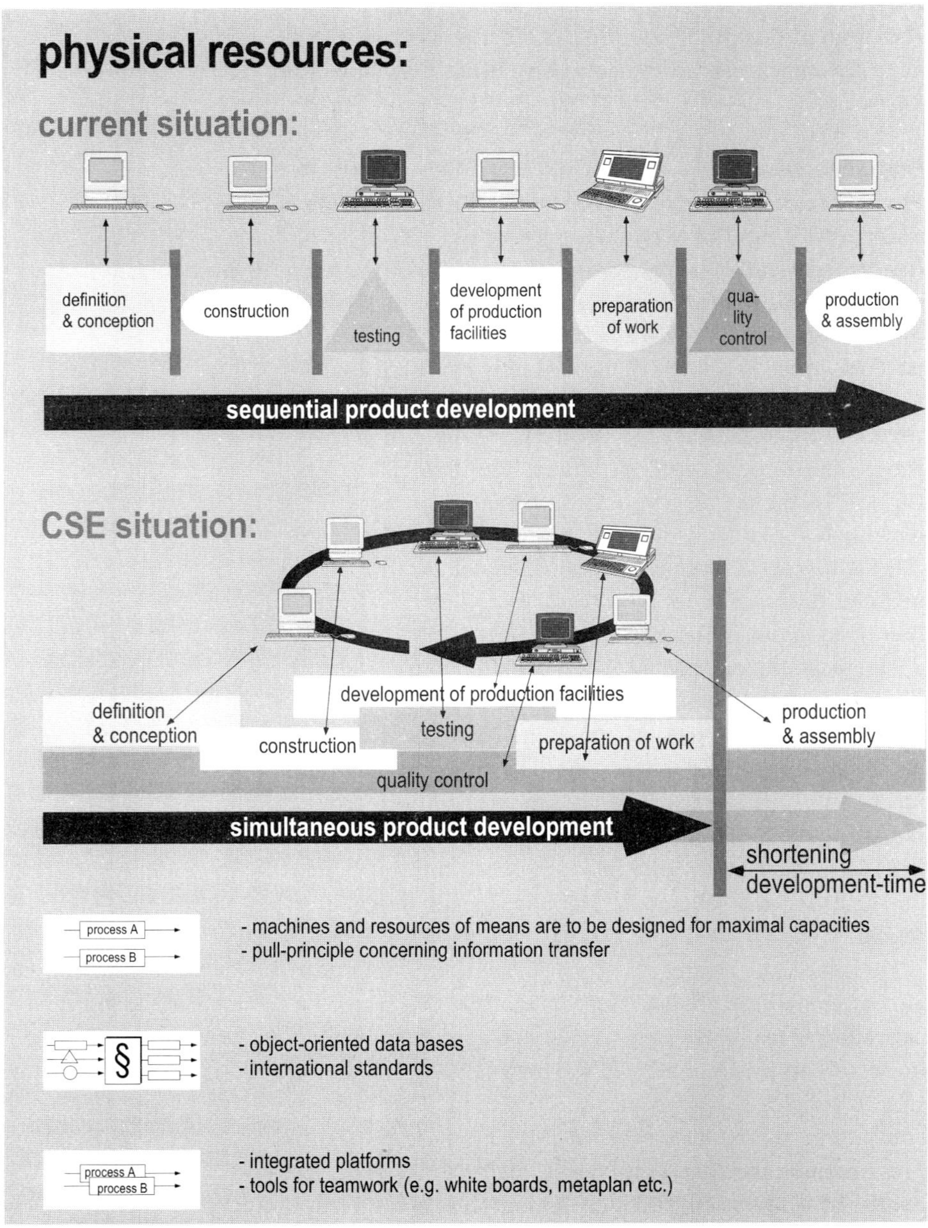

Fig. 2.14: Physical Resources

2.4.2.4 CSE Area: Product

Documentation of the product and the production and their interation in a unified report and information system

The result of the product development process is described with the documentation of the product itself (e.g. product-model, calculations, certificates, circuit layouts, bills of materials, etc.) and its production process (tool-drawings, process-plans, work-plans). In the sense of CSE, it is of importance to pursue the documentation of the product and the production in the earliest phases of the product development process. A documentation has to be integrated in a unified report and information system, as e.g. the duties record book and design specification book. The following figure shows the characteristics of these two tools.

The objective of the report and information system is the support of the information management. Transparency for all employees concerned with the pursued results of the product development process is ensured by the goal-directed documentation of the product and the production process.

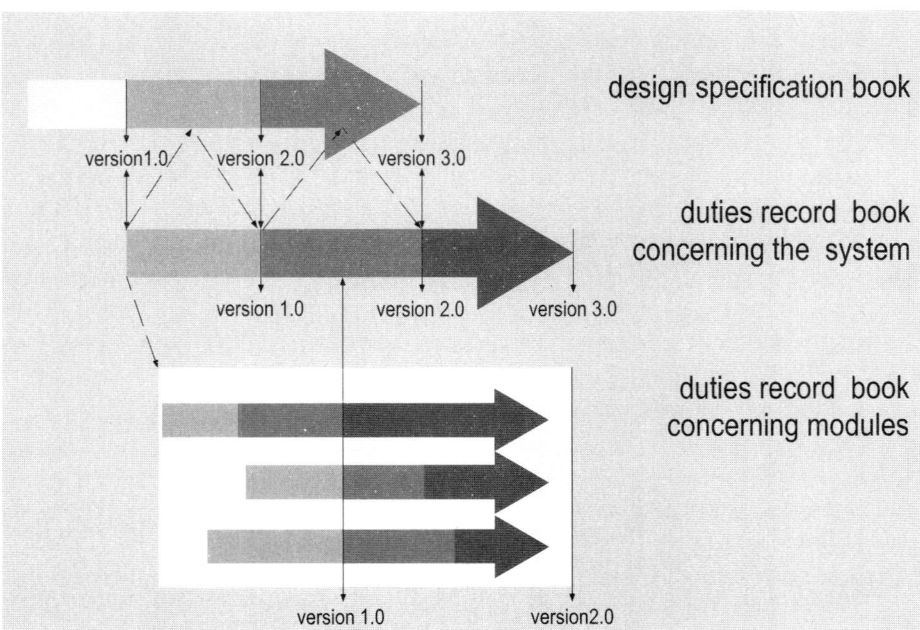

Fig. 2.15: Duties record book and design specification book

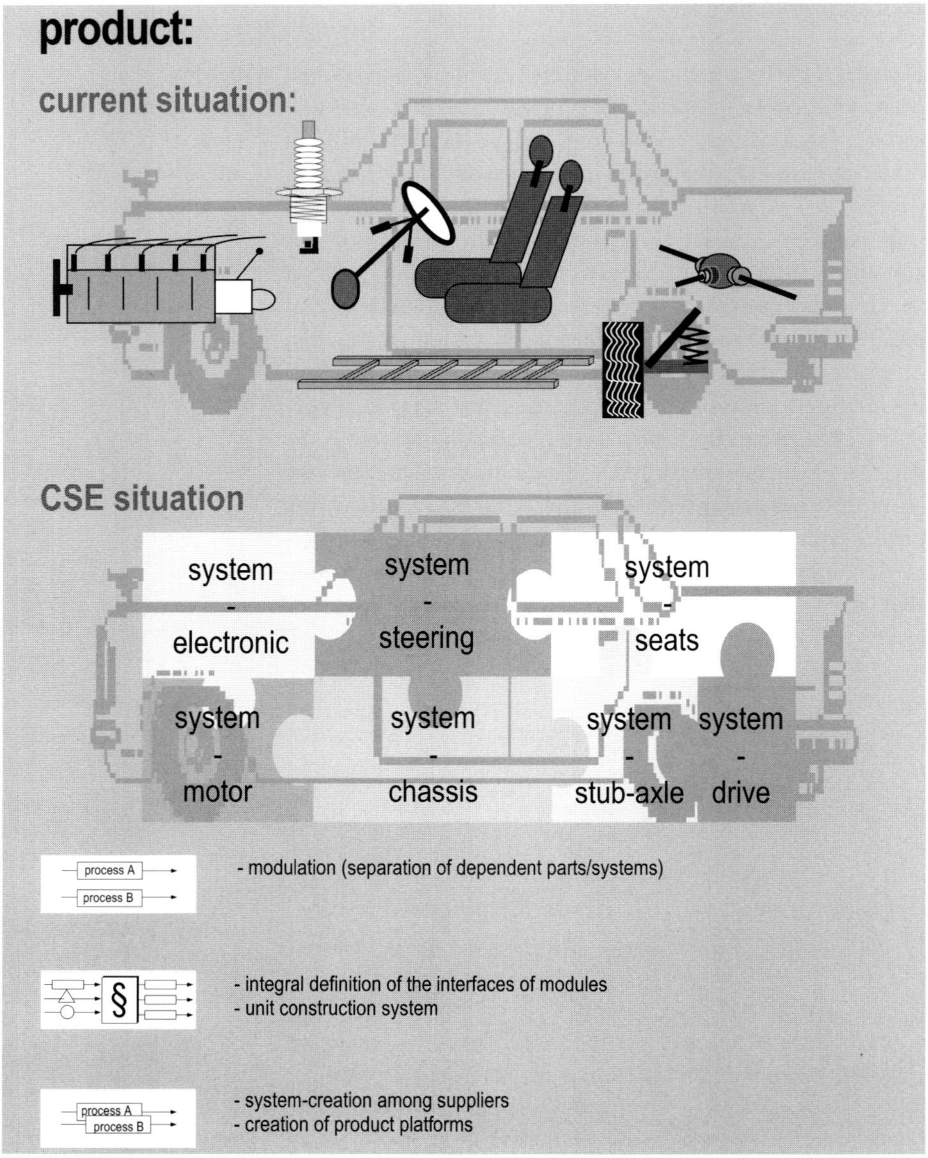

Fig. 2.16: Product

Figure 2.16 illustrates basic differences between the two approaches to the general product.

Consistency of the data

In the general product the most important support provided by tools can be seen in the consistency of the data about the product. The achievement of uniformity has to be reflected in the data and also has to be recognised by the team members. In order to improve a more intensive information flow about the product among the members of staff, tools of Rapid Prototyping [32] have recently been gaining importance. They allow the imitation of product-characteristics at an earlier stage. These prototypes considerably simplify the discussion and problem solution and point out possible contrasting views, early in the process.

Rapid Prototyping in order to ensure an intensive information flow

Regarding the data, big efforts have been made to define clear product structures. Libraries for standard components are an efficient supporting tool for engineers, based on a data model which describes the entire product. They can contain data of standardised items or standard elements of the company. From these libraries Special tools [33] for the creation and the use of the classification of parts are available for the selection of appropriate elements. These tools use the geometric and functional similarity of parts to define standard solutions. Knowledge is always required for generating solutions This is based on the experiences from earlier product development processes. Dependencies among product characteristics can be noted and evaluated. One also speaks about Decision Rationales [34]. They are set up with every important design decision and combined with the documentation of the state of the product data, which enable users to trace back the design history. The history of different products and experience gained during the complete product-life-time can be evaluated It also provides valuable suggestions for the materials used and the possible potentials for the improvement of the products.

Libraries for standard components as product development tools

Decision Rationales

In order to simplify the development and to improve the generation of product-knowledge, it is useful to construct modules within the product structure. Sensible structures of modules can be defined from experience, which, in the following course, can be applied in a similar way or can be adapted to many products. Furthermore, these structures can adopted or adapted in many other products. Techniques of the parametric design can be applied, and it is possible to bind innovative ideas to one module only. These ideas can be transferred to a variety of other products without losing time, after having solved all the implied problems once.

Development of product-modules

Multi-product-development on the basis of CSE

The PSI-strategies as described in this section can be applied for the single product development. In case that the methodology CSE should be applied to the development process for more than one product, the question is how the PSI-straetgies for the development of the whole product spectrum of a company should be applied. A new strategy [35] for the construction of products is required.

Product families

One approach could be to group all products in product families. A product family is a group of products with a common product platform, offering specific characteristics and functions for several groups of customers (niches). The product platform entails components and a specific constructive composition, which are shared by a number of products. Because of this, product platforms are known as the heart of a successful product family. A product family is directed towards a specific market segment while the individual products are targeting special niches in this segment. The objective of this approach are fast and near-market leaps in the innovation and the diversification of product platforms.

2.5 Implementation of Concurrent Simultaneous Engineering – A Business Process Re-Engineering Approach

This chapter deals with the implementation of CSE in companies. Several aspects of the implementation will be discussed such as a general guideline regarding the cultural aspects, changes in management, characteristics, and prerequisites.

Aspects of the CSE-implementation

2.5.1 Culture – The Factor of Success in CSE-Implementation

CSE, as a strategy, was not initially developed as a strategic concept. The first companies to work with previous approaches of CSE had isolated problems for which they found new solutions. In most cases, these activities were initiated by the realisation that developments of products are quite often not finished at the scheduled deadline and with frictional losses. In these cases, people recognised that there were general problem patterns [18]. They analysed these patterns and realized that the underlying reason was a lacking integration of functions within the company. Thus, in many successful companies, the introduction of CSE was not a deliberate process. In general, these companies approached CSE by situatively learning without an explicit CSE-implementation. The consistency and constancy of the solutions were a significant factor for those companies with successful rudimentary improvements. The monitoring of several of these companies resulted in solution-patterns, which could be gathered under the integrated roof of CSE. There is a similarity whenever CSE is applied. Therefore, it is not always regarded as a real process.

Introduction of CSE was not always a deliberate process

From the latest experience and insights into the organisation development [36] it can be derived that the CSE-implementation is a comprehensive and long-term issue. Anyhow, an important chain of a company's value-creation is changed and orientated towards future market challenges.

CSE-implementation is a comprehensive and long-term issue

Major changes in the company go together with the implementation of CSE

It is necessary to change macrocosmic structures and microcosmic processes, leading to a new way of thinking and working [3]. An improvement cannot be achieved just by tremendous changes which are dictated by the top management and which involve only a redrawing of the company's organisational chart. As we know from successful companies, a cultural change and a new working environment have to come along with the implementation. This means in detail that an implementation in the form of detailed methods, behaviour-patterns, attitudes, organisational structures, and the management philosophy is necessary.

The change of a company's culture as a crucial task

The main bottleneck in introducing CSE is the cultural change. The behaviour of the individual within the system is determined by the usual formal and informal processes, the distribution of power and the competence and disposition of that individual towards the organisation. The CSE-implementation is a major intervention in a company. It has to overcome several obstacles. It is the task of the change management to achieve a better understanding for changes within the company and to understand possible doubts of employees. The change management is described in the next section.

2.5.1.1 Change Management

"We never know if something improves, once it has been changed. But we know quite well that we have to make changes in order to improve."

J. Schmidt, management trainer

Permanent adaptation of the organisation to its environment

It is indispensable to adapt organisations or processes to new situations due to the permanent changes in a company's environment. During the past few years, visions and guiding principles, as e.g. Lean Management, Continous Improvement or Business Process Re-Engineering were discussed intensively for lean and competitive companies. Many entrepreneurs and senior managers still have the task to implement elements of these concepts. As it is proved by experience, this realisation of ideas entails in far-reaching consequences with significant difficulties. Changes in a familiar environment are quite often seen as something negative or hazardous. If changes are regarded as a neces-

Changes must be seen as an opportunity

sity, they will onlyresult in obligations. Thus, changes threaten the status quo of the employees. It seems to be difficult to regard changes as individual opportunities. As a result of this, a majority (about 80% [37]) of changes remain unsuccessful. This is when the approach change management is needed.

Change Management refers to the management of changing processes. Changes are not related to the operational business. Instead they concern:

- the company's objective and strategic direction,
- the processes,
- the responsibilities and competencies,
- the company's culture,
- the physical assets.

Two different levels of change management

Change Management can generally take place on two different levels:

- Firstly, there is the basic and profound change of a company's structure, culture, and strategy (radical change in the form of Business Re-Engineering).
- Secondly, the change of a company's segment involves only minor modifications and amendments (normal change management in the form of organisational learning).

Radical changes

Radical changes to the structure and culture lead to significant improvements or even leaps in productivity. To achieve this improvement, a radical change management is foremost the task and the challenge of the senior management. The preparation of the implementation, the company's declaration to adopt the new approach and the promotion during the period of introduction by the top-management are essential to the final success. Nevertheless, it is crucial to determine the right moment for radical changes. A good sense and intuition of the management for the environment and for the own company is required.

Normal changes

Normal changes are a reaction to new conditions within the bounds of existing management approaches. It results in an optimisation of details and the adaptation of the system to slowly changing conditions. Thus, normal changes concern all employees and must be carried out by them. Rules and principles of "Organisational Learning" have to be considered in order to be successful with the implementation of normal changes.

2.5.1.2 Cultural Aspects of Changing Processes and Structures

A company's culture cannot directly be controlled

The change in the personal attitude and conduct reflected in a company's culture is also confronted with the awareness that a company's culture cannot be directly controlled by means. There are interdependencies between the conditions in the areas of CSE and the company culture. Regarding the dynamic point of view, a change in the organisational structure must be followed by a change of the company's culture.

On the other hand, a changed frame of mind as a characteristic of the culture is influencing sub-processes of the product development process. The changes of both sides do not necessarily have to go to the same direction and need not take place simultaneously. Practice has shown that a change of processes, structures, and work techniques can be executed faster than the adjustment of a frame of mind and conduct. An employee, e.g., who has received only orders from his superior will have difficulties when he has to take over responsibility in a team or make decisions of his own. Because of these reasons, the implementation is of considerable importance regarding the interplay between the company culture and the changes in structures and processes. If the team organisation is implemented without the participation of the employees, it will later be difficult to delegate responsibility to the team level.

Opposition of organisational changes by employees

Employees and managers, who want to maintain their status quo, might oppose substantial changes to the organisational structure. Often managers only recognize the personal disadvantages such as the fear of losing personal responsibility, power and reputation or even the loss of their jobs. The potential of working as a coach or as a moderator, are not immediately obvious. When implementing new approaches, this behaviour causes a delay in power struggles. A list of apparent arguments "prove" the disadvantages of the "new approach". Similar barriers can emerge among employees. This is indicated in a rising fear of change and in the possibility of overwork. These barriers can be overcome by giving information and by an intensive

From a functional manager to a moderator or coach

participation of the affected employees when the new approaches are designed.

In order to synchronise changes in structure and culture, it is necessary to make use of a guiding principle in the sense of a CSE vision. Thus, the behaviour and attitude of employees can be changed permanently. A successful implementation can only be achieved if there are no doubts about the direction and the objective of the change.

A guiding principle in order to enable changes

2.5.2 A guideline for CSE-Implementation

2.5.2.1 Principles and Prerequisites for a successful CSE-Implementation

The consequent implementation of CSE is more than just using new tools. It involves a new orientation of the complete product-development process. It is necessary to develop a profound understanding for the product development process and its correlations. It is not enough merely to think in dimensions of methods and tools. In order to meet a successful implementation there are some principles which should be regarded as prerequisites.

CSE-implementation involves a new orientation

CSE, as an integrated concept, has a uniformity on all company levels. Thus, it is important to link the company's strategy closely to CSE. Clear decisions, regarding the objectives and the introduction are required at management level. These decisions can only be taken sensibly, if the performance potentials, opportunities, and risks of CSE are sufficiently known. The management has to believe in the CSE vision to embody it and set an example with its own conduct.The initiation of a project as a form of a temporary organisation is significant to the CSE implementation.

Strict connection between strategic company goals and the implementation of CSE

The management must state what it expects from implementing CSE. There are two alternatives: Either the management has no experience with CSE or the management has some experience and knows about the advantages of CSE and the potentials of its own company. If the management has no clear idea of CSE and its implications for the company's strategic goals, a minor pilot project could be established. This project should be supported by sufficient capacity, budget, and staff with some authority to make decisions. After having completed the pilot project, the results will point out existing potentials. A management decision can be made next. In case that the experience already exists, appropriate CSE goals can be set by the management, e.g. a reduction of development costs by 30%.

Clear definition for the responsibility of CSE implementation – nominating a promoter

CSE-implementation includes changes to the system regarding checks and balances within the company. Managers lose their power and influence; current processes are to be changed. People are afraid of the new situation. They tend to perpetuate the old and familiar structures. In the light of this situation, a person in senior management must be convinced of CSE and must take over responsibility for the CSE-implementation. This person acts as a partner of the employees who are involved with the CSE-implementation. He supports the decisions of the senior management and overrides decisions of opposing managers.

Organisation first – technical support afterwards

The implementational strategy should be marked by the principle "organisation first, technique later". Generally speaking, organisational structures and processes must be changed first. Based on this premise, appropriate methods and tools can be introduced and adapted. Sometimes it is not possible to follow this rule consequently, since the existing infrastructure (e.g. existing and alterable CAD and networks) has to be taken into consideration. If investments made for this are not abandoned, the equipment, requirements, and restrictions for the design of the processes and structures will emerge.

Concentration on strengths

One promising approach is the concentration on strengths during the product development process at the beginning of the implementation. A company which is extending its strengths has better opportunities to increase its competitiveness than a company which is focusing on its weaknesses and thus neglecting its strengths which differentiate it from its competitors. These implementations are in most cases successful, thus the motivation to initiate further steps is increased and opposing employees can be convinced of the advantages. Concerning personnel and cultural aspects, the deployment of process- and power-promoters as well as the intensive participation of employees are of crucial importance when implementing CSE.

Include the employees

Employee orientation is the last but not the least principle for the implementation of CSE. The product development today is a complex process in which pursuing the classical management is not possible. Classical management assumes that the process is deterministic, but it is not.

The people who perform the product development processs should be involved in the definition of supporting tools. They should be allowed to decide what the process should look like in future. The employees implement CSE! The objective of the implementation should be to get the employees involved in bringing about the changes rather than having them simply affected by the changes.

2.5.2.2 Characteristics of the Implementation Process

Cyclical, iterative process as implementation process

The management of the implementation process as a project is based on a cyclical, iterative process. The implementation process cannot be planned in detail in advance because of the complexity of the intention. This complexity is a result of narrowly linked correlations of procedural, structural, and methodical elements of the product development process. The ability to handle details carefully need be strengthened on the basis of a clear guiding-principle of the future product development process. This iterative approach is marked by a short conception phase, which avoids concepts (guidelines, manuals) going into too much detail. Practice has shown that it is an advantage to work out a basic framework to a certain degree (30% concept). Then, the frame will be adapted, evaluated, and improved to the specific project. Similiar to the waterfall-principle (see figure 2.17) more employees will be actively involved with the further progress of the implementation process. This makes full use of the complete experience and knowledge-potential available in a company. Other effects are that a widespread acceptance and a change of the awareness among the employees on all levels of the company can be achieved.

Step-by-step expansion of employees involved in the implementation process

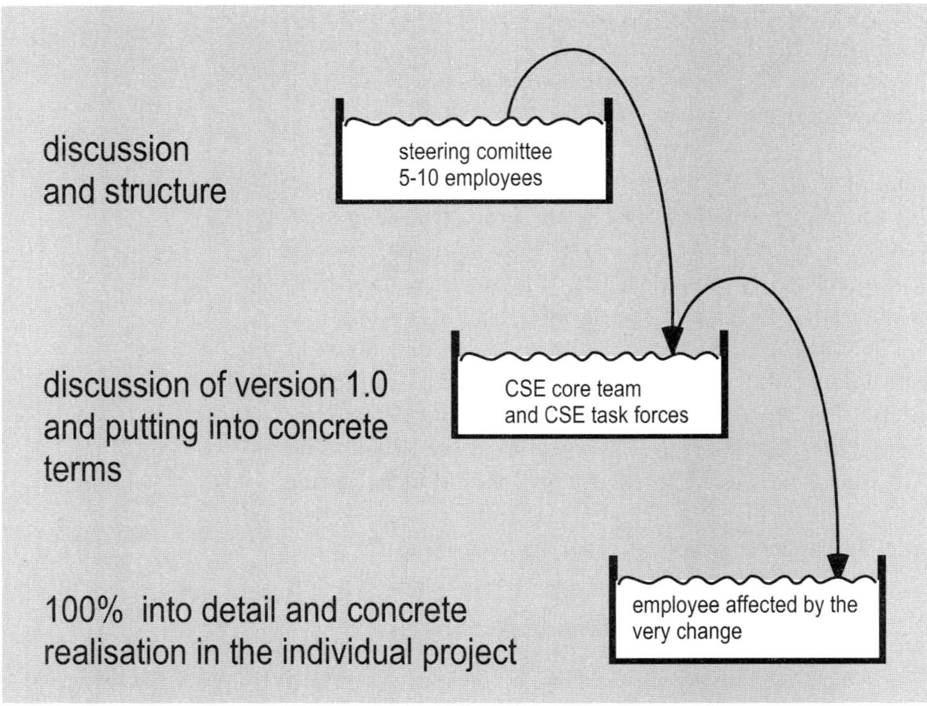

Fig. 2.17: Involvement of employees: The waterfall principle

Possible imbalance of old and new approaches

While implementing CSE, an imbalance of old and new approaches can temporarily emerge. Bigger amendments, for example, can be first introduced to a project which is later used as a model for the whole company. Further concepts and the next steps of the implementation process can be approached on the basis of the experience gained while pursuing the coherence and consistency among all the development activities [3].

2.5.2.3 Elements and Tools of the Implementation Process

The elements described in the following can be derived from the discussed principles and characteristics in the previous two sections. As already mentioned, one does not go for a rigid implementation concept. These elements can be used during the implementation process, depending on specific conditions. Compared to conventional procedures of the organisation development a faster and more profound change of the product development process can be achieved.

The Implementation Teams

In order to ensure a prompt introduction of a CSE oriented process in the product developement the following project organisation is required. Its basic structure is divided into three levels, each with one or more teams. These teams are set up independent to the project progress and pursue different tasks.

1. CSE Steering Committee

The CSE steering committee is composed of senior managers of the company. The tasks are the set-up of targets, the creation of a CSE-vision and the assesment of the progress.

Implementation teams on three levels

2. CSE Core Team

Senior managers, representatives of the employees and directly and indirectly affected employees form the CSE core team. The overall task of the core team is to put the CSE vision, created by the CSE steering committee, into concrete terms. At the beginning of a project, its role is more of an organising nature and at a progressive stage it has a rather guiding role.

3. CSE Task Forces

The task forces form the third level of the project organisation. Each taks force is guided by a member of the core team. The objective of these tasks forces is to work out detailed solutions and their implementation based on results achieved by the CSE core team. The members of the task forces are selected due to specific problems to be solved. It should be achieved that an increasing amount of directly affected employees will be nominated in order to support the acceptance of the achieved out results.

Re-Engineering in the Product Development Process

Process design wants to minimise information deficits, deceleration, bureaucracy and iterations within the actual product development process. The design is concerned with the formal processes, since technical problems remain mostly unpredictable. It is based on an analysis of the information flow of selected former projects. Process design should be carried out at the start of the implementation process and should be repeated after each pilot project. Gained perceptions will be considered when the processes are redesigned.

The Process Programming Tool PACE®:

PACE® as tool for designing processes

A supporting tool for designing processes is the PACE® method. The modelling of the process structures of the product development of DASA and Hidrosorefame was done by using a software platform named PACE®. PACE® is a graphical modelling and programming tool for discrete event systems. It is based on extended, timed and coloured Petri-Nets. This kind of Petri Net-models has been used for decades for the description and evaluation of hard- and software systems. The dynamic behaviour of a system or program can be visualised and analysed with the PACE® simulator. Petri-Nets represent a generalisation of state-transition models. They are especially suited for describing concurrent systems, such as those shown in the case studies.

The PACE® system description language does not refer directly to a specific application field. The mapping of the language elements to the corresponding elements in the application field is usually obvious. Consider, e.g., the dynamic objects moving through a PACE® net. They are called tokens. In a communication system, a token might represent a message. In the model of a manufacturing system, a token can represent a product, or some material. For better acceptance, the graphical standard elements (circles, squares and tokens) can be substituted by user-defined individual icons.

In simulation applications, the language that is used to describe the dynamic flow of the objects through a system need not depend on what the objects represent in the real world. Independently of what an object represents, we need some way of describing its internal state, to express under what conditions it moves from one location to another, how much time such a movement takes, how the object selects between alternative paths, how its internal state is changed, etc. The PACE® system description language combines Petri-Nets and Smalltalk. Hierarchical Petri-Nets are used to describe the structure of a system, as well as its concurrent behaviour. On the other hand, data and sequential behaviour are modelled by Smalltalk objects and code sequences. The tokens flowing through the net contain attributes, which are Smalltalk objects. The net elements can be inscribed with some Smalltalk code, in order to do computations with the objects.

Workshops:
The general function of workshops is to work out design approaches or methods of CSE and to adapt them specifically for each company. One member of the support team is responsible for the preparation, moderation and direction of a workshop. The selection and mix of the workshop participants is done by the CSE support team. Participants should be selected by the choice of subjects and the employees involved. The cross-linking of different topics and workshops has to be taken into consideration. The transfer must be accomplished by the CSE support team and an overlapping manning of the workshops.

Work out of design approaches or methods in workshops

Tools for the First Steps

Pilot Projects:
Newly evolved processes, structures or procedures are applied to pilot projects. The advantages of the changes are demonstrated with the help of practical examples in the context of a company-wide information campaign. The objective of this campaign is to strengthen the attention to CSE among the employees. The implementation of CSE should be pushed by internal discussions. A pilot project should be a typical project for a company with a high probability for success since its failure would have negative consequences on the implementation process. The project team should consist of employees with a high acceptance concerning CSE.

A pilot project should be typical for a company

Coaching:
Coaching is the support of a group which is managed by itself, accompanied by a supervisor. Objectives of coaching are the control of the group process and the support of workshops or pilot projects by project "externals". In ideal circumstances, the coach belongs to the CSE support team. The coach should not take over the project-management but serve the project team like a coach in sports.

2.5.2.4 The Implementation Process

Procedure to implement CSE

In order to achieve a successful implemention of CSE, a specific procedure is suggested which contains the following issues:

- Goal setting,
- Working out a CSE-vision,
- Creation of key topics,
- Coordination of the key topics and coaching the introduction,
- Progress assessment.

These issues are illustrated in figure 2.18.

Goal setting

The steering committee works out a clear and conclusive conception of what should be changed, in which direction and how. Doing this, the objectives of the implementation must be defined, in qualitative or quantitative terms. These objectives will be specified and adapted in the course of the definition process of the guiding principle, as well as in the course of the progress assessment.

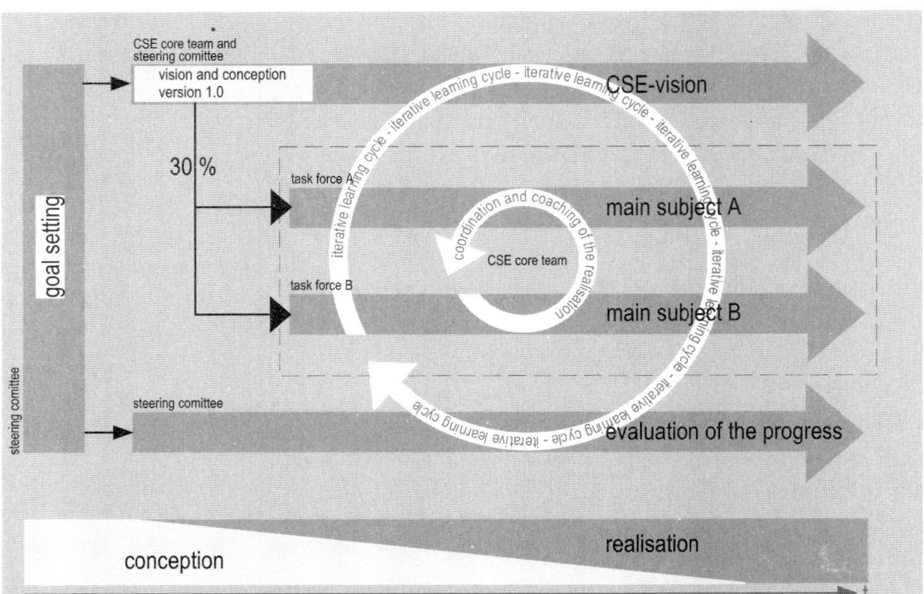

Fig. 2.18: Methodology for CSE implementation

Working out a CSE-vision

A guiding principle which is commonly formulated and which creates CONSENS is the result of this task. The content of this guiding principle is:

- The strategic goal direction for the change on the grounds of future challenges.
- The formulation of principles characterising the new situation.
- The concept and scenario for activities (30% concept).

Therefore, the guiding principle represents an instrument which aligns the senior managers and employees towards a common objective. The guiding principle serves as a guideline for the implementation process and should be accepted by all members of the steering committee. At the beginning of the defintion phase (until the formulation of the principles) the steering committee is involved while the 30% concept is worked out by the CSE core team. At this stage the CSE core team is most significant for the creation of awareness and acceptance.The key factors innovation, quality, time, and productivity are to be integrated into the guiding principle referring to the basic conditions such as customers, employees or productivity.

Creation of key topics by the CSE core team

Key topics are issues or problems which can only be solved with the specific knowledge of the affected employees. These key topics are to be elaborated in this phase, i.e. a 100% solution is the target. Having created the key topics, the activities for their solution will be worked out and introduced. The CSE task forces are responsible for the activities in this phase.

Coordination of the key topics and coaching of the introduction

The results and interim results of the CSE task forces have to be coordinated and harmonised in order to ensure an integrated and interlocked implementation. Thus, disfunctional effects can be avoided. This coordination can be achieved by meetings of the CSE core teams. In these meetings the leaders of the several task forces report on the results of their teams and coordinate their own results with those of other teams. If necessary, the results will be corrected and improved by the task forces.

Another objective is to urge thematically the CSE implementation with a permanent coaching of the process.

Progress assessment

In regulary intervals the progresses of the implementation will beassessed on the basis of the defined goals (ratios) and activity plans. In meetings with the steering committee the following issues will be discussed:

- state of the implementation process,
- state of the key topics,
- comparison with the targeted guiding principle,
- taking of decisions,
- adapting the goals (ratios).

With the application of the procedure desribed above a successful implementation can be achieved.

References

1 Stalk, G.jr.; Hout, T.M.: Zeitwettbewerb: Schnelligkeit entscheidet auf den Märkten der Zukunft. Frankfurt a.M., New York 1990

2 Warschat, J.; Wasserloos, G.: Simultaneous Engineering. In: Fortschrittliche Betriebsführung und Industrial Engineering, Vol.40(1991) No.1, pp. 22

3 Clark, K.B.; Fujimoto, T.: Automobilentwicklung mit System. Strategie, Organisation und Management in Europa, Japan und USA. Translated and published by E.C. Stoko. Frankfurt a.M., New York 1992

4 Bullinger, H.J. (Editor): Neue Produktionsparadigmen als betriebliche Herausforderung. In: IAO Forum - Innovative Unternehmensstrukturen. Berlin, Heidelberg, New York 1992

5 Rommel, G.; Brück, F.; Diederichs, R.; et. al.: Einfach überlegen: Das Unternehmenskonzept, das die Schlanken schlank und die Schnellen schneller macht. Stuttgart 1993

6 Tress, D.W.: Simultaneous Engineering in der Elektroproduktion. In: VDI Berichte No.758, 1989, pp. 205-220

7 Perillieux, R.: Technologietiming. In: Handbuch Technologiemanagement. pp. 268-284. Editor. Erich Zahn. Stuttgart 1995

8 Mansfield, E.; Schwartz, M.; Wagner,S.: Imitation Costs and Patents: An Empirical Study. In: Economic Journal, Vol.91 (1981) December, pp. 907-918

9 Robinson, W.T.; Fornell, C.: Sources of Market Pioneer Advantages in Consumer Goods Industries. In: Journal of Marketing Research, Vol.12 (1985) February, pp. 305-317

10 Buzzel, R.D.; Gale, B.T.: The PIMS-Principles. New York 1987

11 Lambkin, M.: Pioneering New Markets: A Comparison of Market Share Winners and Loosers. In: International Journal of Reserach in Marketing, Vol.9 (1992), March, pp.5-22

12 Cooper, R.G.: The Dimensions of Industrial New Products Success and Failure. In: Journal of Marketing, Vol.43 (1979) No.3, pp.93-103

13 Vanhonacker, W.R.; Day,D.: Cross-sectional Estimation in Marketing: Direct versus Reverse Engineering. In: Marketing Science, Vol.6 (1987) Sommer, pp. 254-267

14 Rosemann, M.; Wild, R.G.: Die Einbeziehung des Total Quality Management in das CIM-Konzept. In: zfo 1/1993, pp. 34-39

15 La Banca, P.: Gegen die Uhr. In: Manager Magazin 6/1993, pp. 168-173

16 Opitz, H.: Moderne Produktionstechnik. Essen 1970

17 Pantele, E.F.; Lacey, Ch.E.: Mit Simultaneous Engineering die Entwicklungszeiten kürzen. In: io Management Zeitschrift, Vol.58 (1989) No.11, pp. 56-58

18 CALS: Applications of Concurrent Engineering to Mechanical Systems Design. In: CALS Technical Report 002, Final Report of the Reliability and Maintainability (R&M) Mechanical Study Design, June 1989

19 Hartley, J.R.: Concurrent Engineering – Shortening Lead Times, Raising Quality, and Lowering Costs. Cambridge, Norwalk 1992

20 Armstrong, Michael: A handbook of management techniques, 2nd ed., 1993

21 Esprit III Project CONSENS: EP 6896. Workpackage 1. Task 1.2, Deliverable H 0.2, CSE-Methodology and Implementation Guideline

22 Fanger, B.; Lacey,E.: Hürdensprint in der Produktentwicklung. In: io Management Zeitschrift, Vol.61 (1992) No.5, pp. 81-84

23 Hill, W.; Fehlbaum, R.; Ulrich, P.: Organisationslehre 1. Ziele, Instrumente und Bedingungen der Organisation sozialer Systeme. Bern, Stuttgart 1974

24 Ulrich, P.: Management: Eine konzentrierte Einführung. 6th edition, Stuttgart 1992

25 Müller, K.: Management für Ingenieure, Grundlagen, Techniken, Instrumente. Berlin, Heidelberg, New York,London 1987

26 Carter, D.E.; Baker, S.B.: CE Concurrent Enginnering. The Product Development Environment for the 1990's Reading, Massachusetts a.o. 1991

27 Dierkes, Meinholf: Unternehmenskultur und Unternehmensführung. Konzeptionelle Ansätze und gesicherte Erkenntnisse. In: Zeitschrift für Betriebswirtschaftslehre. Vol. 58, 5,6 /1988, pp. 554-575

28 Krackhardt, D.; Hanson, J.R.: Was das Organigramm verschweigt. Informelle Netze-die heimlichen Kraftquellen. In: Harvard Business Manager, Vol.16 (1994) No. 1, pp.16 ff.

29 Platz, J.; Schmelzer, H.: Projektmanagment in der industriellen Forschung und Entwicklung. Berlin a.o. 1986

30 Sprenger, R.: Ideen bringen Geld, bringt Geld auch Ideen? Es ist Zeit das betriebliche Vorschlagswesen abzuschaffen. In. Harvard Business Manager, Vol.16 (1994) No. 1, pp. 9 ff.

31 McGrath, M.; Anthony, M.; Shapiro, A.: Success through Product and Cycle Time Excellence. Boston a.o. 1992

32 IMS: Intelligent Manufacturing Systems. International Conference on Rapid Product Development, 31.1.-2.2.1994, Stuttgart 1994

33 Warschat, J.; Koch, D.: Implementing Concurrent Engineering through an Integrated Vehicle Documentation System. In: Proceedings of the 25th ISATA, Florence 1992

34 Lu, St.: Research, Development, and Implementation of Knowledge Processing Tools to support Concurrent Engineering tasks. The 1992 Research Plan for Knowledge-Based Engineering Systems. Research Laboratory University of Illinois at Urban-Champaign 1992

35 Meyer, Marc H.; Utterback, James M.: (1992), Core Competencies, Product Families and Sustained Business Success, Sloan School of Management Working Paper 3410-92, International Center for Research on the Management of Technology, Working Paper 65-92, Cambridge 1992

36 Rieckmann, H.: Organisationsentwicklung – von der Euphoriezu den Grenzen. In: Die lernende Organisation. Konzepte für eine neue Qualität der Unternehmensentwicklung. Edited by Thomas Sattelberger, Wiesbaden 1992

37 No author named: Ein trauriges Beispiel. In: Manager Magazin (1994) 6, pp. 172-173

3 The Organisational Environment of Concurrent Simultaneous Engineering (CSE)

S. Berndes, A. Stanke
supported by J. Piller

Description of the elements of a CSE-organisation

The objective of this chapter is to develop and to describe the elements of a CSE Organisation and to apply the whole model to the CSE area organisation which was described in section 2.2.1.

When organisation hurts!

The background, theory, and the general implementation process of CSE were described in chapter 2. Chapter 3 deals now with questions of how to create an optimal organisational environment in which CSE can be adjusted. In young and small companies the CEO or some other senior executives perform a strong project leadership and the functional managers work together in order to achieve the global goals. Mature companies have extensive specialisation, a huge amount of employees and a number of ongoing operating concerns as well as development project concerns. Rigid organisational structures are typical of these companies and hinder the ability to innovate ([1] and [2]). In this case, organisation hurts and it is worthwhile to ask how R&D should be organised since innovation is crucial for success.

Model for a Team Organisation

As an answer to this problem, a model for a Team Organisation is introduced in this chapter which enables companies to face these challenges. The elements of the model are illustrated in figure 3.1. The assumption is that processes are the core of the organisation. This means in the case of the product development that these processes are represented by projects. In the centre of the model are the projects. Starting with viewing on the relevant aspects of one project, the organisation is systematically developed in this chapter. The first step of the description is the investigation of just one project with its relating issues:

- project-team,
- project-manager.

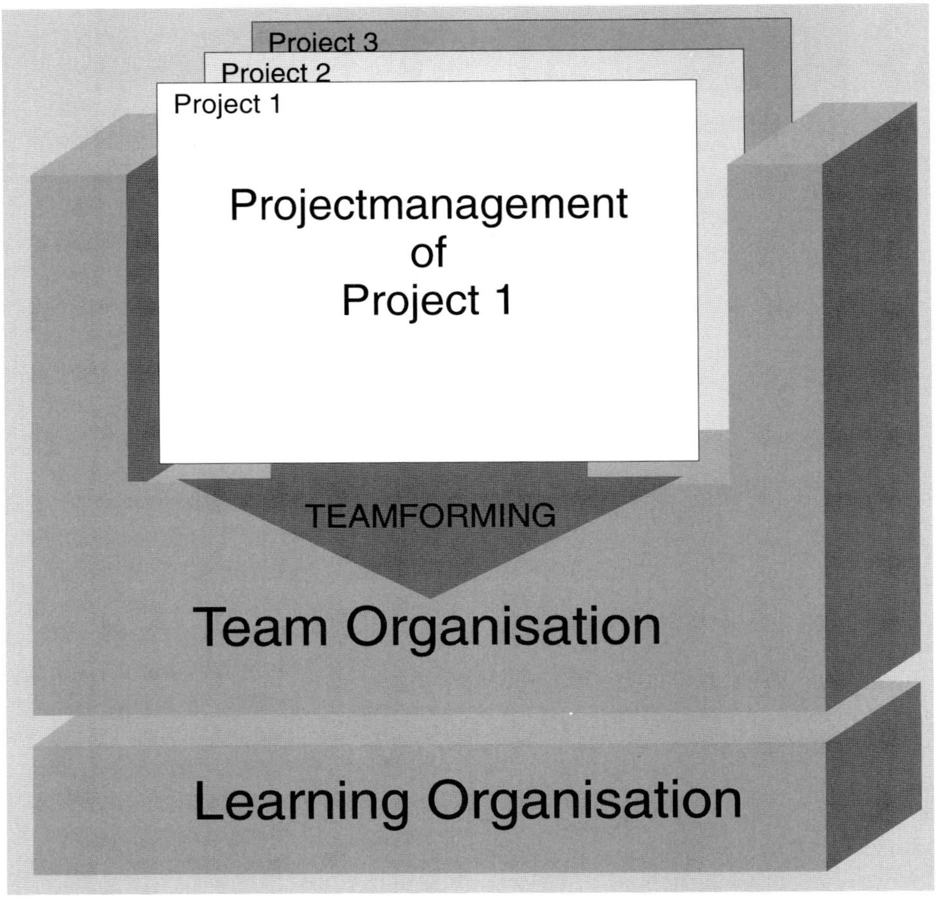

Fig. 3.1: Elements of a CSE organisation

From investigating a single project an organisation for multi-project-management is formed

The next step is the transformation of the above issues to several projects in a company. This is possible since all R&D-tasks from change request developments, derivative developments to the development of breakthrough products, can be carried out in the organisational frame of projects. The difficulty is to manage the project-teams and especially to form them. Therefore, the authors recommend the application of the organisational approach described in this chapter. If this approach is consequently applied, a Team Organisation will emerge that is able to learn and to improve itself. The described approach represents a platform for the specific and individual application in a company.

3.1 Required Characteristics of a CSE Organisation

An organisation that applies CSE has to meet three basic characteristics:

1. Projects are the focus of intention in a company

Process orientation in R&D by focussing on projects

The focus on projects is the consequent implementation of process orientation in the R&D. Accordingly, the orientation on projects is the possibility of achieving a higher degree of integration.

Definition of the term "project"

A project is a unique operation with a complex structure and a determined objective. In addition, it is marked by a fixed final date, by a cost limit and by limited human resources. In the CONSENS-project, a project is defined as the systematic selection and link of autonomous processes and the planning, steering and control of the input and output, as well as of the resources. The structure of a project depends on the prevailing conditions and tasks to be solved. A derivate development requires a different structure than a new product development with external cooperation partners.

Objective of the traditional project-management

To manage a product development process project-management is the appropriate approach. The basic objective of traditional project-management of former years was to get order in a (creative) chaos. Formalised techniques and mature rules were used for achieving this order. Both ensured the ability for repetition and for determination within the project.

The product development process as a solution process in the bounds of an integrated point of view

From a more theoretical point of view [3] the product development process can be seen as a solution process. In order to explain the behaviour of a solution process, the existence of components of problems, of potentials, and of targets is assumed. In an integrated point of view, the solution process is the step by step break down of components of problems and the simultaneous building up of components of potentials. The result is that problems and potentials adopt different states in a specific situation. Thus, they are subject to a dynamic that depends on process. However, three conflicts arise comparing the traditional and systemtheoretical approach:

- The chaos has positive effects on the emerging of order but it should be minimised already at the beginning of a project.
- Processes cannot be determined but should be deterministically controlled.
- Targets are vague in the beginning of a project but should be well known already in the beginning.

As a consequence, a few requirements have to be regarded in order to overcome and control these conflicts:

Requirements in order to overcome these conflicts

- The planned targets within the project have to become softer, more flexible, and more open but not the overall target and the prevailing conditions.
- The limits of the ability to plan must be obvious and taken into account while carrying out the project.
- Planning and executing functions do not have to be separated.
- Plans are developing themselves with the progress of the project and have to be permanently changed and adapted in accordance with the progress.
- The project-team has to take over the controlling tasks and has to manage the project processes in the form of self-organisation.

New kind of project-management

Bringing the traditional project-management approach together with this theoretical approach, a new kind of project-management will emerge. The main objective of applying the new project-management is to allow the flexibility and the creativity, that create changes.

This atmosphere is generating a creative chaos. Simultaneously to this creative chaos, clear targets and limits, as e.g. targets concerning costs, time, and results have to be kept. Thus, a certain degree of necessary order can be enabled.

2. Clear definition of organisational interfaces

Optimal interfaces

A clear definition of organisational interfaces is unavoidable in order to achieve an efficient operation within the project or R&D-department. Interfaces are optimal if they are stable to the evironment even in the case of internal variations e.g. in the case of technological changes of the product during its development i.e. that the technology of a solution is changing and the interfaces remain the same. They must be seen as an opportunity to form the company and its structure. A clear definition of sensible interfaces is an important prerequisite for standardisation. Accordingly, a much faster interdisciplinary communication is possible since the long lasting search for the right communication partner is not necessary any more.

Interfaces as opportunity to form the company

Clearly defined interfaces should have at least two results. On the one hand, easier communication processes in the company and, on the other hand, a definite handing-over of results. Thus, parallel working can be achieved, and easier decision making in a team.

Practical experience proved that the clear definition of interfaces mostly leads to a reduction of interfaces since the existing processes are examined quite exactly. In addition to this, interfaces can be reduced by a strict process orientation and by the integration of all functions in teams that are related to the product development process.

Reduction of interfaces due to a clear definition

3. Flexibility

In the bounds of this book, the term flexibility relates to the characteristics of projects and is synonymous with the capability to react to changing conditions. The adaptation of the organisation to changing prevailing conditions has to be understood in these terms. Two cases of reactions can be differentiated:

Flexibility means capability to react on changing conditions

- reactions concerning the target of a project: This case is not treated here.
- reactions concerning the organisation: The degree of extending this case can vary, from replacing single team-members because of several reasons, to the new structuring of a project already in its course.

In the second case, two aspects have to be taken into account:

Two aspects of reactions concerning the organisation

- The project's specific prevailing conditions have to be taken into account, i.e. for example that the right project structure has to be installed in the right way.
- The changing conditions and disruptive factors in the course of a project must be handled, i.e. that the organisation has to be changed in an appropriate way to enable a successful result from the project.

Individual forming of teams and team-structures for achieving flexibility

The required flexibility can be achieved with an individual forming of teams and team structures. Nevertheless, a certain degree of standardisation is recommended in order to achieve an efficient process.

In addition to these three characteristics a future oriented R&D has to meet several conditions in order to ensure an efficient product development process. These conditions are:

Conditions of a future oriented R&D

- that the R&D must allow optimal processes due to the increasing competition and that these processes have to be adapted permanently,
- the division of labour and the capability to take decisions in order to achieve the required flexibilty mentioned above,
- the reduction and shift of hierarchy,
- the general improvement of the information flow.

3.2 Project-management in a CSE Organisation

It was mentioned above that in the Team organisation projects are the focus of intention. As a consequence of this project orientation project-management as it is understood in this book is introduced in this section.

Project-manager and project-team as main pillars of project-management

3.2.1 The Project-Team and its Characteristics

Beside the project-manager the project-team is the main pillar of project-management. The project-team is a temporary institution that ensures the coordination over organisational interfaces and it should focus all the capacities and resources that are allocated to the project towards the achievement of the target. The team is formed by specialists who are required to fulfil the project. The interdisciplinarity of the teams supports the integration and the flexibility of the project, since different points of views for problems are represented in the team.

Project-team as temporary institution

The task of the project-team-members is to coordinate the work in other different cross-functional teams ([4] p. 84 f.). Their responsibilites can be divided into two parts. The team accountabilities – i.e. those when representing the functional or other engineering team and the functional accountabilities ([2] p. 209 f.), ([4] p. 84 f.):

Team hat accountabilities and functional hat accountabilities as responsibilities of teams

- Functional accountabilities:
  - ensuring functional expertise on the project,
  - representing functional perspective on the project,
  - ensuring that sub-objectives are met that depend on their function,
  - ensuring that functional issues impacting the team are raised proactively within the team.
- Team accountabilities:
  - sharing responsibility for team results,
  - reconstituting tasks and content,
  - establishing reporting and other organisational relationships,
  - participating in monitoring and improving team performance,

- sharing the responsibility for ensuring effective team processes,
- examining issues from an executive point of view,
- understanding, recognising, and responsibly challenging the boundaries of the project ([2] p. 207 f.).

Characteristics of project-team -memebers

The project-team-members must be outstanding representatives of their function or the team where they work. They should have the following characteristics:

- willingness for cooperative work,
- intellectual mobility,
- steadfastness,
- willingness to improvise,
- willingness to take risks,
- ability to generate ideas,
- power of endurance and reliability,
- helpfulness and consideration,
- Knowledge in Project and Time Management,
- Expertise in Presentation and Moderation Techniques.

Acceptance of each other and concentration on the task

"All key participants must have strong personal incentives (...)" ([6] p. 116) in the team and the project. Therefore they must accept each other and be able to concentrate on the task ([2] p. 207 ff.).

The empowerment of teams is a necessary condition but it is sufficient that the combination of abilities collaborate. Scholars state the following characteristics to be outstanding for the performance of the project-teams ([4] p. 61 ff.), ([5] p. 42 ff.), ([6] p. 85 ff.):

Three characteristics influencing the team performance

- communication,
- effective coordination,
- decision making with consensus.

Communication

Communication is a prerequisite "(...) enabling rapid execution and eliminating mistakes all too common in product development" ([4] p 62). The core-team-approach which will be described in the section 3.2.4 inherently hinders any information to be filtered up and down the hierarchy. Also, it avoids mistakes in interpretation by bringing together the deciding team-members in one place and it helps to minimise the cases in which people do not know that they have to communicate. But in addition to the advantages of the core-team-approach, the team-members are obliged to communicate with each other.

Coordination

Coordination means ([4] p. 66 f.):
- knowing the importance of each task,
- knowing who is in charge of which task,
- knowing interdependencies between tasks.

Polemically stated, the project-team supports the project coordination better than technical methods like e.g PERT. It minimizes delays due to unknown interdependencies and hinders activities to be taken out of sequence. How the project-team performs the task of coordinating the working functional teams is presented in chapter 5.3.1.

Decision-making with consensus

Another aspect is effective decision-making. Team-members know better what and when to decide on which issue ([4] p. 67) and they integrate different points of view. Decisions which need to be timely are better made in the empowered team without excessive involvement of hierarchy. To have a fast decision is good but it is better if a decision is made by consensus. Only consensual decisions in the main points will be implemented with the same spirit in all functional teams. Consensus means that each member of the group can honestly agree to the following three statements:

What is consensus?

- "I believe that you understand my point of view,
- I believe that I understand your point of view,
- whether or not I prefer this decision, I will support it because it was arrived at in an open and fair manner." ([5] p. 42)

How to achieve consensus?

People must learn to work together achieving consensus. If they do not achieve it they feel frustrated and say, the meeting was a waste of time ([5] p. 42 f.). Achieving consensus is done by exchange of ideas, definition of a common vocabulary and production of shared meanings, assumptions and priorities. Just to develop the consensus moving team-members to the same location is most important. A collated team has more opportunities to interact and to collaborate ([5] p. 41) and to find its spirit.

Of course these items are supported to a reasonable extent by the project-team approach itself but must also be understood as a commitment of the team to their own way of working together.

3.2.2 The Project-Manager and his Role

Orientation, concentration, and reflection are tasks of the project-manager

His challenging tasks are the orientation in a complex environment besides the concentration on the essential. Accordingly, he has to reflect on the existing situation and the suitable resulting reactions. The project-team should be "(...) headed by a strong and talented leader (...)" ([6] p. 115) and he should:

Characteristics of the project-manager

- drive the team,
- be responsible for time, cost and quality,
- behave more like a team captain,
- have good interpersonal skills,
- have entrepreneurial foresight.

The team-leader as concept champion

The project-team leader is not a primus inter pares. With important projects, where much money and effort is spent, an accountable and even outstanding person is necessary. This person manages, leads and evaluates other members of the project-team which have to report to him throughout the projects duration. The team leader must be concept champion. He does not manage resources, time and cost only. The team leader plays a central role in the project by ensuring the system integrity of the final product or process, therefore he is:

- direct market interpreter – gathers firsthand information from customers, distributor visits, industry shows; has own marketing budget, market study team, direct contact and discussions with customers,
- multilingual translator – fluency in language of customers, engineers, marketers, stylists; translator between customer experience/requirements and engineering specifications,
- "direct" engineering manager – orchestra conductor, evangelist of conceptual integrity and coordinator of component development; direct eye-to-eye discussions with working-level engineers; shows up in drafting room, looks over engineers' shoulders,
- programme manager "in motion" – out of the office, not

too many meetings, not too much paperwork, face-to-face communication, conflict resolution manager,
- concept infuser – concept guardian, confronts conflicts, not only reacts but implements own philosophy; ultimate decision maker, coordination of details and creation of harmony. ([2] p. 209)

Within the project, he has to act as a coach and minimise or eliminate the frictional losses mentioned above. Outside the project he represents an entrepreneur. A project-manager is the stakeholder of the team and the project in one person. Summarising, he is responsible for the integration of the team and leader of the project.

Inside the project as a coach, outside as an entrepreneur

Furthermore, the project-manager takes over personnel responsibility. In the bounds of this responsibility he has the following tasks:
- he is leading employee,
- he is assessing employee.

But his task is not to arrange contracts for employment or to decide on the salary for the project-team-members.

3.2.3 Rules of Teamwork

It was already discussed in previous sections that CSE depends on teamwork. It is not the objective of this book to provide a comprehensive decription of teamwork. Nevertheless, because of the importance of teamwork to CSE two basic classes of rules for successful teamwork are described in this section.

CSE depends on teamwork

Rules Related to the Persons

These rules are concerned with personal aspects. The teams should have the optimal number of members. Experience has proved that five to seven members is most efficient. The different roles in the team like the transcript writer are defined in principle or depending on the situation, but never by accident.

Rules related to personal aspects

Illustrating this class of rules a practical example of a German company belonging to the electronical industry is provided. This example shows how this company lives and realises rules related to persons. The managers and employees identified eight rules as important:

1. Creating confidence

Confidence

Confidence is indispensable. Confidence can be created by a fair behaviour in crucial situations and by the maintaing of the stated views.

2. Demanding criticism

Criticism

The acceptance and active demanding of criticism of one's own behaviour creates a constructive and open work atmosphere. This means in concrete terms that every employee should be able to question himself.

3. Ability for conflicts and constructive compromises

Coping with conflicts

It is not possible to rule out conflicts in an organisation, with teamwork in advance. Actually, one can say that in a flexible organisation such as one which is developed in this chapter, conflicts can occur more often than in rigid structures. The objective is to improve the willingness and capability for coping with conflicts. The existence of conflicts must be seen as a source for a further development of problem formulations.

4. Offering and creating a sense of belonging to a team

Sense of belonging

The sense of belonging to a team of a team-member can be reinforced by an understanding of personal views and the required acceptance therefore. It is the task of the project-manager to support and demand this understanding within his team.

5. Support for experimenting and learning from mistakes

Never make a mistake twice

Creative solutions can only be found if strict rules for behaviour are avoided. The courage to follow new approaches must be rewarded. Doing this, mistakes can be made but these mistakes must be understood as a basis for the development of the aspired solutions. However, the objective is to avoid making a mistake twice and should not be aimless.

6. Supporting one's own initiative

Delegation of responsibility

Initiative of one's own can be supported by the delegation of not just the tasks to be solved but by delegating the responsibility and competence to decide as well.

7. Ensuring clear tasks are set

Clear targets

Effective projectwork can only be achieved if clear targets for the particular tasks are set. This applies to the whole project as well as to work-packages. The setting of the targets should be done in a dialogue between the participating employees, e.g. in a management by objective process.

8. Transparency of decisions

Transparency

The reasons for decisions taken in a team or outside a team must disclosed. Decisions can only be implemented effectively and quickly if they are seen in a larger context.

Rules Related to the Organisation

Rules related to organisational aspects

At the beginning of each project the project-team meets for the kick-off meeting. In this meeting the structure and first plan of the project are discussed and improved. In the course of the project, regular project meetings are held , e.g. at each milestone and the topical problems and the progress of the projects are discussed. The presence at these meetings is compulsory, i.e. no assistant can be sent. Measures for the cross-functional coordination are taken as well as the adaptation of the plan. At the end of each project, the team meets for an audit in which positive and negative experiences of the projects are collected and assessed (compare section 3.5.2 Project audit).

Meeting-culture

For the process of meetings it is compulsory that the members are prepared, only then decisions can be made. Before a decision is made, alternatives are looked for and assessed. The decision should be based on a consensus and not on a voted majority in the team. In the event that opinions differ from each other, the opponents should not be persuaded but be convinced. The people affected by a decision should be informed early and integrated.

Special equipment

Teams require appropriate meeting rooms and specific tools in these rooms in order to work efficiently. These tools are e.g. flip charts, white boards, overhead projectors or material for moderation and are to be provided.

3.2.4 Possible Organisation Arrangements for R&D

Different types of organsiation for R&D

After having introduced an idea to project-management and the related problem and before a Team Organisation is elaborated in the following section, a short "tour de raison" through R&D organisation models is done now. The project-team or core-team-approach is presented and the question of how to expand that approach to the complete R&D is asked.

The most recently published literature on CSE a typology of organisations is given ([2] p. 190 ff.) and ([4] p. 61 ff.). A summary is provided in the following (see figure 3.2):

- functional team structure,
- lightweight team structure,
- matrix management approach,
- autonomous team structure,
- heavyweight team structure.

Functional approaches work well in companies that don't have strong competition

The most common ways of organising the product development process are up to now functional approaches. They work well in companies in which the time competition is not strong. But the disadvantages become stronger if time becomes tougher. Functional approaches are characterised by a special "throw it over the wall attitude" ([4] p. 69), and a complicated communication, coordination and decision structure defined by the hierarchy. These approaches also support a reward system that goes along the functional goals which tends to produce decisions and work results which are against project necessities ([4] p. 68 ff.)

Programme coordinator approach

In order to overcome some shortcomings of the pure functional organisation, derivates of it have been implemented, for example the programme coordinator approach. A programme coordinator manages schedules and coordinates the activities. But these managers "(...) are rarely involved in any actual management (...)" ([4] p. 75) and have no competence. Thus, this leads to plans without belief and almost no positive effect. In many companies that have implemented project-management, a multiple functional project-manager approach ([4] p. 75) has been established. That means many project-managers for one project and in the end a lack of overall accountability when problems arise. Like the functional organisation itself, these

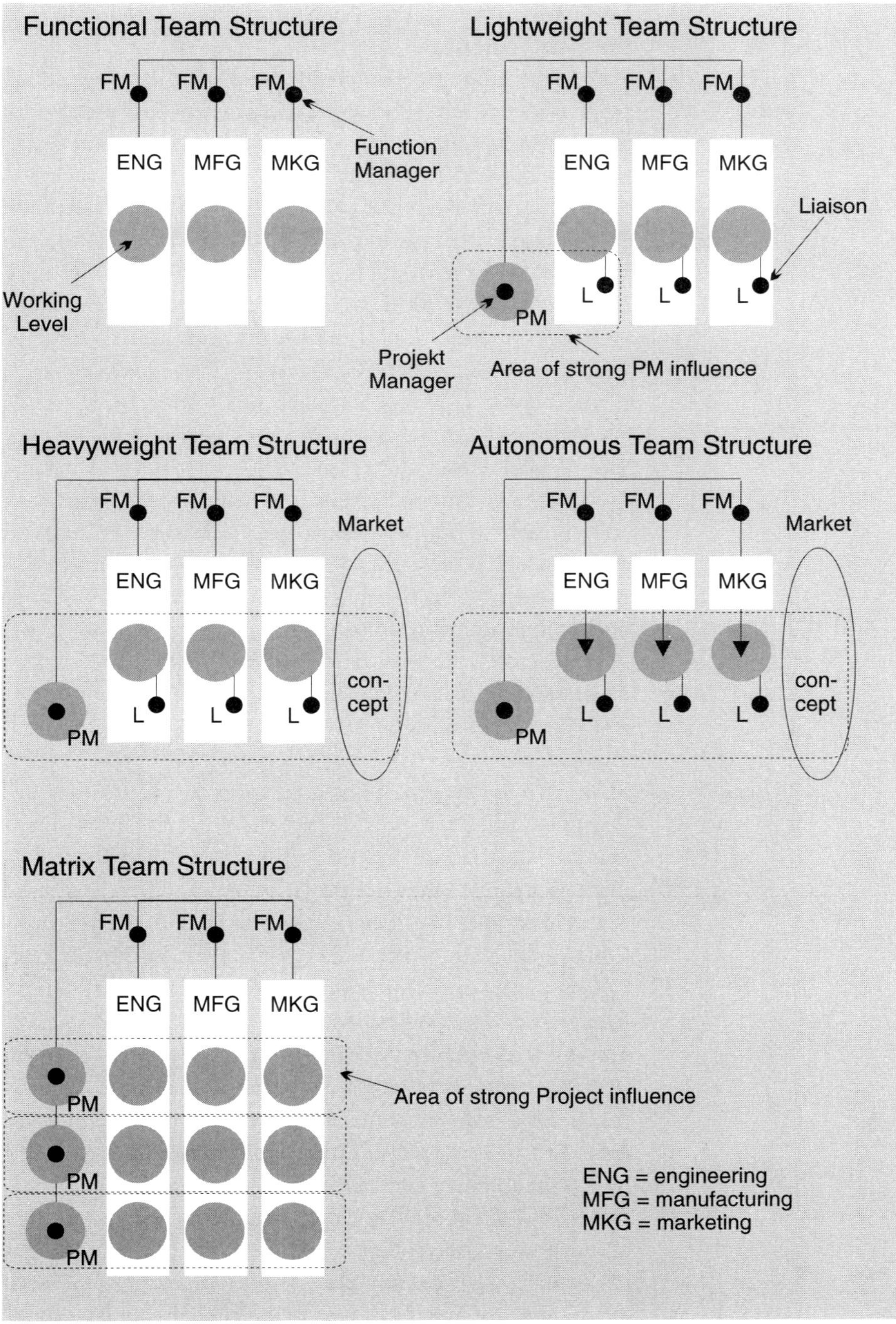

Fig.3.2: Types of development teams ([2] p. 191 and [4])

project-managers tend to optimise their function and not the global optimum.

Lightweight team structure

The lightweight team structure "(...) tends to degenerate when development processes accelerate because much time is spent on communication activities between teams, particularly at hand-off points." ([4] p. 78)

Matrix management approach

The matrix management approach is one of the most fully investigated organisation model. The idea is "(...) to take the best from the functional organisation and combine it with a temporary project focus" ([4] p. 78). As experience shows it "(...) requires a significant amount of executive time to ensure that the project-team is managed fairly" ([4] p. 78). And it "(...) requires clear goals and objectives for each project (...)" ([4] p. 78) and each person must be continuously reviewed. But again, is this approach feasible in a dynamic environment? Project managers must be respected to be able to get their tasks done, as long as they have no direct access to resources. Furthermore, open communication between project-managers and the functional managers is required just to solve those conflicts which are mostly the reason for not communicating openly. So, "matrix management usually fails because an individuals responsibilities and loyalities are in constant conflict" ([4] p. 78) if employees serve two masters.

Autonomous team structure

An extreme approach is the autonomous team structure ([2] p. 191 and [4] p. 80 f.). Here, the product development process is performed by one organisational unit alone. The management is rather simple. Give the team a general sense of direction and a budget, separate them from the organisation and let them work. This approach is feasible for start up companies and new ventures. But there is a total dependence on one project-team. The relatively closed structure may have a negative impact on the implementation of an overall company strategy. An example could be that the team may follow another make-or-buy-strategy and nobody knows it or that autonomous teams tend to duplicate skills and may drift away from the company's core customers and technologies ([4] p. 81).

3.2.5 Core Team – Heavyweight Team Structure

In many cases, the heavyweight team structure or the core-team-approach is the best one ([2] p. ff. and [4]p 82 ff.). "Teams offer improved communication, stronger identification with and commitment to the project, and a focus for cross-functional problem solving" ([2] p. 200).

The heavyweight or core-team-approach is best in many cases

In order to achieve the advantages of this approach the project-team has to be empowered and trusted. Empowerment means to give the team the power to decide on implementation level issues, to give them resources for each phase and clear guidelines by targets while strategic decisions must be made from the management or a product approval committee ([6] p. 115 ff. and [4] p. 90 f.).

Empower the team!

Empowerment has disadvantages. The disadvantages of strong teams are that ownership and commitment may lead to expansion of the teams role and purpose, especially if defined poorly by the senior management. Often it happens that the remaining part of the organisation is viewed as "second class" and a distinction between haves and have-not-teams regarding resources and management attention may arise. Top teams demand top-priority for support by the rest of the organisation which leads to conflicts ([2] p. 202 ff.). The question arises of how to balance the individual project and the needs of the rest of the organisation?

This balance is reached by defining the roles of the team and its members and by introducing clear goal definitions. A project charter and a contract book are instruments to make this happen. Before selecting the team, a project charter expressing a clear mission is decided by the senior management. Each project-team-member must before joining this team obey the idea stated in the charter ([2] p. 204 f.). In order to form the project-team and to control whether it understands the mission defined by the top management, the team is given time to write a so called contract book ([2] p. 205 f.). The time should be as short as possible – one week should be enough. A possible structure of contents of the contract book propose Wheelwright/Clark ([2] p. 205):

Project charter

Contracts books

- Executive Summary,
- Business Plan and Purposes,
- Development Plan: Schedule, Materials, Resources,
- Product design Plan,

- Quality Plan,
- Manufacturing Plan,
- Project Deliverables,
- Performance Measurement and Incentives.

The roles of the core team-members and the project manager have been investigated in previous sections. Further roles necessary to establish a working core team are introduced in the following:

Roles in the Play – Executive Sponsor

The executive sponsor is client and mentor of the project-team. While empowering the project-team and its leader he must "(...) retain the ability to guide the project and its leader (...)" ([2] p. 211). The executive sponsor has "(...) to oversee the team actions so that they stay in line with the company's vision" ([5] p. 41). In order to give clear advice to the team, one vision and one interpretation of the senior managements in communication with other executive staff is done through this person. The executive sponsor has the task of taking control over ([5] p. 40 ff.):

- Resource commitment (...),
- Pricing for major customers and major accounts,
- Potential slips in major milestone dates (...),
- Plans for the transition from development to operating status,
- Thorough reviews at major milestones (...),
- Review of incentives or rewards that have company-wide implications for consistency and equity,
- Cross-project issues such as resource optimisation, prioritisation, and balance." ([2] p. 211 f.)

These issues may serve as boundary issues to limit the project-team.

Roles in the Play – Functional managers

Functional managers are persons which lead functional or other teams which have to be coordinated in the product development process with help from the project-team. They cannot take part in all project-teams, because of missing time. They should focus on functional excellence rather than contributing to day-to-day problems. Their task will move from an operative to a more consultative one ([4] p. 87 f.).

Roles in the Play – Facilitators

Facilitators help the project-team throughout the project runtime. They have process improvement focus and assist during planning, scheduling and coordinating tasks and measure them ([4] p. 86). They service the project-team with knowledge from other projects.

3.2.6 Portfolio of Approaches

The core-team-approach is not feasible for all projects - a portfolio of approaches is necessary

The core-team-approach should be used for top priority projects. It is not feasible for small change requests or product modification projects. But companies have all sizes of product development processes. Therefore, a mix of different approaches or a portfolio may be the solution. But organisations tend to be attracted by a dominant approach. This leads to conflicts if the dominant approach and the one used to perform distinct tasks do not fit. This is the case when in a functional organisation some product development projects shall be managed using a lightweight team approach. While companies that have adopted the core-team approach as their dominant orientation "(...) will find it relativly easy to implement autonomous tiger teams", lightweight and functional approaches as well ([2] p. 215 ff.). Due to the fact that the project-team task is to manage other engineering teams which carry out detailed work, it is easy to give these teams either more freedom or autonomy for smaller tasks. It does involve just "adjustments in the standard approach (...) instead of adding new activities or capabilities that conflict with the established system." ([2] p. 216). But, nevertheless, the question is, how to integrate these different kinds of team-work into one organisational approach. There will arise conflicts if the core-team-approach is dominant and functional departments are in power. Successful project-teams produce self-conscious members which build together a strong informal network. The network and its members are a challenge to the functional departments and their managers. If the core-team-approach should be dominant and implemented by adapting training efforts, communication structures and patterns of leadership ([2] p. 216), what should be done with the former functional departments? Or formulated positively – how might such a new organisation look? The following model may give an answer to this question.

How to integrate different kinds of team work into one organsiational apporach?

3.3 Team Organisation

Teamwork as basis of a Team Organsiation

The basis of the Team Organisation is team-work. People working in teams on the most challenging tasks show the ability to stand on their own feet. A self-controlled team is an intact group of employees who are responsible for a whole work process and who supply the result of their work or their service to an internal or external customer. According to their function, the members of the team are working in order to improve their performance as well, to solve day-to-day problems, plan their work and monitor the results ([7] p. 23).

3.3.1 The Concept of the Team Organisation

Team organisation is made out of teams

A team organisation consists of self-controlled teams which have the right to establish further teams. Each team has a name which should characterise the task of the team. The project charter and contract book contain the team definition (compare chapter 3.3.2). Both replace job descriptions. Here the aim of the team must be stated, its competences, resources, budget and the way of doing business (or relevant process definitions) and the performance evaluation criteria and the way of evaluation. The list of team-members contains information on persons and their position or role in the team. The Team Organisation consists of two kinds of teams: temporary and permanent teams.

Temporary teams

Temporary teams are defined to perform a task which is unique e.g. a special product development project or a task-force to solve a single problem. These teams will be terminated after finishing their work and achievement of the targets. If these teams need additional resources, they must negotiate with the senior management. Typically, such teams report to the establishing teams and have to deliver the result to them.

Permanent teams

Permanent teams perform tasks which have to be done more than once in the same manner for a long time (e.g. documentation, modification projects). These teams are established by the senior management of the company. They report to and have relations with those teams as stated in their team definition.

Each employee in the company has the right to enter teams. Each person should stay in the teams he entered until the task has been finished or the team has released him from being a member.

Joining teams

The team as a whole is evaluated by comparing their actual achieved results with those defined in the project plan. The team-members should be evaluated in the team only. This may be done by giving the team the competence to divide the incentives or rewards for achieving the goals. In case of differences, the management or coaches may help to solve the conflict. If team-members think that the team is too slow, not motivated or does not understand them, they may try to direct their main effort on other tasks. On the other hand people with a bad reputation will not become members of interesting teams. The tasks of the classical human resource management concerning personnel management and the career planning will shift to a coach-like one.

Evaluation of the team and evaluation of team-members

Teams have a manager. This person is either elected from the team-members or appointed by an authority outside the team. This depends on the importance of the task and is part of the team definition.

Teams with an elected or appointed leader

The important information about the teams can be expressed by defining team objects with attributes, this is shown in the figure 3.3.

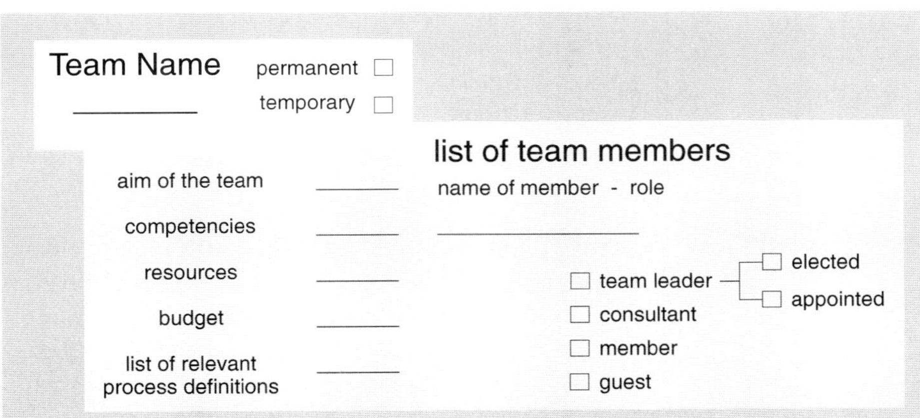

Fig. 3.3: Team objects with attributes

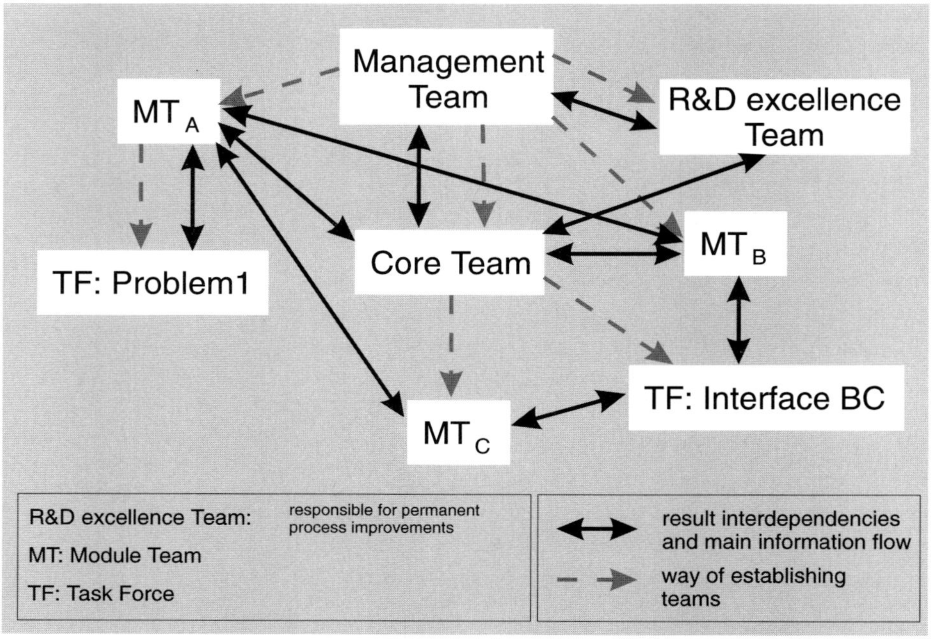

Fig. 3.4: Visualisation of a team structure for a R&D-team organisation

Network of teams

The definition of each team expresses the importance and the position of the team in the company. Together with dependencies between teams which express the establishment and termination of teams, a non-hierarchical networklike organisation is constructed. The team structure itself states (see figure 3.4):

Team structure

- number of teams,
- persons in the teams,
- rough dependencies between the teams:
  - appointment,
  - result interdependencies,
  - reporting.

Personal structure

Another important aspect is the personal structure. Here is documented which tasks in which teams are performed by each person compare for this figure 3.5.

Name	Abbreviation of Name	Team Members in	Function
Jarson	J	Module Team A Task Force II Core Team	member member member
Lang	L	R+D excellence Module Team A FEM-Team Task Force II	leader member member member
Miller	M	Management Team	CEO
Norton	N	R+D excellence Module Team B	member member
O´Ryan	O	Management Team R+D excellence	member member
Pascal	P	Module Team A	leader
Victor	V	R+D excellence FEM-Team Module Team C	member leader member

Fig. 3.5: Personal structure for a R&D-team organisation

3.3.2 Open Questions Arising with the Team Organisation

What can be done with functional managers?

If the functional departments are disbanded, the former functional managers can serve as consultants to the teams and to the senior management. To make it clear, they should work like facilitators. Functional managers will consult over questions of functional excellence, the engineering teams, – the former functional departments – and the senior management too. Maybe they are co-opted in most teams and their task is to implement new methods. It is like today, where young and less experienced engineers tend to consult the outstanding engineer or functional manager in case of problems. Of course, method training with e.g. work shops will remain. It is desirable to have the former functional managers also working on normal engineering tasks. They would know about problems and would keep in touch with the basis.

How to save the knowledge of teams?

Teams or persons will work together not only once. But if people work together in temporary units, they have the problem of saving knowledge. The task for today is to define and to take care of the process knowledge. Permanent teams establish their process definitions. Project-teams and task forces evaluate their solution and the selected process and are obliged to document. Documented knowledge must be evaluated by facilitators if it describes repeated processes and filled in standard procedures. The interpersonal skills grow and team working abilities become a core competence in the company. (compare section 3.5 Organisational Learning in a Team Organisation)

How may we manage and coordinate the resources?

The distribution of tasks was the main job of the functional manager. There is only one, maybe a few, top priority tasks and a huge number of smaller tasks. The project-team manages deals with the top priority project and tries to get the resources needed from the engineering teams. The conflict in distributing resources between the top priority and the less important projects remains. Today, the conflict is between the functional manager and the project-team leader. Within a Team Organisation, the conflict will be between project-team leader and the representatives of different engineering teams. The conflict remains but will be solved faster and better than in a functional organisation due to the fact that resources and teams can be more easily shifted. Negotiations to solve the problems are the job of the involved team managers. If they fail to find a solution, a management decision is necessary. All this can be done on the basis of the Engineering Process Manager (EPM) method (Compare section 5.3.1 Engineering Process Manager).

Are engineers or members of teams fixed to one team?

No, each engineer may engage in a number of teams. He is responsible for himself and has to manage his resource. In each team he brings a defined amount of man power and the team decides what to do with this portion. In doing so, the engineer is not fixed to one team or department. He may search for his optimal role in the company on his own, supported by a coach or mentor. The team decides whether to employ him or not. This may depend on his personal reputation. Engineers will try to enter teams with a high probability of success and big reward, but wouldn't be found in cases where a difficult task is to be done, or a task where nobody can win. How would one get a person to work on it? In cases of a permanent team with an management-appointed leader there is not a problem. In case of a temporary, the task for the senior management is to deploy enough resources and to present interesting work environments for such "horrible" tasks.

Can people work on different project with the same personal commitment?

No, but personal commitment is important for high risk tasks only, while it needs not to be high for teams which handle day-to-day problems. Of course, each person should be aware of that problem and should try to concentrate on its challenge. I think people will learn from it.

Who evaluates the engineer and steers/controls careers?

The team – and the internal client of the task – evaluates the engineer. Part of the team definition are goals and rewards in case of success. This is negotiated between client and the team. The distribution of rewards inside the team depends on the team itself. Many options occur which are well known from private life. People disagree, are unfair to each other , do not want to work but want to get the reward. But reputation of teams and persons will be known in the company and people will react.

Who organises the equipment?

The method engineering team with the consultant on top organises the equipment investment. They know best what they need. Different process teams will have good ideas to improve the productivity. In the end it will result in budget negotiations like those we know today.

3.3.3 Advantages of a Team Organisation

Expansion of the idea of the core-team-approach to the rest of R&D

We stated that it is possible to expand the team idea of the core-team-approach to the whole R&D organisation. Here we want to summarize the advantages of a Team Organisation. We can distinguish three main arguments concerning the:

- dynamic structure,
- dynamic information processing,
- personal management.

3.3.3.1 Dynamic Structure

A dynamic organisation is an ever lasting compromise but the best what we can have

Teams make up a dynamic organisation. That means to implement a dynamic perspective of organisation rather than a static one. A static organisation perspective means to find an optimal solution for the distribution of work and one of the best ways of installing information and reporting channels. But this perspective seems not to be adequate in a fast changing environment. A dynamic organisation is from the static aspect an ever-lasting compromise without one consistent principle. But in a complex business, this required optimum is changing so fast that the organisation cannot be adapted to it. So the consequence is to regard the organisation as being permanently being changed to some extent. The Team Organisation makes it easy by defining new teams, task forces, by re-allocation of resources, by changing project priorities to adapt to new situations more easily. And this is done with the help of two methods:

- organisation done by the management and
- self-organisation by teams and their members.

Organisation done by the managment

Organisation is the task of the management. The idea is to define a global set of rules for teams to guide future teamwork, to establish permanent teams and a couple of important temporary teams, to give them clear aims and responsibilities. They also define performance goals and distribute rewards among the teams. They serve also as conflict solvers. And this game leaves room for self-organisation. Self-organisation is the action within distinct rules, to manage and process the organisational issues of an enterprise by employees and teams which have no genuine leadership responsibility. The teams will start to organise themselves and employees will try to find the best position according to their abilities.

Self-organisation

Of course self-organisation will lead to conflict, we suggest resource conflicts mostly. But most of the resource conflicts can be solved by simply using the EPM method together with clear priorities of the management. Teams decide to postpone the one or other process, to run down not needed teams or stop processes. Important jobs, of course, must have management attention – this was the task of the executive sponsor, all other tasks can be assessed by employees – trust them! An advantage is that teams can be changed more easily than departments, because heads of departments have a strong stake in their department and try to hinder people being deployed in other units.

Change of principles in order to solve problems

If conflicts remain, the task of the management is to change the principles in order to solve them. The principles need to be company-specific. Due to the fact that the senior management must remain accountable for the company to the outside, they hold the ultimate authority to issue directives. They have the opportunity to give special competences to distinct teams. Therefore, the organisation keeps the ability to manage a possible crisis effectively.

3.3.3.2 Dynamic Information Processing

Information in a Team Organisation is processed parallel and teams are given the power to decide

The way of information processing and decision making in hierarchies has two main problems. In hierarchies information is filtered walking up and down and the assumption that the superior will come to a better decision than the guys below is wrong if the time is too short to collect all necessary information ([8] p. 66). The Team Organisation does not have these problems.

Better information processing performance and faster reactions

Information in a Team Organisation is processed parallel and teams are given the power to decide. This leads to a better information processing performance and faster reactions ([8] p. 67). Clear strategic missions defined by the senior management are the frame for many decisions made by teams and employees in parallel. If more people immediately take part in discussions on the effect of changing conditions and start to change their business after getting a clear mission, the actions in the whole organisation will be better rectified. The network which is a non-hierarchical self-organising type of organisation ([8] p. 67) adapts itself to changing conditions. Important is that the relations in the network are not pre-defined. The openness enables learning from experience. Instead of waiting for the management to teach and change the organisation, this is done in a self-organised way. The duration of decision-making may be longer than in a classical organisation but the decision is immediately known in the relevant parts of the organisation without interpretation bias and with a bigger consensus. This leads to a faster and more consequent implementation of the decision.This parallel processing relieves the senior management and gives them time for strategic thinking.

Once more, – people are responsible for their jobs, trust them!

3.3.3.3 Human Resource Management

A more individual employment of employees

Last but not least, the Team Organisation allows the workplaces and career options of employees to multiply. People can be employed in a more individual way and that this is done is the responsibility of each of the individuals. Employees are self-responsible. They organise their work-environment.

The newly defined status of employees

The status of a person is related to their role in different teams, the number of teams in which he works and the kind of teams in which he serves. It is not so easy any more to say – this person is more important, has a greater status than that person. It means the true integration of expert-, managerial-, and project career. This fact gives rise to the development of the personality of employees which is the most valuable capital of a company. The actual situation is as follows: the functional career is due to the fact that the functions have the most power, most desired independently from the qualification of the job-aspirants. Additionally, career-bottlenecks occur often. The opportunities to make such a career are not enough ([9] p. 160 f.). That means as long as there is a predominant career model, this problem cannot be solved and people not having the right attributes will either become bad leaders or frustrated. To dislink the status and hierarchical position is a precondition of organising the R&D along the project requirements and also by taking into account the individual.

Managers and employees as entrepreneurs

This kind of organisation is shocking for managers and the employees too! They are more responsible for their job, career and outcome than they were before. The organisation pushes management and employees into the role of entrepreneurs. Summarising, conditions for the Team Organisation are:

Conditions for a Team Organisation

- clear and enterprise success-oriented rules of the game,
- an open information policy,
- trust the employees.

3.4 Incentive & Motivation in a Team Organisation

Reduction of the importance of hierarchy

One condition of a future oriented R&D in terms of a Team Organisation is the reduction of the importance of hierarchy. As a result of this, different incentive and motivation systems are required since the position in the classical organisation plan is not appropriate to assign the payment and the status anymore. One approach to motivate the employees in a Team Organisation is described in this section and is related to the issues career, planning of the career, assessment of managers and employees, and payment.

Career

New understanding of career

A new understanding of career and its relevance is necessary in a Team Organisation. The objective for employees is now to develop competencies [10] and not to climb up the organisational ladder. Development of competencies is related to

- special competencies on a subject,
- methodological competencies, e.g. reducing complexity in the product development process or handle occurring conflicts,
- social competence, i.e. skills for communication and cooperation,
- competence of personality which is e.g. creativity, the behaviour of somebody or his values which influence a project.

Career is life-long learning

Consequently, career means life-long learning and involves taking over new tasks in changing teams and to break permanently new ground within projects. It is a permanent process of gaining experience and of experimenting. One has to be willing to take over further responsibility and to develop personal skills.

Planning of the Career

Career planning in accordance with planning of life

Planning the career means the individual planning of the development of an employee in accordance with the planning of his life. Planning of life contents the targets of the life, e.g. a break and the later come-back to work. Each employee is responsible for his own life-long learning but with the support of his company. Responsibility for specific subjects and for leadership are equal in a Team Organisation.

The concept of career planning is divided into the three sections training on the job, coach approach, and flexible working time:

Training on the job

- Training on the job, i.e. the purposeful take-over of several tasks in projects as a continuous process. The tasks increase systematically in their difficulty, complexity, and extent of responsibilty. Thus, it can be avoided that the project members will be burned out but that the employees can develop their competencies.

Managers as coaches

- The managers as coaches for the planning and development of competencies. The objective is change from leading to professional competencies and the other way round. However, this leading to an aimless job-hopping has to be avoided.

Flexible working time

- Flexible working time, i.e. a working hours scheme for weekly, monthly, yearly, and lifetime working time is required. This enables an individual planning of the working time and its coordination with the life planning, e.g. to take off for a half year after finishing a project and to come back to the company after this time off.

Competencies as platform of the career

The basis of the approach are the competencies of each employee. These competencies have to be continuously developed appropriate to the existing task. The different competencies are summarised in a circle which is shown in figure 3.6.

The whole circle is divided into three sub-circles and every subcircle into several elements. Each element represents a specific competence. The inner circle represents the specific competencies on a subject, the middle circle the methodological competencies and the exterior circle leading & managing competencies.

Application of the competencies in several fields

The competencies have now to be applied in several working fields. Working fields can be related to different species of projects. For example the derivative development project of product group 1 or the breakthrough develop-

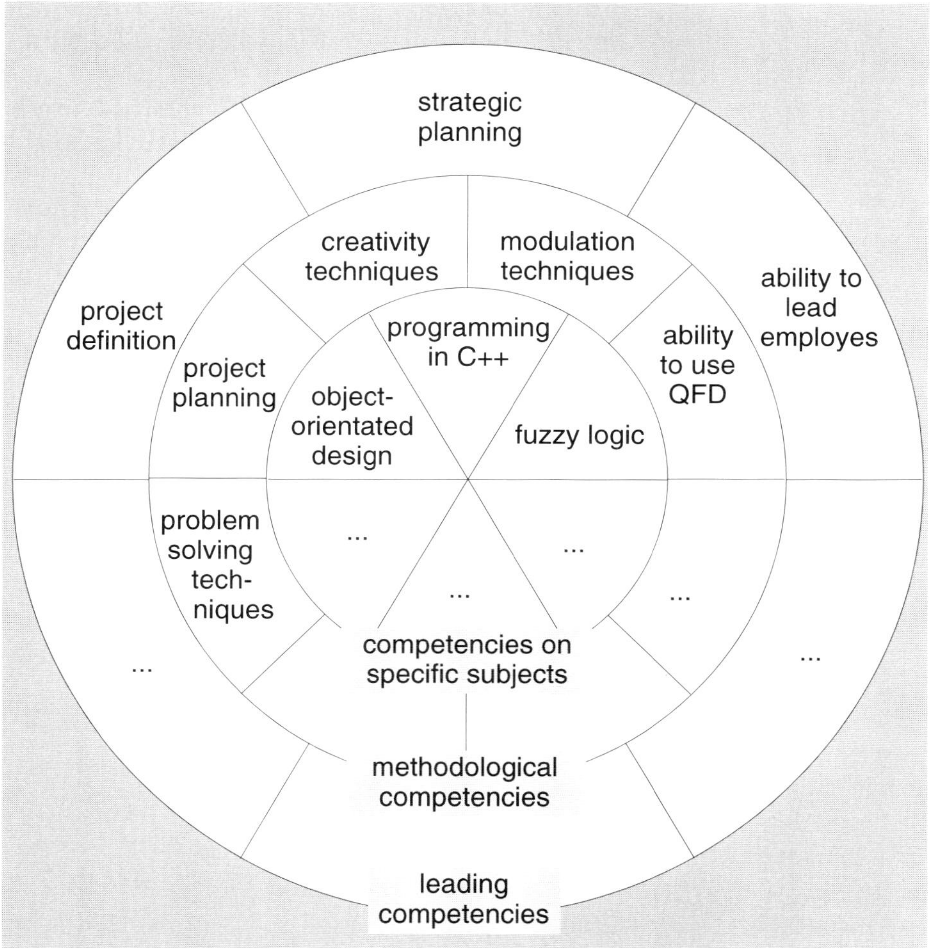

Fig. 3.6: Circle of competencies in a company belonging to electronical branch

ment project in product group 2. These working fields are represented as several circles in the figure 3.7. The circle as a whole represents the sum of required competencies for fulfilling the target. In the course of a career, several fields are passed through. With every change new competencies should be developed. The objective is that the employees gain more and more competencies. Two developments are possible:

Management career

- To move systematically from the inner circle to the exterior circles and take over step by step the functions of project member, an assistant project-manager, a project-manager, and of a senior manager. This represents the classical career,

Career on a specific subject

- or to gain specific competencies on a subject and move in the figure in a horizontal way instead of a vertical way.

Both possibilities are absolutely equal. The sort and forming of the functions on the vertical axis depends on the existing structure of the Team Organisation and is specific for each company.

The described approach is illustrated in the figure 3.7 career portfolio.

An example of a career

An example of a career is given on the base of a software engineer. He starts his career as project member in the inner circle and develops competencies in programming in "C + +" and "object-oriented-design" in product group 1.

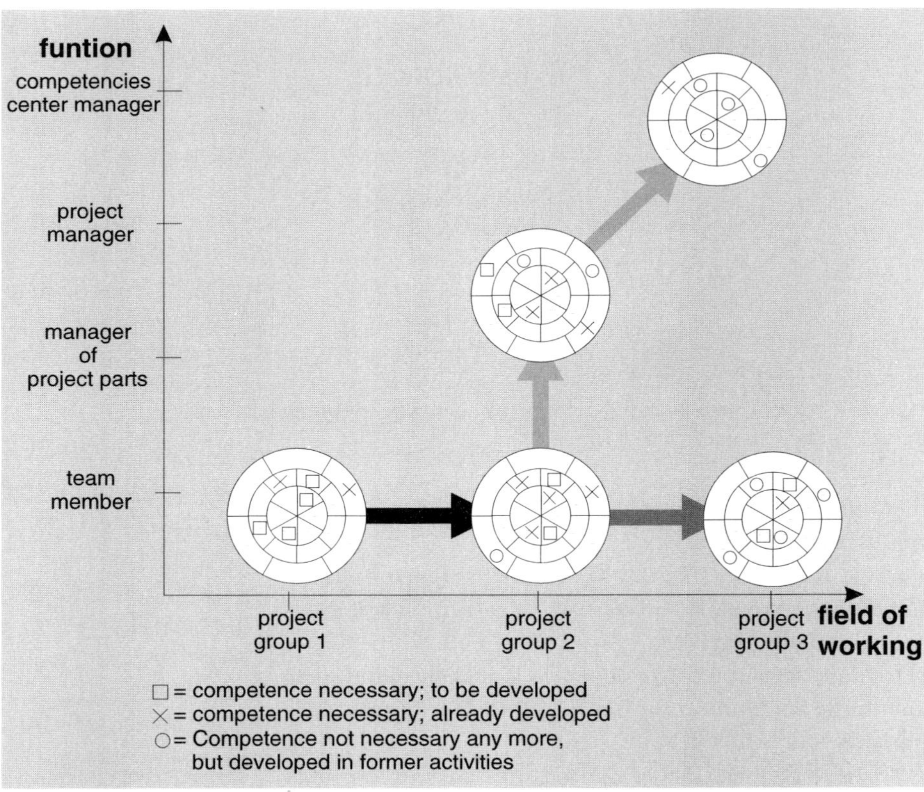

Fig. 3.7: Career portfolio

After a certain period of time he will change into the product group 2 and will develop the additional competence in "fuzzy logic & neuronal networks". The competencies from the first product group are partly still used. The next change could be a vertical one back to the product group 1 as assistant project-manager or a horizontal one to product group 3 in the same circle of competence. In both cases he needs new competencies for fulfilling the job. For the vertical change, some of these competencies are in the middle such as the usage of QFD or moderation and problem solving techniques. Taking over tasks as project-manager or senior manager requires different competencies than the developing of the career in a specific subject. In both cases, these competencies will be consequently developed.

Assessment of Managers and Employees

The success of a team is decisive for the assessment of its members

Employees are assessed by the success of the team-results. The assessment is done by different persons within the company, e.g. by the project-manager or by a functional manager. It takes on the one hand the development of competencies i.e. the achievement of objectives which are individually agreed and the quality of the achieved results into account. On the other hand, the work within the team, such as the willingness to cooperate, creativity, and ability for compromises.

Project managers are assessed by their service towards the team-members or employees and the project. The assessment is carried out by superiors who assess the project success achieved by management competencies and additionally by team-members who assess the managing-success achieved by social competencies.

Payment

Break between payment and hierarchical status

The payment of an individual follows the principle that those have a good payment who have made great contributions to the customers, colleagues, and the company. A break between the payment and hierarchical status is to be achieved. The salary is oriented on the success of the work and not by the status of employees. This should not be a levelling down but an equal treatment.

The payment is divided into the components basic payment and flexible success oriented bonusses:

- the basic payment depends on the individual professional, social, and personal degree of competence,
- the flexible success oriented bonuses. The success of projects is measured by the assessement of the project targets concerning the time, performance requirements, costs and their keeping. Additionally, the success on the market of the project is taken into account.

3.5 Organisational Learning in a Team Organisation

What is a Learning Organsiation ?

This section explains what a Learning Organisation in a CSE environment means and how it can be achieved that an organisation learns.

The starting point of the investigation is the Team Organisation derived and described in the previous sections. This Team Organisation is not fixed or static but it is subject to permanent change due to the ongoing temporary projects. However, as one can see in the following, the Team Organisation has got the potential to optimise and develop itself. The objective is that the organisation is able to adapt itself optimally to changing prevailing conditions. This adaptation should not take place as a result of a central order by the management but because of the experience and understanding of the employees. If this state is achieved and can be maintained, the organisation as a whole is learning.

Organisational Learning is based on individual learning which can be defined as in the follows:

> Learning is the reception of information and the derivation of insights out of them concerning their own thinking and behaviour[11].

3.5.1 Characteristics of a Learning Organisation

All employees participate in change processes

Changes of structure, of strategy, and of culture are resolved in the hierarchical organisation by the senior management. In a Learning Organisation all employees participate in these processes [12]. Exactly, this is understood under the term "Learning Organisation." Normal change management (compare chapter 2.5), is permanently carried out by all employees without any specific initiation. That is the reason why Learning Organisations are more successful at carrying out smaller changes than hierarchical organisations.

The following abilities must be developed in a Learning Organisation in order to gain and to pass on knowledge on the one hand and on the other hand to amend their own behaviour due to new generated knowledge [13]:

Systematical solving of problems

Set of problem solving instruments

The company XEROX introduced a set of instruments in order to solve problems in a systematic way. This set is introduced to the employees in company-wide seminars lasting several days and are compulsory to be used afterwards by every employee. Elements of this set are:

- Generating ideas and collecting information,
- Achieving consensus,
- Analysing and representing data,
- Planning of measures.

Experimenting with new approaches

Understanding of the backgrounds

This involves the deployment of scientific methods with the objectives of gaining new knowledge by means of problems and doing this under appropriate conditions. The result should be that the employees not only know how to do something but that they understand why things happen in specific ways. Two approaches can be identified:

Ongoing test-programmes

Ongoing test-programmes are series of small experiments on which processes of continuous improvement in the production are based. It must be possible to send employees from all levels in the company to scientists and other organisations in order to exchange information and to generate new ideas. Accordingly, abilities to experiment and to analyse the data must exist or have to be developed in seminars.

Demonstration projects

These projects are in most cases more complex and of larger scale than ongoing programmes. Quite often they represent the first project in which new methods and principles are tested (learning by doing). Projects of this kind should be developed by multi-functional teams.

Learning from experience

Mistakes are the best teacher

Mistakes are the best teachers. Companies have to analyse reasons of success or of failure and to record the results in a way that is perceived as open and accessible by the employees. A basic prerequisite of learning is the ability for criticism from everybody in a company. This is leading to the following two demands for an organisation:

- Projectaudits must consequently be carried out,
- The "lassons learned" have to be perceived and announced.

Learning from others

It is much too expensive to ignore the knowledge of others. This knowledge should not be wasted and therefore has to be taken into account. One has to break with the attitude of "not invented here". Methods for learning from others are:

Do not ignore the knowledge of others!

- Benchmarking,
- contacts to customers, i.e. customer surveys, discussions with customers, analysis of how the customers are using the product, or comparisons with competitors.

Fast and effective transfer of knowledge within the whole organisation.

After the knowledge is gained it must be quickly transferred otherwise it is used only for local applications. This transformation can be achieved by:

Transformation of knowledge within the organisation

- reports: orally, in writing or in the form of figures,
- checklists,
- job rotation,
- visits to other companies,
- training and other further educational programmes,
- programmes for standardisation.

3.5.2 A Concept for Organisational Learning in a Team Organisation

As already mentioned, organisational learning is based on individual learning but must be clearly differentiated from it. Organisational Learning is not a logical or natural consequence of former projects, it must be worked out. The following difficulties are to be overcome:

Organisational learning must be worked out

- The expenses that should not be underestimated for observing, analysing, and assessing of the past projects.
- The employees are already needed in the next project after having finished the former one. Time pressure and tasks of the new project prevent the reflection of the former project by the team-members.

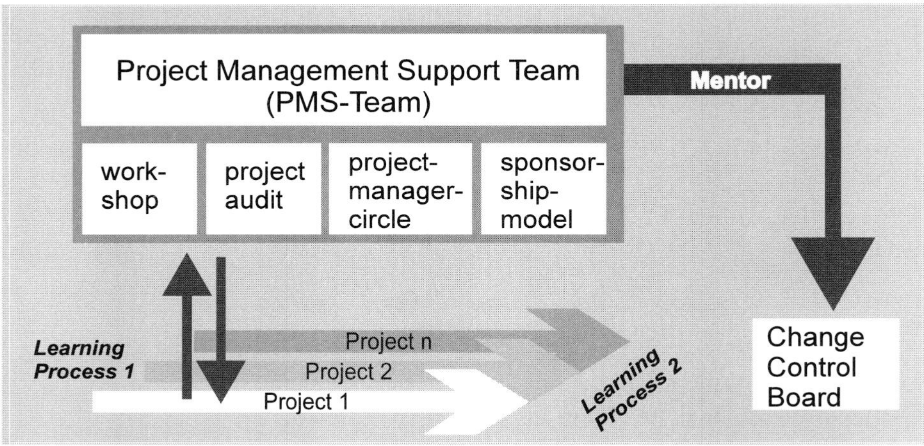

Fig. 3.8: Concept for Organisational Learning in a Team Organisation

A concept is now introduced for successful learning in a Team Organisation that applies CSE. This approach is shown in figure 3.8.

Projects as basis of the concept

The basis of the concept is again the projects of a company. New knowledge is mostly generated in these projects, thus the learning processes start on a project level. This learning by project work is the so-called "learning process 1".

Learning process 1

At this stage, "learning by doing" and the "experimentation with new approaches" is in the focus of intention of the learning process. At the end of each project, an audit, as a systematic project-review, must be carried out. Compare for this audit the building block "project audit".

Summarising, it can be said that the learning process 1 is carried out only within one project and its project-team-members. The next step must now be to provide this knowledge gained in the latest projects to other employees of the organisation. This step is called "learning process 2". In order to enable this transfer of knowledge to the whole organisation, four building blocks in the concept for a Learning Organisation are introduced. These building blocks of the concept are instruments which should be applied within the learning processes depending on the situation.

Learning process 2

Workshops

The objective of regular workshops is to work out improvements and amendments of the structure and the operations. Thus, the CSE environment can be systematically optimised and adapted to changed prevailing conditions. A workshop has to be carried out if obstacles are arising that cannot be solved within a project.

Optimisation of the organisation by means of workshops

Concerning the participants of a workshop, it can be said that generally all directly affected employees should participate. A workshop can be initiated due to the results of a project audit or by the project-manager-circle.

The proceeding of a workshop can be that it is initiated by the "Project-management-support-team" (PMS-Team) in accordance with the mentor. This initiation is based on results either by the project-manager-circle or by a project audit. Ifthe objective of the workshop is quite comprehensive, the senior management (change control board) should be involved from the beginning of the workshop. The general course of a workshop could be like it is described in the following frame.

Analysis of structures, processes, methods, and instruments and the search of the reasons and backgrounds of existing problems

The formulation of targets for improvements and the demonstration of first approaches for solutions

The building up of a first basic concept, i.e. to work it out only up to 30%. Depending on the degree of change, the senior management has to be involved. Additionally, the effects on other processes have to be checked.

The basic concept has to be adapted,worked out, and tested in cooperation with the directly affected employees. A plan of action for the implementation has to be set up as well.

After a defined period of time (mostly six weeks), there should be a review in order to control the implementation. This can take place accompanied by a pilot project, by the project audit, and/or by the project-manager-circle.

The worked out results should be quickly introduced in the project work.

Project audit

Assessment of projects

Project audits are carried out to assess finished projects and to derivate improvements for future projects. If it is necessary, ongoing projects can also be analysed and assessed. The assessment can be based on a simple form-sheed, an example for a form-sheed is shown in figure 3.9.

In the bounds of this meeting, the project-team, the project-manager and a member of the Project-management-support-team are assessing the project with this checklist and additionally with discussions and analysis of project specific data. Thus, possibilities for improvements can be identified and recorded. It is important to state that these project audits are not a control of the project-team-members but the basis for individual and organisational learning processes. The results of the audits are to be integrated in general improvements.

Project-manager-circle

Exchange of experience

The project-manager-circle is formed by all project-managers and the PMS-Team with the objective of a deliberate exchange of experience between the projects and for the

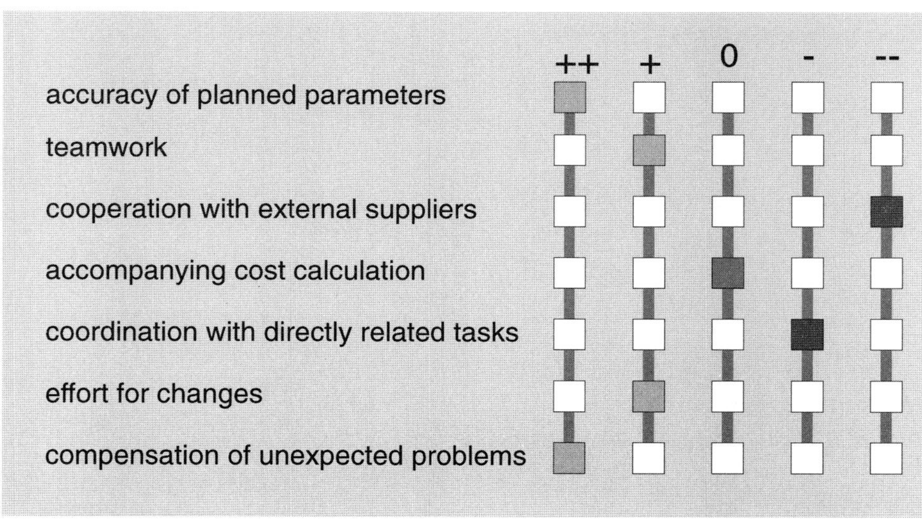

Fig. 3.9: Form-sheed for a project audit

identification of project-wide obstacles. These circles are prepared, moderated, and assessed by the PMS-Team and are held every six weeks without a duty of attendance for the project-managers. The topics of the circles result out of:

- the discussion of current problems within the projects,
- a defined main topic ahead of the circle,
- required information for the improvement of the processes in CSE.

Sponsorship-model

Active support of project-managers

Under the sponsorship-model the active support of a new announced project-manager by an experienced one (sponsor) can be understood. This sponsor is nominated by the PMS-Team after a new project and thus a project-manager is announced. The project-manager can ask his sponsor for support with specific issues (coaching). The build up and carrying out of the whole sponsorship-model is coordinated by the PMS-Team but it is not involved in the coaching.

Possible interactions in the concept

Two different kinds of interactions between the projects and the four building blocks can be identified. Firstly, the interactions towards the projects from the sponsorships and project-manager-circles. That means that the projects are accompanied by these two building blocks. The general passing on of all kinds of measures for improvements is the objective. Secondly, interactions targeting all the building blocks from the projects. Here, the projects are assessed and the gained knowledge and experience is recorded.

The project-management-support-team

The Project-management-support-team which was several times mentioned above acts as a catalyst of the learning processes. It is a cross-functional circle of subject mentors that is responsible for the maintainance and further development of CSE after its introduction. Additionally, it coordinates the deployment of the four building blocks of the learning concept and works out suggestions of improvements for the change control board.

The change control board

Head of the PMS-Team is the mentor. He represents the connection to the senior management and is responsible for the overall improvement of CSE. The change control board is an institution of the senior management in which all the decisions concerning more than one project are taken. Only in extraordinary cases, such as changes of the structure, or its adaptation is the change control board to be involved. Generally, the transfer of experience and knowledge should have a non-bureaucratic and effective character.

3.6 Conclusion

Empowerment of project teams as core of the Team Organisation

In this chapter the elements of a CSE organisation are presented. The core of the organisation is the empowered project team with a strong leader. This enables optimal product development processes by overcoming the problem of insufficient coordination between functional departments. Prerequisites for successful team work are communication, effective coordination and decision making with consensus. This should be supported by following rules of team work, establishing a clear framework of meetings and forms of cooperation which are reliable. In doing so, projects as a whole gain the importance they need.

Core team approach as best solution for integration

The core team approach is, as described in the text, the best solution for integration of project-management in the modern functional architecture of a company. It should be the dominant approach in companies. But nevertheless, many product development processes are too small or even too big for organising them in a core team. Therefore this approach, the dominant one, gives the possiblility to organise small projects in a lightweight structure or bigger ones in an autonomous project team structure. But the question is how to integrate teams for different tasks and the functional back bone. We propose a so called Team Organisation which is built up of teams.

The Team Organisation allows a dynamic adaptation of the organisation

We show that this solution has the attributes required from an CSE organisation. A Team Organisation allows a dynamic adaptation of the whole organisation to changing conditions. This is possible by the mere fact that teams can be founded and resolved in a rather informal way – normal employees are organising themselves in teams. Of course conflicts will be presented in this kind of a dynamic organisation and it will be an everlasting compromise and an equilibrium between management's organisation and employees self-organisation must be found. Another important aspect is the Team Organisation's ability to improve the information processing, decision making and decision implementation by placing in parallel the activities of different teams and giving rise to direct formal communication between them. Of course, to steer the set of teams, clear missions must be agreed upon in the company. And last but not least – the Team Organisation enables the organisation to multiply the career options of the employees who will be able therefore to follow their specific personal abilities and performance rather than a carreer model which is one-dimensional and limited to a hierarchical notion of status. The whole Team Organisation is only feasable if the executives trust their employees.

New incentive & motivation system in a Team Organisation

An incentive and motivation system for the team organisaton is presented in which instruments are given to manage the career. The hierarchical status of an employee will not be the main attribute to distinguish the salary. This is done by use of the circle of competences in which three categories of competence – subject competences, methodological competences and managing competences – are put in. A second source for the salary is the success of the person in the team and the team as a whole.

Organisational Learning in a Team Organisation

The organisational learning will be ensured in the Team Organisation firstly by helping it not to petrify and secondly by introduction of a number of methods to learn. Also by establishing a project manager circle or a sponsorship model whose task is to ensure that in project audits documented experience is reflected in process or organisational changes.

Projects are the focus of attention

We hope that these arguments together describe an approach which fulfils the requirements stated by CSE. Projects are the focus of attention. The core team approach and the Team Organisation as the organisational model to integrate different projects enable this to be realizable.

Team Organisation supports changes in the organisation

Also, the team organisaton improves the clear definiton of organisational interfaces by giving clear tasks and missions to the teams and giving rise to a densely webbed information network between the teams. The Team Organisation inherently supports changes in the organisation. It does not know the optimal status of the organisation and thus supports flexibility that is the ability to react to changing boundary conditions.

References

1 Brockhoff, K.: Stärken und Schwächen industrieller Forschung und Entwicklung. Results of a survey in the Federal Republic of Germany, p. 61, Stuttgart: Poeschel 1990
2 Wheelwright, S.C.; Clark, K.B.: Revolutionizing Product Development. Quantum Leaps in Speed, Efficiency, and Quality, p. ff 188, New York: The Free Press 1992
3 Balck, H.: Projektmanagement im Wandel – Wandel im Projektmanagement. In: Zeitschrift für Organisation ZfO 6/89, p. 396-404
4 McGrath, M.E.; Anthony, M.T.; Shapiro, A.R.: Product Development Success through Product and Cycle Time Excellence, Stoneham MA: Butterworth Heinemann 1992
5 Carter, D.E., Baker, B.S.: CE Concurrent Engineering. The Product Development Environment for the 1990's, Reading MA and others: Addison Wesley 1992
6 Rosenthal, S.R.: Effective Product Design and Development. How to cut lead time and increase customer satisfaction, Homewood Illinois: R.D. Irwin 1992
7 Byham, W.C.; Wellins, R.S.; Wilson, J.M.: Power Teams. Spitzenleistungen mit autonomen Arbeitsgruppen, Landsberg/Lech: Verlag Moderne Industrie 1992
8 Schmidt, J.: Die sanfte Organisationsrevolution. Von der Hierarchie zu selbststeuernden Systemen, Frankfurt/Main, New York: Campus 1993
9 Domsch, M.: Laufbahnentwicklung für Industrieforscher, in: Domsch, M.; Sabisch, H.; Siemers, S.H. (editor): F&E-Management, Stuttgart: Schäffer-Poeschel 1993, S. 153 – 178
10 Compare for this approach: Bleicher, Knut. Haben heutige Organisationen noch Zukunft? In: Das biokybernetische Modell, p. 161-181, edited by Jürgen Fuchs, 1992
11 Sommerlatte, Tom: Lernende Organisation. In: Das biokybernetische Modell, p. 114-122, edited by Jürgen Fuchs, 1992
12 Dierkes, M; Raske, B.: Wie Unternehmen lernen. Erfahrungen und Einsichten von Managern. In: Manager Magazin 7/1994, p. 142-154
13 Garvin, D.: Das Lernende Unternehmen I: Nicht schöne Worte – Taten zählen. In: Harvard Business Manager 16 (1994) 1, p. 74-85

4 Enabling Technologies I: The CONSENS Platform

4.1 Support of CSE by the CONSENS Platform

S. Kessler

In chapters two and three we have discussed CSE implementation and introduction in a company from a strategic point of view. Major issues are to improve, change and concentrate on processes, people, organisations and structures. To define and well improve the structure, all these new processes and organisations form one aspect of deciding on CSE for a company. The realisation and implementation afterwards, the changes of the daily work in the company and its suppliers is another. This is the real challenge for today's companies! Not only the processes and structures of the new environment which has to conform with the product development are important, they also need to be supported by a convenient information technology infrastructure (IT-infrastructure), matching the new requirements. This new IT-infrastructure should also help the users to find their way through the newly created and process oriented environment.

This new infrastructure, therefore, not only has to concentrate on the basic functionality of exchanging and transferring data/information from one company to another or within the company itself. Rather, there is a need for an integration platform to master the new issues of CSE. This is mainly to carry out the new CSE concepts in a convenient and supporting IT environment.

Easy to use environment

User-friendliness is of main importance, allowing the user to concentrate on his/her initial and major tasks. To free him from administrative tasks such as searching for the latest version, not knowing how to implement complex tools in the best way. The users should be supported in order to make the use of the new working guidelines easier, i.e. following a defined, well structured process rather than structuring the work along their own preferences.

The environment must integrate and use all its existing tools and applications, which have proven themselves useful during the product devlopment. New IT-platforms also have to make concurrent and simultaneous teamwork possible. They must allow quick access to the required data, avoiding to wait until the complete task is finished.

The data and information management systems should not only admit easy access, but also allow parallel access and the exchange of data for all members of the product development process. To meet these challenges, the CONSENS consortium developed not only a methodology but also an IT-platform.

The CONSENS Platform

This IT-platform should support CSE in a practicable and convenient way. To reach this goal, numerous different methods and tools are necessary. The requirements consider the whole product development process in a company and all of ist needs for CSE.

Parallelisation, standardisation and integration are the keys to a CSE oriented product development process.

Key issues of CSE

To reach parallelisation, standardisation and integration and to introduce and support CSE during the entire product development process the platform itself consists of three major components:

- a Framework, to model, support, control and integrate processes and teams; to produce within these processes and itssteps all the data necessary for the product,
- an Information Management System, to manage, change, release and store metadata related to the product,
- a Product Information Archive, to store and access a common product data model.

Each component of the CONSENS Platform has to support parallelisation, standardisation and integration in its own manner, independent of the basic functionality behind each component and only from the viewpoint of CSE in a company. To fulfill the requirements of CSE, all components of the platform have to work together in a close and user-friendly way.

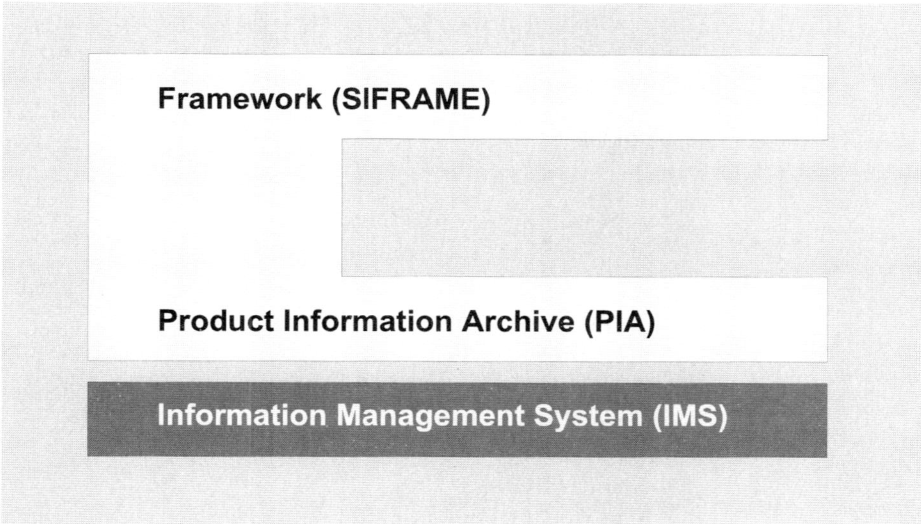

Fig. 4.1: The CONSENS Platform

4.1.1 Support of CSE by the Framework SIFRAME

For the CONSENS Framework the consortium selected SIFRAME, which is developed from the JESSI Common framework, as the basis for their project driven enhancements.

The framework SIFRAME

SIFRAME supports parallelisation, standardisation and integration in a CSE oriented environment with the following features:

Parallelisation:

- controlled and concurrent access to distributed data and information,
- multi project-management,
- client/server architecture, multiple desktops,
- Interactions between different work packages and their interdependence,
- modelling of independent and dependent processes to enable parallel and simultaneous work,
- flexible reaction to changes in the product development process as well as in the organisation of the project,
- support the user in adapting the installed project according to new requirements,
- distributed database, shared schema,

- common method for executing tasks via the user interface,
- to visualise the information/data flow from and to another process,
- structuring the project/product into distinct interrelated or independent work packages which can be worked in parallel,
- possibility to divide the project into its components and a flexible management in work packages,
- to support team work and parallel access of team members to different tasks of an installed project,
- to provide mechanisms to free information/data in time for other users to enable concurrent and simultaneous work in this project.

Standardisation:
- to run something on different hardware platforms in a heterogeneous network, multi-platform capability,
- distributed databases,
- standardised interfaces to exchange data between different tools and frameworks,
- support of standardisation with the facility to reuse results in multiple projects and to ensure their consistency,
- to allow the reuse of results and work packages,
- to prepare libraries for the reuse of processes and project tasks,
- to support the installation and reuse of standardised processes and projects,
- to provide functionality to model processes and their data interdependence,
- reuse of existing components by an interface connected to external documents handling systems and archives.

Integration:
- to integrate different kinds of users (supplier, designer, project manager, ...) to provide each of user with his customised version of the Integration Platform,
- to provide a common graphical user interface for each user of the system,
- to allow the integration of demo-site-tools to enable the user to work in familiar environment, supported by the framework,

- to allow the integration of domain neutral tools to support the user in controlling the processes,
- to offer interfaces for communication and integration,
- access to different tools through the user interface of the Integration Platform,
- to run the Integration Platform as a distributed system to support different user locations,
- to use standardised interfaces and mechanisms to allow the exchange of data between different design tools,
- use of standards for communication and integration (IPC, ITC, CFI, ...),
- to manage the status of results and keep track of their consistency so that entirely specified information can be destinguished from partially specified information,
- to manage all interdependences between work packages and to inform the participants of the effects of their task,
- to provide the possibility to exchange information in a controlled and defined way between different processes,
- to execute tools necessary to fulfil tasks in a convenient and controlled way,
- to check the data transfer between processes,
- multiple schema, interrelated data, shared data model,
- object-oriented (structuring of the "real world"),
- consistency control and storage in the data handling component provides a defined status of the data,
- to support a project/process oriented organisation and changes in the organisation of the currently running project.

4.1.2 Support of CSE by the Information Management System

The information management system IMS

As the management information system, the consortium selected the PDM system Sherpa, hereinafter called IMS as the basis for the project driven improvements. To support a successful introduction of CSE to a company, IMS must work together with other CONSENS modules: the Engineering Project Manager (EPM) and SIFRAME. The integration between the three modules allows not only the setup of a common project environment, it also support the automatic sharing of significant information and events.

The general concepts regarding the domains of each module are:

- IMS is concerned with the storage and access control of the official documentation of both products and projects, from the very beginning of the product development process, to the end of its life cycle,
- EPM is concerned with the progressive planning of projects, the progress control and audit,
- SIFRAME is concerned with the management of processes and thetools to work out activities, related to the product development process.

Integration and data exchange of IMS

The general concepts regarding the integration and data exchange of IMS with EPM and SIFRAME, are the following:

- the information generated using one of the modules, is made available to the concerned module(s) as soon as it is ready, avoiding the need of polling,
- each information is passed on to the interested consumers only, with regard to their project and role,
- the information exchanged between modules is automatically transferred, processed and stored. If there is a need to intervene, the appropriate user will be informed.

The requirements related to a successful introduction of CSE are divided into three major issues. How IMS supports Parallelisation, Standardisation and Integration is shown below:

Parallelisation:

- strong and detailed I/O control of official documents in a multiproject environment by means of selective access rights to the same information,
- the generated by the availability of too much information,
- freeing users from worrying about the status and reliability of documents.

Standardisation:

- projects, that can be set up using templates,
- product documentation, that consists of pre-defined types of official documents, grouped into sets,

- organisation, providing functions to define standard roles for people with different tasks and privileges related to the official documentation and granting the control of access rights and document's evolution,
- the creation of resources and culture, by providing and controlling company-specific guidelines to produce the product documentation.

Integration:

- a common way to access official documents, available for different departments, from the very beginning of the product development process,
- a link between the enterprise level organisation and people who work on the framework level,
- automatic data-exchange regarding the expected official documents in a project, to be taken into account during the planning of the project and defining processes,
- to give sufficient notice of significant events to the project-management module,
- controlled storage of expected documents, planned and produced using other CONSENS modules, as a result of activities.

IMS can fully support the transition period, when CSE is installed in a company:

- if a "CSE project" is set up, it can coexist with many traditional ones,
- the same person can be part of different teams of the traditional organisation. He can also be part of different projects, as well as the document types and behaviours which can coexist,
- if an experimental project can help a company to find the more adequate way to adopt CSE, IMS provides the facilities required for the change. It helps to customise project templates and to amend and re-use previous experience.

4.1.3 Support of CSE by the Product Information Archive

The common product data model PIA

The third component of a CSE oriented IT-Platform is the Product Information Archive, hereinafter called PIA, to store and access data in a common product model. The consortium decided to take the opportunity to design and develop a product information archive to fit the needs of a CSE-oriented environment concerning the product data.

PIA efficiently manages non-redundant, up-to-date product information and improves the flow of information between teams during the product development process. It is based on a generic product model, including time versions and customer variants of products as well as process data, and is used for company-specific sub-models.

As the central database for product information, PIA will be the integrative component in CONSENS for sharing such information.

PIA support the requirements of parallelisation, standardisation and integration in a CSE oriented environment as described below:

Parallelisation:

- shared information model containing product information with parallel access,
- multiple tools can access the data model.

Standardisation:

- supports detection of similarities between products,
- avoids the development of the same product several times.

Integration:

- common data model, accessible by multiple teams/tasks,
- to access and manipulate data in diverse external formats and systems,
- enterprise wide database for the entire product development process.

STEP standard will be taken into consideration

CONSENS strives to be a system which is open to information exchange with existing environments and new applications. Openness with respect to the exchange of product information means that the STEP standard will be taken into consideration. Since the product data in CONSENS is stored in PIA, this system module will be built in such a way that at least the kernel product information of STEP can be exchanged according to the STEP standard.

4.2 The main Components of the CONSENS Platform

S. Kessler

Major aspects of the CONSENS Platform

The CONSENS platform, besides the requirements concerning parallelisation, standardisation and integration, is mainly based on the following major aspects:

- use of international standards,
- modules and components,
- object-orientation,
- user-friendliness,
- flexibility,
- open interfaces.

The CONSENS Platform allows the productive side-by-side existence of components and modules. They serve all the requirements of an IT-technology, supporting the newly designed environment for the product development process of a CSE oriented company.

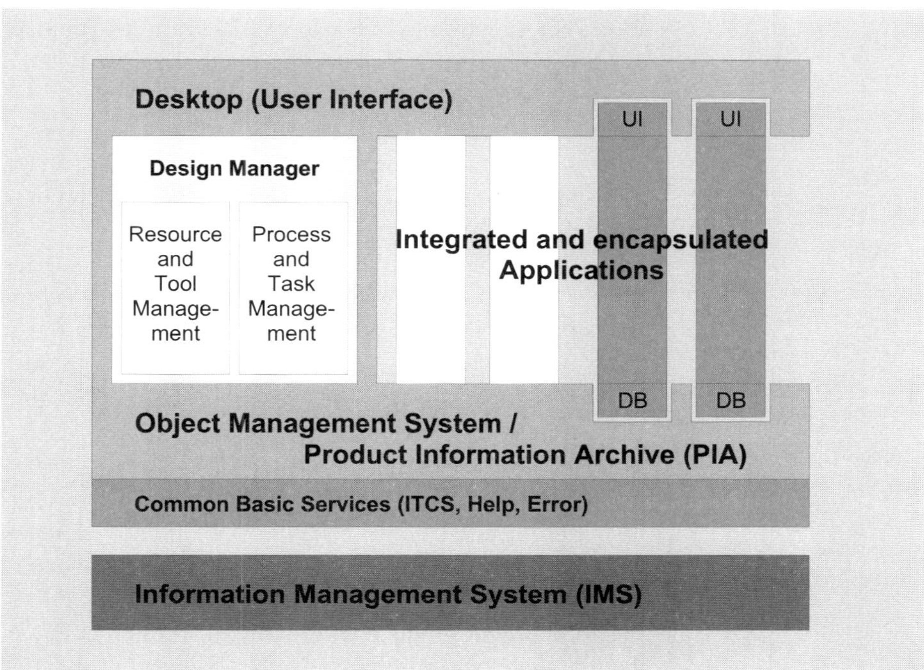

Fig. 4.2: Architecture of the CONSENS Platform

Main components of the CONSENS Platform

According to the requirements described above and the necessary support of parallelisation, standardisation and integration, a CSE-oriented IT Platform consists of the following major components:

- Desktop (User Interface),
- Data Handling System,
- Common Basic Services,
- Design Management,
- Tool Integration and Encapsulation,
- Information Management System,
- Product Information Archive.

Each of these is necessary to fulfil the needs of a CSE-oriented environment in a company. How these components work together is shown in the figure below.

4.2.1 The Framework SIFRAME

The framework SIFRAME is a general environment providing services independent of a specific domain. It is, therefore, a solution from which a domain-specific environment can be built up, using domain specific tools, such as demo site tools or CONSENS Tools.

Services provided by SIFRAME

The services provided by SIFRAME are based on a set of common basic components, such as user interface, design management and data handling. These services ensure the consistent handling and problem free data exchange between the various product components and tools. They further support interdisciplinary teamwork and the organised use of available resources such as teams and tools. They also support the modelling and work along predefined processes, allowing the execution of integrated and encapsulated tools in a user-friendly and easily to accessible way. Processes can be modelled with all their interdependences from one task to another, from the responsibility of one team to the responsibility of another. Within this environment a framework has two clear advantages: firstly, the software driven management helps to control the complexity of the product, and secondly, the design environment provides an appropriate working milieu for the entire product development process during which the best tools can be made available. Having these features, it covers nearly all aspects and requirements an IT-Platform for a CSE oriented environment should have.

Desktop

The Desktop of the CONSENS Platform is the standard graphic user interface provided by SIFRAME. It is based on OSF/Motif and is designed in accordance with the guidelines contained in the OSF/Motif Style Guide and the SNI Style Guide.

The desktop consists of dialogue boxes, windows and browsers in a graphical user interface environment, as well as a built-in help system. These modules have a uniform structure and they use the OSF/Motif dialogue elements. They can also be used by all the integrated tools. One of the tasks of the Desktop is to display objects, relations and attributes in a graphical way for the user. It also presents the methods applicable to objects. The Desktop accepts any input from the user, (i.e., it handles all "events"). It also performs session management. Consequently, at the end of a session the session status (e.g. selected and open objects/windows) will be registered and, at the next start-up by the same user, the old session status can be re-installed.

It is possible to enhance the Desktop is functionaling by adding new global or domain tools and extending the appropriate menus during start-up without re-compilation.

Data Handling Component

The data handling and storage component of the Integration Platform is called "Object Management System" (OMS), which is a structural and object-oriented data handling system. It offers services for data modelling and data-manipulation, as well as for the integrated transaction management and event trigger mechanisms, thus ensuring data consistency and security at low levels. Furthermore, it provides a procedural interface for schema-dependent access to the data which is known as the Object Manipulation Language.

Services of the OMS are of particular importance for all forms of tool integration, from tool encapsulation to deep tool integration.

Common Basic Services

The Common Basic Services (CBS) provide basic functions for the system environment. They can be used by every component of the framework and all the integrated tools. For example, Inter Process Communication (used to control tool handling) is such a Common Basic Service.

Design Management

The Design Management is the central component of the framework. It supports the definition, administration and execution of design projects. It is concerned with the management of project related data and project structures.

Tool Integration

The CONSENS Framework supports two different forms of integration: thorough tool integration and tool encapsulation. Both focus, more or less, on the following four aspects of integration:

- Data Integration:
  Data integrated tools share the same schema. There is only a single data entry point. Such data can be re-used by other tools.
- Desktop Integration:
  Desktop integrated tools share the same style of user interface.
- Control Integration:
  In order to preserve consistency, changes caused by one tool must be reflected by other tools, at any stage of the process.
  Common control flows must therefore be considered. This is especially important for parallel processes.
- Communication Integration:
  Communication channels are established between the tools.

4.2.2 Information Management System (IMS)

IMS is the underlying metadata management

From a functional point of view, four main components represents the Information Management System which works as the underlying metadata management component in CONSENS.

Information Sharing

This component manages all of the information or metadata necessary for the design, development, manufacturing and support of the product. In addition, it handles the flow of information in the company's organisation.

The shared information must be relevant to the enterprise level. It is useless and confusing to share minor changes between departments. These changes are held in documents or parts of documents while the work is in progress are known to a single department or small team.

Document Handling

During the life-cycle of the product, each phase is accompanied by the production of official documents. The typology of these documents is very variable because each document is associated with the product type and the specific phase of the development process and depends on the company's organisation and standard process.

Release Control

During the product life cycle, the information is subject to constant change as the product evolves; this change must reflect the work flow in a company for that specific product. The aim of the Release Control module is to support a fully controlled environment for the collection and development of product information.

Change Management

This component manages the product changing process; it keeps track of what needs to be changed, and why and when the modifications have to be applied.

4.2.3 Product Information Archive (PIA)

PIA, the archive for common product data

PIA contains the product data model for the products produced by the user companies. The product model is described in terms of complete products with their internal structures and their evolution in time. It is an object-oriented model describing the product in terms of objects which are classified according to their qualities. The objects will be products, assemblies, subassemblies, parts, materials, etc.

Through the object-oriented structure of PIA, the data-management is simplified because the complexity of the data can be modelled in very expressive terms. The data can be stored without redundancies and inconsistencies and changes can be made quickly.

PIA provides product data for the CONSENS tools so that they can perform their functions. For example bills for materials of different types can easily be generated from the product data model within PIA. There will be no processing of data retrieved the PIA database other than the functioning of those tools. The definition and changes of the database schema will be made via the integration platform (e.g. SIFRAME).

The generic product model will be built by analysing the users' product data, comparing them with other products' structures and abstracting the generic features from the generic product model. This product model is matched with the kernel information in the STEP Resource models.

The data is stored can be seen in different ways depending on the CONSENS tools which will be using them.

4.3 Modules of the CONSENS Platform

4.3.1 SIFRAME – The CONSENS Framework

S. Kessler

4.3.1.1 Overview

SIFRAME is based on the exploitation of the JESSI-COMMON-FRAME project. It is a flexible modular system for creating customised solutions. For CONSENS, SIFRAME 3.1 is customised to the specific needs of Concurrent and Simultaneous Engineering (CSE) defined by the CONSENS Consortium. Customising in this context means to make use of the existing functioning of SIFRAME 3.1, to adapt it to its specific needs and to develop a new functioning.

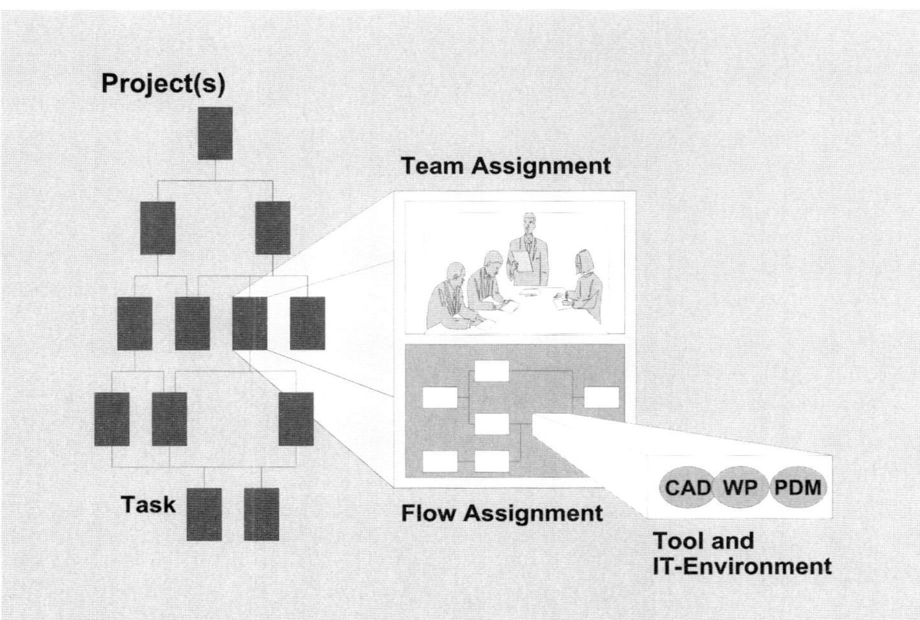

Fig. 4.3: SIFRAME – The Concept

Components of a framework

SIFRAME, the framework in CONSENS, consists of five major components:

- User Interface (Desktop),
- Object Management System (OMS),
- Common Basic Services (CBS),
- Design Management (DM),
- Interfaces with other Products and Procedures.

To deal with new processes and new organisations in the CSE-oriented environments, SIFRAME provides, based on the above described components, groundbreaking functioning to support the user in putting his new working environment to work:

SIFRAME – The Concept

- Framework Isle:
  A framework isle contains resources (the administrative data pool) and projects. Some of the resources, which are contained in the administrative data pool, are users, teams and (process) flows.
- Users and teams:
  In the context of users we distinguish single users from teams. Teams consist of users with the same membership.
- Viewtype:
  A viewtype is a description of the input or output produced by a tool. It can also refer to its subfunctions, e.g. text files, C-source files, drawings.
- Activity:
  An Activity represents a specific tool or a sub-function of a tool which is necessary to perform the defined process step. To execute activities, one or a set of input data (via viewtypes) must be available. The results of activities are known as design objects. Dependencies between activities are specified due to their required input or created output data.
- Flows:
  A flow describes specific time and data dependencies between activities. It contains a set of activities which have to be accomplished either in sequential and/or parallel order.
- Projects:
  Each framework isle can contain projects. Projects can use the resources contained in the administrative data pool. Projects are necessary for the definition of the engineering environment in an organisation.

- Team Assignment:
  Multiple users as part of a team can work on one project and each user can work in multiple projects. The access to the projects is through cells.
- Flow Assignment:
  Multiple flows can be assigned to one project and one specific flow can be assigned to multiple projects. In any event, it is possible to assign a default flow to a project. A project contains cells, which allow the decomposition (work-breakdown structure) of the overall engineering task.
- Cell (Task):
  Each cell has its own procedure and a team of users who can work on this cell. A cell itself is composed of design objects. Design objects are the results of the execution of activities, in the form of data files and directories. To manage, control and model the above listed features of SIFRAME, it is necessary to scribe the functioning. This follows below.

4.3.1.2 Functionality

User Interface

The Desktop controls the graphical user interface provided by SIFRAME for the CONSENS framework environment. It is based on OSF/Motif and is designed in accordance with the guidelines contained in the OSF/Motif Style Guide and the SNI Style Guide. The desktop itself is made up of several different components necessary to provide the specified user functioning. The built-in graphical user interface provides a very user-friendly way of communication with the central functions of the CONSENS framework. The Desktop uses the concept of the OSF/Motif object action mechanism (the user can browse through the selected objects and apply methods to them). It manages user sessions, accepts user input and displays the required objects in the desktop windows. The Desktop is intended to support the user by supplying mechanisms which make the tasks required during framework usage easier to understand and to accomplish. Therefore, it uses the concept of an object-oriented user interface. By using the Desktop, the user browses through the database, selects objects and applies methods to them.

The desktop works as the user interface

The role of the user is evaluated when the set of applicable methods is determined. The Desktop supplies multiple mechanisms for the completion of a task, so that the user can select the most convenient style. For example, the same method may be invoked by using either menus, pop-up or pull-down, or typed-in commands.

The windows of the Desktop have a uniform structure and use the OSF/Motif dialogue elements, i.e. window menu button, minimise button, maximise button, re-size border, scroll bars and pull-down menus. The pull-down menus can be called-up from the menu bar. With these windows, the user is able to control the activities of the framework and find his way through the database.

The Desktop uses dialogue boxes to display information, selection possibilities, error messages, short help information, etc. There are always several response buttons in the dialogue box available to the user, namely to confirm, to cancel or to call the help system and to execute the selected method. In addition to that, there are several boxes to inform the user about the status of the executed methods, e.g. error messages or short information. With the graphical user interface of the Desktop, the users are also supported with mechanisms for selecting objects and methods, windows, menu items, dialogue boxes.

The browser of the Desktop displays the objects of the current database in the main window. The user can use it to work in the database and to apply methods to objects. The browser uses icons to symbolise the objects of the database.

Customisation of the user interface

The Desktop is customisable and consists of three parts:

- the graphical part performs the graphical output. It allows object and menu selection and invokes the Scenarios,
- the Scenarios – in order to perform a user request via the selected object-method combination, a so called "Scenario" is executed by the Scheme Engine. A Scenario is a Scheme Script, combining user interface actions with the application functioning,
- the Scheme Engine – the framework architecture sufficiently supports the concept of an "extension language". Being closely related to the CAD Framework Initiative (CFI) strategy, "Scheme" had been chosen as a core language. The Scheme Engine is the interpreter accepting the

extension language. It may be fed by the user via textual in-out sub-window or by the graphical part to execute a scenario.

Object Management System

Object-oriented data handling

The data handling system of CONSENS is based on the Object Management System (OMS). It is an object-oriented database kernel system offering the database handling capacity for the whole framework.

The OMS guarantees high data integrity via its transaction mechanism and enables high speed access to the engineering data. Moreover, the OMS is an active database; it provides an event-trigger mechanism, which enables the framework to react to certain occurences in the database. The OMS is also equipped with a function for checking the environment, start-up and termination of the database and user administration.

Database Servers

The Object Management System is the implementation of the Internal Data Model (IDM) with emphasis on the representation of engineering design data. The architecture of the OMS is designed according to the client-server model. The server-function is split across three separate, specialised processes:

- Object Manager (OM):
  This is the main server managing object and attribute types, their interrelations and the graph of complex objects. The OM also synchronises the concurrent access to objects and the transformation from main memory data structures to their representation in the disk file system.
- Event Manager (EM):
  This server implements the event-trigger mechanism. Clients may set events on object and attribute types, their interrelations, and the graph of complex objects. This information is stored by the event manager and compared against the information about the current OMS activity.
- Name Server (NS):
  This server implements a message mechanism for the exchange of messages among OMS clients similar to the

message concept in UNIX System V. It also supports remote clients. Remote clients connect to their OMS servers with the help of the NS, which establishes a communication server, representing the client in the server host. The implementation of this concept is based on Internet sockets using the TCP (TCP/IP is a protocol suite standardised by the US Department of Defence in 1978).

Internal Data Model (IDM)

The Internal Data Model (IDM) facilitates the design of a complex object-oriented data model through the following of objects:

- Container Objects (COs):
  These objects are divided into two parts, the object hull and the object contents. The hull consists of the object properties such as OID, name, attributes and edges. The contents are used to store Primitive Objects (POs).
- Primitive Objects (POs):
  These objects consist of the object hull only. They cannot contain any other POs, but may be related via an edge.
- Structured Container Objects (SCOs):
  An SCO contains a set of Primitive Objects. The nesting of Container Objects can be modelled by using edges between COs and POs.
- Unstructured Container Objects (UCOs):
  An UCO contains a byte-stream and has the same characteristics as a file.
- External Container Objects (XCOs):
  An XCO contains a UNIX path name. Data stored in the file system can thus be administered by the OMS in an uninterpreted way.
- Locking Mechanism:
  Locking mechanisms are performed implicitly when an OMS operation is performed. However, objects can also be locked explicitly. The locking mechanism is basically a two phase lock. Shared transactions do not exist. For locking object hulls (attributes and edges attached to an object), the OMS provides a read-lock mode and a write-lock mode.
- Concurrency Control:
  For concurrency control, the OMS offers different lock modes which are based on a two level multi-granul lock-

ing protocol. These two levels correspond to the different levels of granulation (primitive objects, POs vs. structured container objects, SCOs) in the underlying data model.

The advantage of read or write locking of containers is that individual object interfaces need not be locked, thus reducing the administration overheads necessary for locking. Locking the contents of an SCO in read-lock or write-lock mode locks all POs within the SCOs in the corresponding mode. However, the object hull is not locked. There may exist several read-locks for the contents of an SCO at the same time, whereas multiple write-locks are not possible.

Assurance consistency in a CSE environment

The intention lock modes, in contrast, offer a higher degree of concurrency, since transactions which hold an intention lock synchronise themselves at PO level. Every PO, which is accessed during a transaction, is locked. This locking modes are well suited for shared SCO.

The container lock modes are tightly coupled to the different load modes offered by the OMS. A tool can access the containers stored in an OMS database in several modes, which have a strong impact on the performance of the OMS. Containers can be loaded directly into the storage area of the tool. In this case, the tool administeres the data in its object cache. This mode is very efficient since the tool has direct access to the data without further communication with the OM (when in operation, the OMS starts two servers: the Object Manager (OM) and Event Manager (EM).

Data Definition Language (DDL)

In any database system there is the data definition language which depicts the conceptual schema in the database in a format peculiar to the system. For OMS, we have the Object Definition Language ODL). The ODL is part of the Tool Integration Definition Language (TIDL). ODL is used to extend the schema definition part of the data model and store the schema information in the Framework Dictionary (FWD).The FWD stores data such as the user defined schema, tool integration details, encapsulation information, desktop specific information etc. The user controls the FWD via a textual input to the TIDL compiler, a normal

ASCII file holding the schema definition according to the TIDL syntax. The FWD is stored at a special place in the database, thus utilising all the facilities of an object-oriented database system.

Data Manipulation Language (DML) the OML is a procedural interface of the OMS, allowing navigation in the database and manipulation of objects according to the schema defined via TIDL. The OML is structured according to the different classes which are arranged in a class hierarchy. The methods are grouped according to the different classes, e.g. SCO, PO, Object.

Database Editor and Browser

In order to automate the integration process and to verify the database schema through user inspection, the framework offers a Database Editor and Browser. The Database editor is a support tool for the administration of the environment of the framework and the integrated tools. It is supplied with the following functions:

- General administration of the database (e.g. start, stop, create, remove, copy the database),
- Support of tool integration and schema modification,
- Inspection of database contents,
- Modification of database contents (e.g. create, delete and backup objects).

Common Basic Services

The Common Basic Services provide general services for CONSENS. These services include the Help System and the handling of error messages and communication services.

Interface of IPC Mechanism

The Inter Process Communication Service (IPC) supplies the basic objects and methods in order to establish and handle communication connections (or channels) between processes running on the same host ("local connections") or on different hosts ("remote connections"). The IPC service is the basis for the Inter-Tool Communication System (ITCS) as well as for other kinds of communication (e.g. communication between SIFRAME and other frameworks and databases). The IPC service provides basic mechanisms to:

- build up local and remote connections,

- handle connections (e.g. read/write from/to connections, etc.),
- close connections.

Interface of ITC Mechanisms

Interfaces for the Inter-Tool communication

The Inter-Tool Communication System (ITC-System) of the Design Management, an application of IPC, is concerned with the mechanisms for the exchange of data and the control of information in order to support the co-operation between design tools and framework services.

These ITC mechanisms supply the objects and methods in order to:

- establish communication between two tools/services via " point-to-point messages" by using a particular connection between them,
- establish communication between multiple tools/services via " multi-cast messages" which are dispatched and delivered by a " Multi Cast Message Server" (MCMS),
- support " tool handling" , i.e. calling-up and controlling tools on a local or remote system,
- support " remote procedure calls" (RPC).

The interface of the ITC-System conforms with the " Inter-Tool Communication Procedural Interface" guideline defined by the CAD Framework Initiative (CFI) Inter-Tool Communication Technical Subcommittee.

Help Service

The Help Service is intended to facilitate the development of context sensitive and on-line help. It provides the user with a comfortable help facility, which supports the display of information in hypertextform. The Help Service is based on the UNIX file system. It considers different levels of experience users may have (e.g. novices, casual users, experts), and can be called-up in all windows or dialogue boxes with the Help button.

Mailing Services and Message Propagation

The mailing services of the framework are an extension of the Common Basic Services. The objective of having such a facility is to bring all users closer together by means of better communication channels. The messaging can be point to point or broadcast. Messages transmitted would be, for example the notification the of the status of a project, the explicit action required by a user, approval, verification and the exchange of information in real time.

Notification of changes

In certain cases, the activation of certain cells (a cell is an object type defined in SIFRAME which may contain process stages) may trigger messages to be sent to other cells. Subsequently, the recipient can store such messages. The path of messaging can be defined in a similar fashion to the existing data-dependencies in SIFRAME.

Design Management

All global services with user interaction and are not related to a specific domain or company but are combined in the Design Management. The Design Management is the central component of SIFRAME as far as the application of specific functions is concerned. Design Management functions support creation and administration of every object involved in a products life cycle: projects, processes, resources.

All the functions described below are available via selection of objects and methods. Objects are presented as icons on the desktop. The methods applicable to these objects are selected via menu items in the pull-down menus of the objects. The presentation of object methods in the pull-down menus of the icons is context-sensitive, i.e. SIFRAME offers only those methods, that are applicable to the object by the current user, reflecting his access rights.

User Roles

Authorisation of users

Every user in the framework has to take on roles which determine his level of authorisation. This level of authorisation is represented by a set of methods the user may apply to different types of objects. Roles are thus related to methods and determine the access of users to the objects within the framework.

The following roles are provided:

- Framework Administrator:
  A user who has the role Framework Administrator may install tools and register users and teams in the framework.

- Design Methodology Manager:
  The Design Methodology Manager defines activities and processes.
- Project Manager:
  A user with the role Project Manager is responsible for the design environment (tools, activities, flows), creating and deleting rojects and authorising users and teams for objects.
- Design Engineer:
  The design engineer works on cells and carries out any other tasks for which he has authorisation.

Tool Management

Launching related tools

The tool Management administers the different design tools needed for the design engineer's daily work. To be more precise, the tool Management administers those tools that constitute the resources for a design process. These are tools used to accomplish part of a design process (Editors, FEM-tools, CAD-tools etc.). Tools which are not executed inside a design process, but are required to control such processes (as for example the Engineering Process Manager) are administered in the Tool Box. All tools registered under global resources can be used for the execution of activities.

Toolbox Facility

In CONSENS, we distinguish between tools required to perform a step in a process (i.e. to execute an activity inside a project) and tools that may be used outside projects. Tools of the first group usually are referred to as design tools or demo site tools, whereas tools of the other group are called domain neutral tools. A tool may belong to both groups. Design tools are registered under Global Resources of the SIFRAME for later use inside projects. SIFRAME administers the domain neutral tools in the so-called Toolbox, which actually is a container for these tools. Sub functions of the domain neutral tools may be used to define (domain neutral) activities. Access to domain neutral tools as well as domain neutral activities is possible via the user interface: each tool and each activity will be presented as an icon. The Toolbox itself (that is the container object) will be represented as a top level icon. Its contents (tools and activities) are accessible via the common methods offered by the Browser.

In CONSENS, the Toolbox contains a collection of globally available services such as the Product Information Archive (PIA), Management Decision Support (MDS) and the Engineering Process Manager (EPM). The activities which use sub functions of these tools, are for example: initialisation of a project, planning of a project, execution and control of a project, information about the product structure, preparation of a decision -upporting report etc.

Flow Management

Controlled execution of process steps

Activities used in a design process cannot be invoked in an arbitrary order, as data and time dependencies of activities have to be taken into account. In order to give the user the possibility to define a set of activities in a specific order flows are introduced. A flow defines time and data interdependences between activities. As a base for the flow definition, the information on data dependences on activities (-defined via input/output relations of viewtypes) is taken into account. Dynamically, activities and thus tools can be invoked only according to the paths of the flow graphs. The flow management ensures that the order in which activities are executed, as defined in the flow is adhered to.

Flow Definition and Flow Verification

Defined relations and dependencies

Specific features of CAD, CASE or any other tools – hence referred to as activities – that are used in a design or software engineering process cannot generally be invoked in an arbitrary order but in a certain sequence, resulting from the activities' specific functioning. For example, the output data of one activity might be needed as input for another. In software engineering, a specification has to be written before the implementation takes place. To model such interdependences in the framework, the definition of activities' execution in a distinct order has to follow in a so-called flow. These flows are graphically represented at the Desktop. Flow Management functions enable the user to define and edit these flows.

The figure below depicts the different forms of flow implementation which can all be presented simultaneously. The rectangular elements represent activities (named A, B,

C, D, E), their temporal interrelations are symbolised by an arrow which presupposes that the activity left of the arrow has to be executed before the one on the right.

A flow, of course, can contain all of the above shown implementations in any combination at any time.

Project Resource Management

Assigning resources to projects

In the framework-isle the resource-teams and flows will be defined and administered by Access Control and Flow Management functions. These resources can be assigned to separate projects in the isle. In a further step, the project related resources can be assigned to cell versions of the project. The administration of the use of the resources in a project is done by the functional block Project Resource Management PRM.

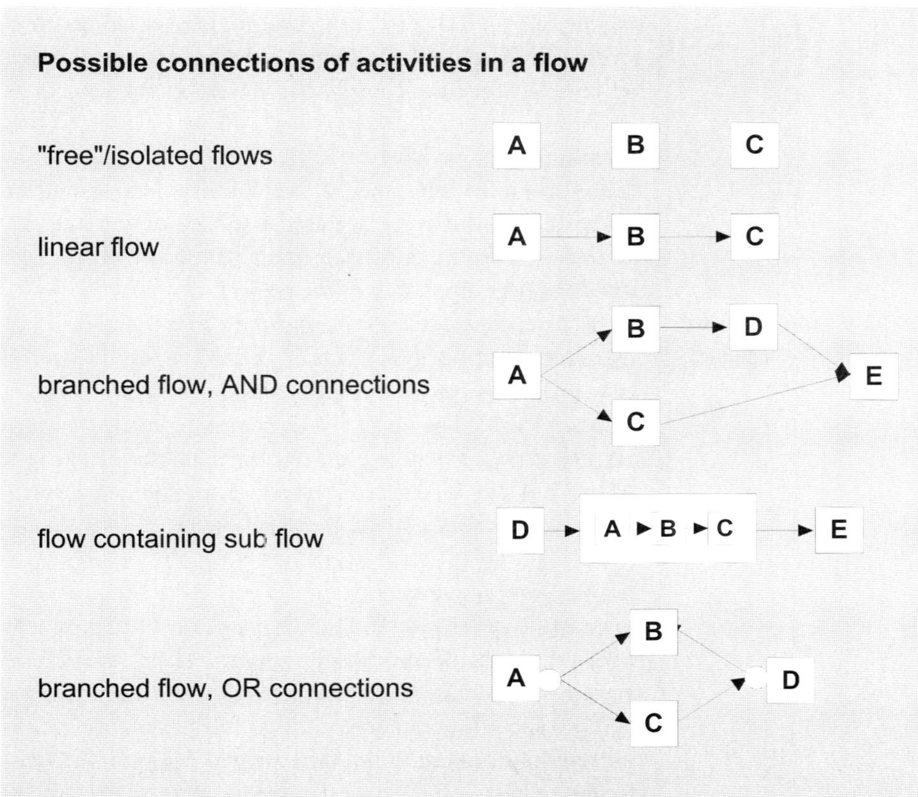

Fig. 4.4: Possible Implementation of a Flow

PRM administers the assignment of teams and flows to projects, cells and cell versions. Furthermore, PRM provides functionalism for getting information on the assignment of resources and administers the relationships between SIFRAME results and EPM results.

Configurations

In the framework of CONSENS, a project contains configurations and configuration versions. A configuration version contains cell versions. A configuration is a consistent set of cells and subcells. If a cell is added to a configuration, then the sub-cells are added automatically; if the cell has parent cells, these are not added automatically. The system provides the functions for configurations as well as configuration versions. This allows the user to individually group cells together in one configuration in order to distinguish between different cases and cell structures. A configuration may be released. Once a configuration is released, writing-access will be refused. To release a configuration, all member cells have to have completed a development state.

Decomposition of Cells

Cells can be interconnected in a hierarchical or netlike structure. A cell may be declared a subcell of another cell. All those connections between cells within the project are called the decomposition of the project.

Work-breakdown or task structure of projects

The interdependences among data in general are not restricted to the design objects of one cell version; an activity to produce a design object may also need design objects from another cell as input. To define such data dependencies, cell versions may be connected in a netlike structure; a cell version may be declared subcell of another cell version. Of course, a subcell can also have subcells.

Cell Variant Management

To give the user the possibility to try out various approaches for the specific technical solution for one cell, the variant concept is introduced. The idea is to allow the user to repeat the flow-specific sequence of the activity execution. He is assured that the results of all the attempts are saved. Within each cell version the user can create as many variants as needed. When the user has finished his work he must decide on one variant only when publishing the results. This is necessary because the results must be unique to complete cell versions. Every other variant of this cell version is implicitly deleted.

Object Version Management

The Version Management functioning is used for supporting the creation of different cell versions, configurations, design objects, viewtypes, tools, flows, activities and activity executors.

Flexibility in the product development process

It is also used to control versions and their connections with objects and with each other. Versions of the same object are used in order to track and record changes made to it over time. It is essential to record the history of changes by tracking version successors and predecessors. Thus, a version graph is built up representing the derivation hierarchy. A new version may be created as an entry to the version graph or derived from some other version.

Workspace Management

A workspace is the private working domain of a user or a team. For each project and user who is a member of a team attached to the project, a workspace is automatically created.

The users workspace environment

We distinguish two classes of workspaces: user workspaces in the global resources and team workspaces in a project. In the global resources, there is one user workspace for each registered user. In a project there is one team workspace for each team attached to the project. These workspaces are created automatically when creating a new user or attaching a team to a project. A user may reserve objects from the global resources for his private workspace or from the project in the workspace of one of the teams attached to the project. In a long transaction, he processes these objects and returns them afterwards. Objects reserved in a workspace in the global resources are accessible only by one particular user, whereas objects reserved in a team workspace are accessible to all members of the team. Reserving objects in a user's workspace is necessary to access some of the objects methods. For example, the method Modify Attributes (applicable to flow versions, activity versions, tool versions) can be applied only if the object is reserved in a work-

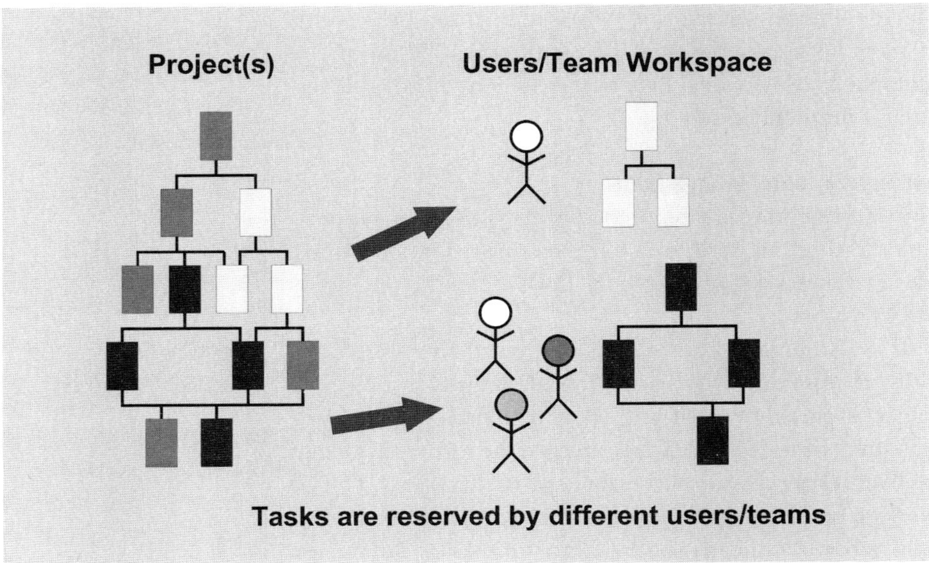

Fig. 4.5: Working in Project(s)

space within the global resources. Thus Workspace Management ensures data consistency in the OMS. When reserving an object in a user workspace, a lock is placed on the object. The object contained in this workspace is the original object. It is not a working copy. The development state of the reserved object is set to "in work".

Consistency Management

In the CONSENS framework terms, consistency control administers and controls:

- the design of object versions
- the consistency of the design object versions within cell versions
- the consistency of the design object versions from different cell versions of the Design Decomposition.

All the objects described above which are managed within consistency control are related to the execution of activities within flows. Each activity needs parts called inputs in order to be executed. These are created by previous activities. The parts, following a successful run of an activity are called outputs. As soon as an already executed activity is executed once more, all other activities, which depend on this activity, are reflected by this change. In order to inform the user at any time about the situation of activities or cells several different possibilities exist within the framework.

Input/output checks before execution

The development progress during the execution of individual activities is represented by the execution states of activities and development states of design objects. Depending on these states, development states for cells are set. Additionally, development states are used to indicate the state of development of global resources and configurations.

The initial development state of a design object is simply the intention to of arrive at the state to be published. This state is also called empty.

States of results of activities

- **empty:** The design object has not yet been worked on. This is the initial development state.
- **invalid:** The state is changed to invalid if a preceding design object, upon which this design object dependends, is executed again.
- **valid:** The design object has been worked on at least once.
- **published:** This is the final state of a design object, and no further execution of an activity on this design object is possible.
- **transferred:** The status of a design object (results) will be

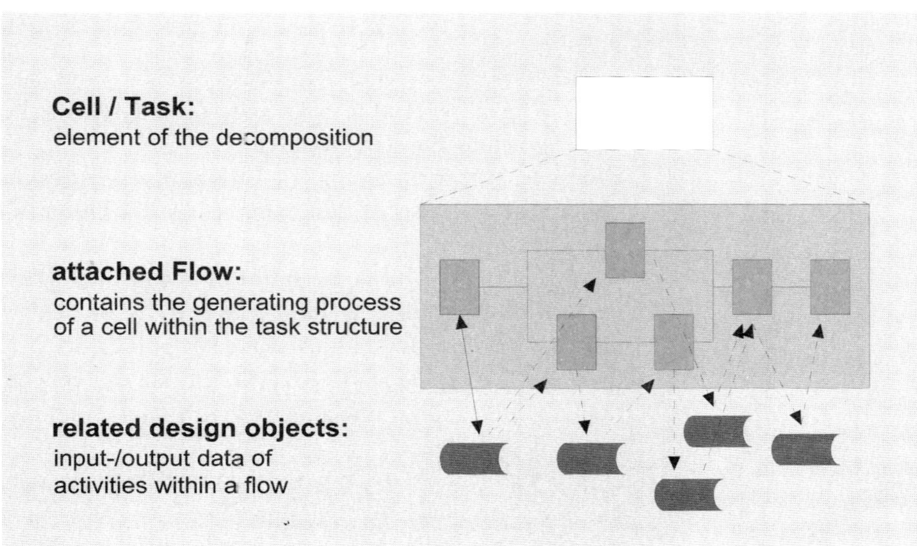

Fig. 4.6: Data consistency in Flows and Cells

changed to transferred if the method „external publish" can be executed successfully and the object has been transferred to IMS. The previous state of the object must have been published.
- **accepted:** The status of a design object (results) will be changed to accepted when the object has been accepted formally by IMS functions. The previous state of the object must have been transferred.
- **documented:** The status of a design object (results) will be changed to documented if the object has been released from IMS functions. The previous state of the object must have been accepted.

The initial execution state of activities is notExecuted with the aim of arriving at the finished state.

States of activities

- **notExecuted:** This activity has never been executed successfully. There are no results.
- **invalidated:** The execution of a preceding activity resets the state of the current activity to invalidated, if data dependencies exist between these activities.
- **executed:** The activity has been executed successfully. Again, the execution again changes the state to executed (successful execution).
- **finished:** The results of this activity have been published for other users. Further execution of this activity is not possible.

States of cells

- **inUse:** During its execution, an activity will have the state inUse. The initial state of cell versions is empty with the aim of reaching the state complete.
- **complete:** This is the status of a variant of a cell version containing activities or design objects whith the status finished res. published.
- **partlypublished:** This is the status of a variant of a cell version that contains at least one activity or design objects which has the status finished res. published.
- **in Work:** This is the status of a variant of a cell version that contains at least one activity which once produced once a design object.
- **empty:** This is the status of a variant of a cell version that has never produced once a design object.
- **released:** This is the status a cell version when the corresponding configuration version has been released.

- **transferred:** This is the status of a variant of a cell version that contains activities which all have the status finished and design objects which all have the status transferred in case they are marked as to be transferred.
- **partlytransferred:** This is the status of a variant of a cell version that contains at least one design object which has the status transferred.
- **documented:** This is the status of a variant of a cell version that contains activities which all have the status finished and design objects which all have the status documented in case they are marked as to be transferred.
- **partlydocumented:** This is the status of a variant of a cell version that contains at least one design object which has the status documented.

Representation of Product Structure in PIA

When of the framework is running, the user should be able to see the product configuration (similar to the desktop display of the existing project configuration). The configuration is represented in a tree structure on the desktop. A limited representation in terms of the hierarchy of the product is feasible, as it would be meaningless for the user to scroll through a product with thousands of items from which to make a selection. Each entity of the tree structure would be an icon representation of the object (or component in the user's term, very similar to a cell in the existing Project configuration). Earmarking an object in the tree structure is possible and, thus, saving the object. When closing and reopening the tree and reopening, the user can return to the last position on the tree.

Interfaces with other Products and Procedures

Interfaces between the Integration Platform and the Information Management System

The typical information carried across the platform and the Information Management System (IMS) is at a more generic level: object types, type specific attributes of objects; links between objects, status of objects and user classes, representing the departments or roles. In addition, there is a need for the platform and IMS to establish some form of conventions or rule for coding objects (parts, document), standard object type structures (for example, a set of documents, draft product structure) and standard procedures (- step or level of evolution of object types significant for the enterprise; requested approval; expected notifications) for new products or for managing product changes, other data to be exchanged with external systems or users, metadata of objects: (physical parts, documents, changes) in detail at enterprise level. Nevertheless, files such as drawings or documents would be treated as a black box. The contents of such files would not be interpreted in any way.

Interfaces between the Integration Platform and the Encapsulated Tools

Tool Encapsulation

The encapsulated tools adopt the standard Black Box Server (BBS) mechanism in OMS for the calling-up of tools. The BBS also allows the execution of a predefined procedure in Scheme, C/C++. Nevertheless, the exchange of data would be via the Step Data Access Interface SDAI. An SDAI program implemented on the tool side database allows the extraction of data from and storage of data in OMS. This is regardless of the SDAI program being relational or object-oriented. The SDAI program on the OMS side is bound to the OML function calls.

Interfaces between the Integration Platform and Integrated Tools

Tool Integration

The Interfaces between the Integration Platform and the integrated tools are essentially the same as the interfaces between the Design Management and the other components of SIFRAME.

STEP Data Access Interface (SDAI)

SDAI – the interface for standard data exchange

SDAI now exists as a Working Draft document within the ISO 10303 Part 22 at the point of writing. The SDAI provides a standard Application Program Interface to the STEP data representations. The data has been defined using ISO 10303-11 (the Express language document). The objective of the SDAI is to provide application-programmers with a consistent environment to be used in the development of software. In fact, it gives application programmers an illusion that the data being manipulated through the interface is represented in the underlying data storage exactly as it is described in a conceptual schema. This illusion should be the same regardless of the data storage technology which the SDAI is providing access to. The SDAI has been described in the ISO documents as an implementation language-independent form. Some language specific bindings are to be expected in the future. A manifestation of the standard in a particular computer language would be a language binding to SDAI. Language bindings for C, C++ and FORTRAN would be available through the ISO.

The system architecture depicts the possibilities of data exchange between encapsulated and integrated tools via the SDAI. The SDAI can be seen as a higher abstraction of the OML in OMS. The implementation methodology is in accordance with established procedures in STEP implementations as outlined in ISO 10303 documents.

In order to use the SDAI interface, the tool has to specify its schema in EXPRESS. Schema information and instances are then stored in the FWD. SDAI will constitute a set of functions written in C or C++ which can be used inside the

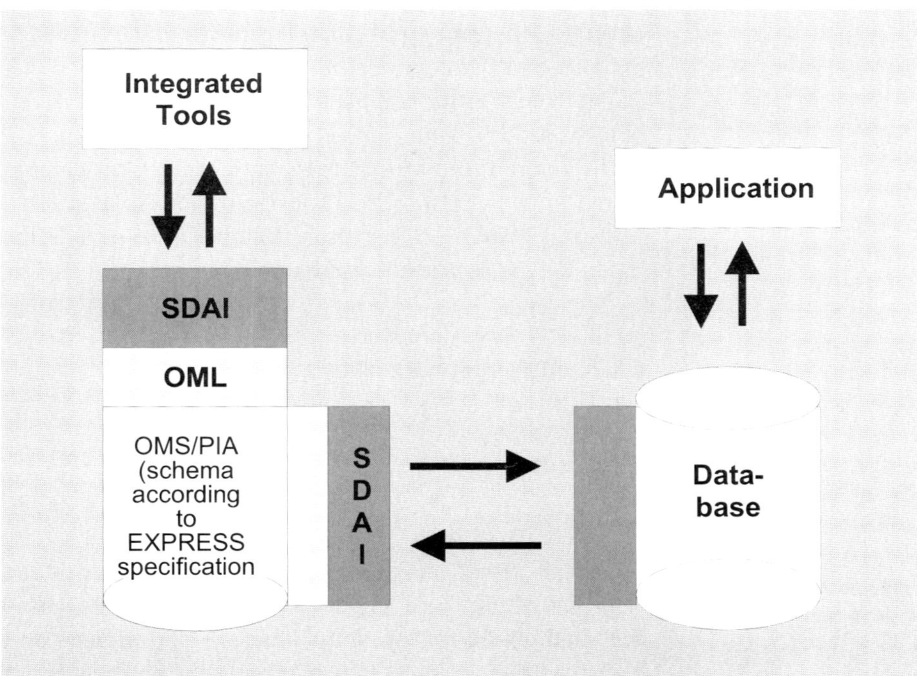

Fig. 4.7: CONSENS system architecture with a breakdown of a STEP interface

tool in order to create, delete and manipulate objects according to the STEP conformance schema.

The SDAI implementation in OMS is a module which implements a language-binding to the SDAI specification. The SDAI constitutes a set of functions and methods (perhaps in C language header files) for adoption by the tool developers. With a STEP-compliant schema definition and a STEP interface, using OML function calls on the OMS, the tool developers are given the flexibility and efficiency for designing software which communicates directly in CONSENS and eliminates the hassle of data exchanges between different formats which are commonplace in an integration platform.

4.3.1.3 Outlook

With the CONSENS Framework SIFRAME, the company, all its employees and, therefore the users of the system work in an open and flexible environment. This enables them to promtly act and react to market requirements, reducing costs and improving quality. The client-server architecture of SIFRAME, working in a heterogeneous environment, is enhanced in the next release of SIFRAME V4.0 by a PC-Client. It enables the company to manage its product development process within all its facilities and departments.

The possible step-by-step exploitation allows the company to start with CONSENS in one distinct area of the overall process, providing all necessary facilities to to expand the use to the entire process or the whole company.

4.3.2 IMS – Information Management System

L. Miotti

4.3.2.1 Overview

In recent years, scores of CAx applications have been offered to design and engineering departments, boosting the productivity of individual engineers, designers and managers. Despite these improvements, the productivity of the organisation as a whole has not improved.

The reason is simple. Thanks to CAx tools, designers and engineers are producing a lot of drawings, sketches, specifications and documents: It's becoming harder and harder to obtain the right information, to find the latest revision, to ensure proper checking etc. As a matter of fact, many companies have now the ability to design engineering changes in less time than what it takes to approve and route those changes.

This inability can hamper the competitiveness of a company. For companies, striving to be competitive in the current tough markets, the accurate, rapid and controlled communication of technical information throughout the organisation (and also its suppliers) is becoming a critical success factor, in order to provide a timely, cost effective and quality product.

This chapter focuses on the documentation issue. The relationships and interdependencies between documents are complex; significant amount of time and costs are associated to document control and monitoring functions such as collating, controlling, collecting, revising, storing and distributing the information. The typical questions a user would ask himself, accessing a document archive are:

- Where is the document I need?
- Is this the latest version?
- Is someone using and modifying it?
- Are there other documents related to it?
- May I make a copy?
- How can I distribute it?.....

As a matter of fact, poor documentation control is one of the greatest factors that contribute to project overruns, a late introduction to the market, excessive manufacturing cycle times. All this must be in place if an organisation wants to successfully adopt concurrent engineering methodologies.

Inside the CONSENS platform, IMS – Information Management System – is the component which takes care of the official project documentation. At the same time, it assures user-friendliness and the flexibility necessary to match a company's standards and its evolution. Thus, IMS contributes to the improvement of time-to-market, cost and quality of the products.

The main aim of IMS is to manage the official documents that describe a product and the project set-up to develop it. Each company has its own standards to establish which types of documents are defined as official, in which way the organisation acts to create, store, transmit, develop and promote the documents: In other words: which kind of control each document must undergo, which group of documents are expected to be produced in each phase of the project, to whom and subject to which events documents must be distributed etc. So, IMS has to fulfil two main objectives, related to two different kinds of users: the Application Manager and the Application Users.

- The first objective is to allow the customisation of the objects to be managed, to grant the compliance to the company's standards and the capability to follow its evolution. This objective is achieved through the Application Management Function, that the right information is delivered to the right person at the right time and for the right reason.
- The second objective is to provide functions for the management and control of information and documents in specific projects, with proper access to Application Users. These are people with different roles in the project. They constitute the User Functions.

4.3.2.2 Functionality

Here the focus is on IMS provided functions: Other chapters cover the issue related to the relationship of IMS with other CONSENS components. Despite the division of IMS functions between Application Management Functions and User Function, it's better to adopt, for presentation purposes, another grouping rule:

- Information Sharing functions,
- Document Handling functions,
- Change Management functions,
- Release Control functions.

This is the grouping one can find in the related deliverable of CONSENS project.

To give a better understanding of IMS functions, a short description in the next paragraphs will be followed by the presentation of an operative example (creation of a new project).

Information Sharing Functions

Information Sharing functions manage all metadata necessary for the design, development, manufacturing and support of the product, and the flow of information through the company. The shared information must be relevant at enterprise level. It is useless and also confusing to share minor changes between departments: Minor changes held in documents or parts of documents, while the work is in progress are known to a single department or a small team. Abiding to the engineering process standard established by the company, Information Sharing will report significant events, i.e. events that change the status of a project's metadata:

- identifiers of products, parts, official documents,
- their structure or configurations,
- their references to other parts or documents,
- document contents.

Information Sharing is not focused on the technical data creation or processing. Therefore, the contents of documents are not checked. Instead, it is focused on the distribution and exchange of information between different departments and users, in a controlled, secure and efficient way (data integrity), based on the users' access rights. Here is a list of main functions:

- Generation of meta data for products, parts or official documents of a definite type, either from scratch or by making copies from existing documents.
- Creation, maintenance and revision of references between documents, parts and products.
- Browsing through products, parts, documents and pieces of information, their history, their references.
- Getting reports.
- Getting copies of a document's contents (without changing access rights).
- Transparently handling the distribution of messages (alerts), informing users about significant events or requesting them to act.
- Recording the evolution of information, keeping track of its history.

Document Handling Functions

During the life cycle of a product, each phase is characterised by the issue of official documents. The typology of these documents varies widely because each document is associated with the product type and the specific phase of the development process. It also depends on the company's organisation and engineering process standards. The document can be individually associated to a product or a part or it can be linked with other documents to characterise a product. For each document, it is possible to define the stages of a document's issue and evolution until it finally becomes obsolete. A product or a part is released when all supporting documents are released.Document Handling functions manage the document with regard to document meta data (title, code, author, status, ...) and document contents (graphic file, text file, ...). The document Handling also manages the flow of documents to different departments or sections of the project team, using Information Sharing functions.Basic Document handling functions are:

- The issue of a new document of a defined type, generating the metadata.
- To get document's contents, using the proper tool, in order to read it.
- To check a document's contents in order to update the document, thus preventing it from being modified by others.
- To check a document after modification, generating new versions.
- Getting messages about a document's issue or modification.
- Making a document obsolete.

Change Management Functions

Change Management functions manage the product change process. They keep track of what needs to be changed, why and when the modifications have to be applied.

When the need to change a product arises, a change request is generated, related to a part or document. The change request describes the reason why the change is needed: It is distributed to responsible reviewers, notifying people involved in the approval process. In order to identify the extent of the change and to analyse its impact, the

change request refers to the affected parts or documents, so that all interested users are aware of the change and can contribute to the decision making process. Once the change has been completely analysed, a change order (if a change is needed) is generated and distributed to the interested people who are responsible for the modification of the product structure and the related documentation. Closing the change order (and the associated change request) terminates the change process, while track is maintained of both the old and the new situation.
Main functions are:

- Generation of change requests, change orders, other documents related to the change process.
- Collating comments, problem descriptions and other supporting pieces of information to change requests.
- Associating (linking) the change requests, change orders related objects.
- Approval or rejection of a change request and the related change order, to advance the change process.
- Warning interested users of significant events during the change process.
- Browsing and reporting through all the information related to the change process.
- Recording the history of change processes.

Release Control Functions

The complete product information is very complex and consists of many pieces. These represent parts of the product or official documents describing it or changes to be done on a part etc. Such a piece of information constitutes an object from the system point of view. Each object, is subject to constant changes as the product evolves during ist cycle. This evolution is governed by a set of rules, the Release Procedure, that depends on the organisation's unique design methodology. The Release Procedure specifies for each object:

- the significant stages of the evolution, represented as a sequence of promotion levels,
- the set of conditions (approvals, checks) that are required to promote the object to the next stage,
- at each promotional level, the access rights each user has on the object,

- the significant events that must be monitored and who should be notified of the event.

On the one hand, Release Control functions provides the design process management with the necessary tools for the accurate modelling of the Release Procedures:

- defining and managing users and users' classes,
- defining and managing Release Procedures,
- setting up and managing projects, as collection of teams and object types,
- defining and managing alerts.On the other hand, they allow each authorised user, during the product development process, to move a specific object up and down its predefined path of evolution.

User Interface

Both the Application Manager and the Application User interacts with IMS, via a graphical interface based on windows of several types.There are two different kinds of approaches applied to set up the user interface. The first one provides to give users the entry on the functions (create, delete, ...) and then requires the selection of the instances of the objects the function will run on.The second one is the vice versa; the user selects the instances of the objects she/he has to operate with and, after selection, the interface will give entry to the appropriate functions.The Application Managers use mainly the first kind of approach, while Application Users' interaction is a mix of both of them.The different windows used by the user interface are listed in the following table.

BAR
LIST
FORM
TABLE
BAR-FORM/TABLE

- Bar Windows:
  This is usually the initial window displayed: Bar windows contain horizontal menus; these menus usually contain options that, when selected, activate other windows or execute functions.

- List Windows:
  List windows frequently contain vertical menus. They are used either to execute functions on selected data or to support the user with list of acceptable data to insert in input fields.
- Form Windows:
  Form windows can be designed to look like a data-entry form. Each form window consists of a heading and a main body section. The heading section is optional; if present, it is usually used for title or description of form's purpose. The main body is mandatory; it is the section that contains the input fields to the object for data entry.
- Table Windows:
  Table windows are constituted of rows and columns of data entry fields. A table window looks like a spreadsheet; it can present both scrolling and non-scrolling areas in a unique form. As form windows, table windows consists of a heading and a main body section.
- Bar-Form/Table Windows:
  These windows combine the features of the bar windows with those of the form/table windows; they are used to customise the methods on selected objects, so that any object can have its allowed functions.

The user can operate on windows via the keyboard or using the mouse. Under the currently active window there are always eight buttons, that can be operated clicking the mouse, used to perform actions (SELECT, DO, QUIT) or to obtain information (HELP) or to move in a list or a table (HOME, END, PGUP, PGDN). On the bottom of every form there are two lines reserved for system messages One line is devoted to trace the function and to messages of successful completion or error on it. The second line is dedicated to give information about specific fields or items included in the form (enter part number, ...).

4.3.3 Product Information Archive (PIA)

D. Koch

4.3.3.1 Overview

PIA is (like SIFRAME and IMS) one of the three essential parts of the integration platform. Its specification has been described in [1] and [2]. [3] gives an overview of its relevance for Concurrent Engineering. The coordinated work of team members demands a frequent exchange of information, so that the impact of decisions made at one stage is not allowed to remain unchecked by those whom it might affect until a later stage in the project.

A shared information model is a central component of a CSE framework

A shared information model, hence, is a central component of a CSE fra-mework. At the core of any CSE implementation must be a common model of the product data because one of the fundamental aspects of concurrent/simultaneous engineering is the ability of all tasks/teams to access an information model which is commonly understood and shared. In product development, most of the information that needs to be shared concerns the product to be developed and the existing ones. An integra-ted product model, therefore, must be at the kernel of such an information model for the product development. Through the system`s components, it must be possible to access and manipulate data which resides in diverse external formats and systems. The components should also present the data as specified by the user or application.

The Product Information Archive thus concentrates on the management of product information in a CSE environment, using a generic product data model implemented in an object-oriented database system.

PIA provides the sharing of up-to-date product information to all CONSENS components. Its main function is to support the information flow between parallel processes, i.e. in CONSENS the information exchange between the CSE tools, In addition, a human user interface allows the manual retrieval and update of product information.

PIA stores the temporal steps (versions) of the product development, thus allowing the CONSENS Tools and Advanced Engineering Services con-current access to the most recent, up-to-date version of the product status, as well as to earlier development stages. This enables a parallel execution of the different processes while sharing the same product information.

PIA supports the information flow between the processes and tools

The following example illustrates how the concurrency of processes is supported by the usage of PIA. Let us assume the case of an existing product for which a new variant is to be developed. The appropriate production facilities for the original product exist already. It is desirable to avoid a major change in the production facilities for the new variant. During the construction process, several tools (like the Design to Cost tool) can be used to generate a first version of the new variant. This information is stored in PIA where another tool – Design of Production Facilities – can access it and therefrom derive a first layout of the required production facilities. If it turns out that the first version causes an unacceptable change in the facilities, feedback is given to the construction process that the first version of the new variant should be further optimized. This results in the development of another approach to a new product structure. The construction and design of production facilities in this example are executed in parallel, thus granting earlier feedback on unviable approaches and reducing time to correct errors.

A database system is essential for handling the documentation, especially of complex products. All the data must be in a consistent state, and there should be no unnecessary redundancies. All processes requiring product data should have efficient access only to the data they need, and this data has to be represented according to the respective application views (like bills of materials, process plans, etc.). The more complex the products are, the higher are the challenges for the database system. Nowadays, relational database systems are industrially accepted for administrative applications. Their user-friendliness and simple data model present definite advantages over the previously used hierarchical and network databases.

Requirement: a powerfull database

However, for the innovative and challenging areas of technical data management, additional features are desirable. The object-oriented database technology was developed with the goal to reduce the shortcomings in modelling power and efficiency in traditional database technology, and to specifically address problems of engineering data applications.

Object-oriented database

Object-oriented database management systems provide a user-friendly data model that involves the actual objects the users are concerned with. Objects can be represented in the way the users think about them, and the translation into other data structures, like tables, is not necessary. This means that the system development and maintenance are simplified.

In the object-oriented data model, complex objects nesting within other objects can be represented intuitively, i.e. conforming very much to the way a designer or work planner thinks about a product he or she deals with. It is an object that contains other objects (e.g. subassemblies) which, in turn, consist of further objects (e.g. parts), etc. In relational data-bases these nested interdependences are split up to be represented in different relations (tables), with the consequence that the complex structure needs to be reassembled for queries concerning structural information.

Complex objects not only have the advantage of a close modelling of the users` perception of application objects. They also enable fast responses to queries that retrieve them. There is no trade-off between performance and non-redundant data storage (as in relational databases) because the same object can be contained within different other objects. In CONSENS, the object-oriented database system OMS (Object Management System) that is part of SIFRAME is used for the implementation of PIA. This has the advantage that the same database can be used to store product and project information.

Object-oriented system: OMS

The interface between PIA and the different CONSENS software modules consists of a number of standardized functions (implemented as object-oriented methods) that enable the manipulation of product data in the form of database queries. The set of functions has been identified by the needs of the different CONSENS modules. It can, however, be enhanced by additional functions, for instance when new tools with new information requirements are added to the system platform. The PIA product model is flexible enough to allow new views of the model to be generated by additional queries.

During the product development process, PIA is used as a tool for mana-ging and controlling product evolution: Several CONSENS tools or other systems can create a first version of the structure of a new product in the database. Teams working on the improvement and further development of the first draft can retrieve this information from PIA and use it for further development. The results of each development stage are written back to PIA and can later be traced for reference or for pursuing a different branch of development, e.g. if it turns out that one design decision has led to an undesirable product design.

Extensibility of PIA

The product data stored in PIA includes structural information of the products, as well as master part data for the products and all their components (subassemblies, parts), and process plans. PIA keeps track of the time validity of versions and different variants of the same component.

Managing and controlling product evolution

In addition to providing the facility for the creation and retrieval of product structures and inserting master part data, PIA has a graphical user interface that allows browsing through products and their structures, creating process plans, and retrieving various kinds of other product information.

The main advantage of PIA in comparison to the currently used EDMSs (Engineering Data Management Systems) is that it does not keep the product information as unstructured documents – which, as such, could be redundant and only reflect partial views of the product. Instead, PIA keeps the product information as an integrated model which comes as close to the reality of the product as possible. Any kind of required view can be generated from this integrated model, since it represents the structure of the objects, thus granting the system greater flexibility as well as the potential for future enhancements. The generation of views is performed by queries to the object-oriented database which are coded as methods stored in the database along with the respective object classes.

The PIA product model consists of two parts: a generic product model that carries the basic structural and temporal features of any product, and a company-specific model that presents a customization of the generic model to the requirements of a particular product in a particular company. The model is described in object-oriented terms as classes with instance variables.

Generic product model and company-specific models

CONSENS can be used efficiently with just the generic product model. However, a company-specific model used in addition to it presents further advantages. The classification of products, assemblies, and parts accor-ding to product-specific features facilitates the management of the product data and improves the system's performance. The company-specific models consist of subclasses that inherit their features from the classes describing the generic product model.

The instances of the respective classes are nested within each other, according to the class definitions.

4.3.3.2 Support of Standards

Compability with STEP

In order to ensure the openness of the system, the PIA data model has been adapted to comply with core entities of the STEP Resource Models (in particular Parts 41 and 44), so that the data exchange with other systems can follow the STEP standard. However, PIA is not intended to be an implementation of STEP. It addresses only a part of the STEP features and goes beyond the scope of STEP in others. A major difference between PIA and STEP is that the link between the generic product model and the company-specific models is achieved in PIA directly by letting the company-specific classes inherit all the properties of the generic model. The relationship between the generic model and company-specific ones is out of the scope of STEP, however. Another difference is the differentiation between versions and variants in PIA which provides semantic knowledge about product variants that logically belong together as well as the temporal development of products and their components. This allows, for instance, the generation of bills of materials, listing variants of the component in question. Since STEP makes no difference between variants and versions (they are both reflected in the entity ProductVersion), a generation of bills of materials, representing related variants, is not possible in the way it is by the PIA model. A third important difference is that STEP does not use the object-oriented possibility of classifying products and their components in a class hierarchy. PIA makes use of this object-oriented feature, thus providing a modular approach to modelling categorization of similar objects.

Last but not least, STEP does not include an integrated model for process data yet. STEP, on the other hand, covers a variety of aspects of product data that PIA is not concerned with (for instance, the representation of geometric data which is simply a reference to a CAD file in PIA).

4.3.3.3 Functionality

The main function of PIA is providing an integrated product information model for ensuring consistent product information flows between the engineering processes. Hence, the Tool Interface via which the CONSENS components exchange product data is the kernel functionality of PIA. The SDAI Interface provides STEP compliant data exchange with systems external to the CONSENS platform, and the graphical User Interface enables manual control of the product information.

Tool Interface

For an industrial company, the usage of the Tool Interface is important for software developers who want to integrate existing systems with the CONSENS platform. The Tool Interface provides a library of C functions which allow the input, retrieval and update of product information in the database. The functions concern:

Integration with CONSENS tools and pre-existing environments

- input,
- retrieval,
- changes,

of

- master part data,
- product structure,
- process plans,

for

- customer variants of a given component,
- different development or release versions of a given variant.

This means that every piece of product information is time dependent. The retrieval and update of the most recent version in the database is the most common and, therefore, default case. However, it is possible also to access information of previous versions. This may be necessary for going back to a previous development version if a later version, for instance, turns out to be inconsistent with previously unknown restrictions in the development.

Temporal versions and customer variants of product data

In order to connect an existing system to PIA and use the Tool Interface Library, a C-program has to be written that will reformat the output of the given system for input. The output will be reformatted as the parameters of the Tool Interface Functions in case of the input of data from the system into PIA, as well as reformatting the output of the Tool Interface Functions for input into the given system, in case of reading data from PIA.

A more frequent case will be the input from existing systems into PIA, since the different CONSENS tools are already designed to support the development process and once the product data is part of PIA it should remain there for further management.

Using the Tool Interface Library for connecting preexisting systems with PIA requires a certain amount of effort but ensures the most expressive way of exchanging product information between PIA and other systems since the library functions have a high level of semantics (for instance, there is a function for reading a bill of materials with versions and variants).

The Tool Interface is mostly used by the CONSENS Tools themselves to guarantee the smooth, parallel, and consistent interaction with each other. This activity is not (and should not) be seen by a human user. Itis the way data integration is achieved in CONSENS.

SDAI Interface

For the integration of the CONSENS platform into an existing environment, a partial SDAI Interface has been implemented. SDAI is the Standard Data Access Interface of STEP (Standard for the Exchange of Product Model Data). It is defined as a functional specification, i.e. headers of functions that shall perform certain basic activities for product data exchange, like opening a session, retrieving an object, or reading the value of an attribute.

STEP compatibility

The semantics of the SDAI functions are on a relatively low level compared to the Tool Interface Functions.

SDAI is not a software that can be bought but a specification document. Implementations of this specification depend strongly on the systems used. This means that in order to use it, most applications have to take care of their own implementation. This has been done also in the case of CONSENS, where SDAI is implemented using the database query language OML of OMS and C++.

User Interface

A graphic user interface based on X Motif enables human users to mani-pulate product information just like the Tool Interface. The same functions are provided.

Usage

Graphic access to product data

A company which uses CONSENS and thus PIA has different possibilities for usage:

- loading the first version of product data manually, i.e. using the graphical User Interface, after which further manipulations will mostly be car-ried out by the CONSENS Tools,
- integrating existing systems that are product data sources in the com-pany (like CAD systems) with PIA by using either the Tool Interface or the SDAI Interface,
- using CONSENS and PIA as it is without integrating other systems in the company. Then, the source of the first version of product data will be provided by KnobieCAD which is an intelligent interface from a CAD system to PIA.

The usage of PIA is mostly hidden since it provides the way different tools and systems exchange their product data and can access historical data.

Integration of product information flows

However, human users can equally retrieve, manipulate, input, and check the product information in the database. It can serve as a product infor-mation archive for all the data of newly developed products.

4.3.3.4 Outlook

PIA has been implemented as a software prototype for a product data-base that enables the integration of product information flows in a CSE environment.

Customisation

It is an integrated part of the Integration Platform of CONSENS. It has been shown that by means of company-specific sub-models the database can be customised to serve particular needs that are not included in the generic model because they depend on the given company. This makes PIA very flexible with respect to the integration into existing systems and adaptation to new requirements.

Future enhancements may include the integration of production-relevant data into the model, i.e. product information that pertains individual, produced products or assemblies, like, for instance, individual test reports.

The completion of the partial SDAI Interface to a full one is likewise for future work.

References

1 D. Koch, T. Diepold, O. Schumacher: Architectural Specification of the Product Information Archive, ESPRIT III 6896 CONSENS, Deliverable D3.2, Edited by: IAO, Stuttgart, October 1994
2 D. Koch: Functional Specification of the Product Information Archive, ESPRIT III 6896 CONSENS, Deliverable B3.2, Edited by IAO, January 1994
3 D. Koch, J. Warschat: Object-Oriented Product Information Management for Concurrent/Simultaneous Engineering; Proceedings of the CEEDA '94, Bournemouth, UK

4.4 Summary

S. Kessler

The CONSENS Platform is an innovative software solution to support the enter product development process in a CSE oriented environment.

This platform enables the newly defined organisations to describe, monitor, manage and execute in real-time the newly defined processes in a company. With CONSENS as a self-documenting system organisations are able to enhance quality by having a clear view of how their processes work in the new CSE environment. The users are "free" from administrative overhead such as searching for the latest versions of data and information. The tools necessary to perform single process steps are launched automatically, together with the right information/data when ever the user executes one of these steps. The graphical representation of process flows ease keeping the newly introduced processes for product development.The deep integrated interfaces between the platform SIFRAME and the information management system IMS allows the user to store and retrieve all metadata related to the current product development in an easy and convenient way. PIA as the product information archive serves for all integrated and encapsulated tool as the kernel storage for product data. Through interfaces, such as SDAI, the tools used within the product development process can access easily data necessary to work concurrent and simultaneous in the projects.

The CONSENS platform does not replace any of the already existing tools and applications in a company. Instead it serves as an umbrella under which all these tools are integrated in order to support the new environment in a CSE oriented company. Resources and process flows can be changed at any time. They allow a flexible reaction on changes in the requirement during the product development. All this ensures that the right information is delivered to the right person at the right time and for the right reason.

5 Enabling Technologies II: CONSENS Tools

5.1 Introduction

E. Sleeckx

The transformation from a sequential working enterprise to a CSE oriented enterprise requires not only a fundamental restructuring of the enterprise but also the installation of supporting information systems. Within the CONSENS project such information systems have been elaborated and implemented in the companies of the end-users. Similar to the transition methodology, the information system must fit into the overall strategic of CSE, i.e. contribute to Par-

Information system must fit in the PSI-strategy

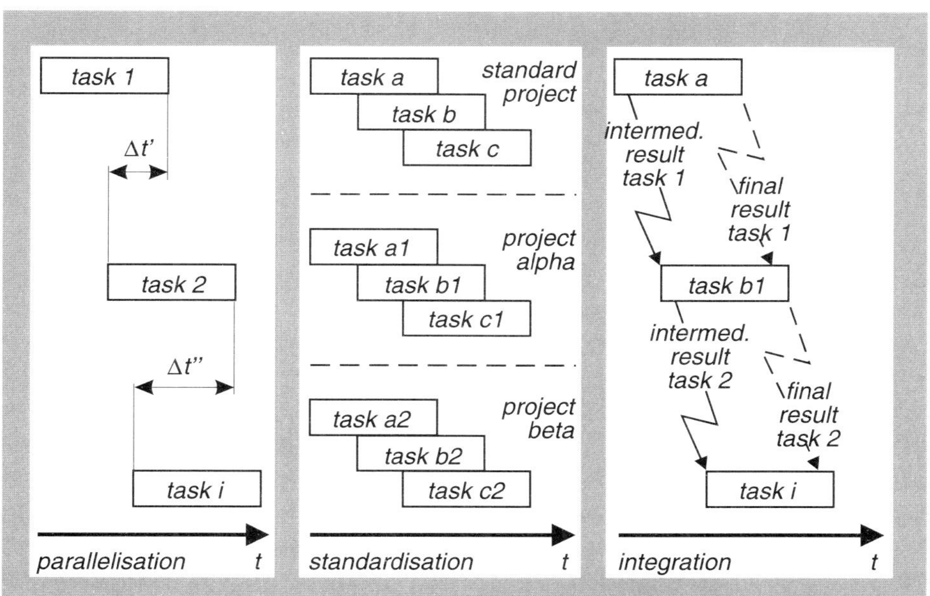

Fig. 5.1: The PSI strategies

allelisation, Standardisation and Integration. In the areas of organisation, process, product and resources, the CONSENS information system fully supports the strategies. For the sake of clarity, the meaning of these strategy factors is recapitulated before the relation of basic modules of a CSE information and CSE strategy is described. Parallelisation means the capability to carry out interrelated processes simultaneously. Standardisation is the unification of processes which show a high degree of similarity or the possibility of repetition. Integration is summoning and carrying out simultaneously different aspects related to product development. For integration, the product and the development process must be regarded as an entirety (see figure 5.1).

Firstly a justification is given for the selection of the basic modules of the CONSENS information system. An answer will be given to questions like: What have been the main reasons for selecting some modules to be integrated? Is there a logical connection between these modules?

5.2 Support of CSE by Product-oriented tools

E. Sleeckx

Product and Process oriented tools

However, before we can answer these questions let's take a look at what modules have been selected and at how they fit into the CSE strategy. In the CONSENS platform two types of tools have been integrated: project (or management) oriented tools and product (or design) oriented tools. The project-oriented tools are: EPM (Engineering Process Management) and MDS (Management Decision Support system). The product-oriented tools are: DtC (Design to Cost), DfMA (Design for Manufacturing and Assembly), DfQ (Design for Quality), DoPF (Design of Production Facilities), ICAD (Intelligent CAD-system) and finally GT (Group Technology). In the following paragraphs, we will describe the relationship between the product-oriented tools and the PSI-strategies. In addition we will take a look at the two project-oriented tools.

Design to Cost (DtC)

Don't overlook the product costs

Controlling the cost of products has always been a major concern for the product development department. Due to the increasing pressure to reduce lead-time, less explicit attention can be given to cost control. Therefore, tools must be developed and integrated in the product development chain to continuously monitor and control the cost of products, to avoid the reduction of lead-time having a negative impact on product costs. The relation of DTC with the PSI strategy of CSE can be found in the following areas: Standard procedure in most companies is to make a cost calculation only when design is completed. With DTC, the calculation can be updated anytime and done in parallel to other tasks. DTC can hence contribute to parallellisation when the designer is taking into account the manufacturing cost and the assembly cost during the functional design. DTC contributes to a better structuring of cost data, specially when the module is used in conjunction with a product information archive. This will make the processing, storage, retrieval and demonstration of cost data easier. This leads to an improved integration of all company activities requiring or using cost data. The DTC module also causes a natural integration of design and manufacturing, and has a strong relation with other modules like DfM and DfA.

Design for Manufacturing (DfM) & Design for Assembly (DfA)

Think manufacturing

The stringent focus on reduced lead-times puts the slogan "Do it right the first time" in the spot-light. Continuous disturbances of regular production processes are due to engineering change orders. These are caused by improper design for manufacturing which distinguishes a sick and ready-to-die company from a healthy and expanding company. The main problem with designing a product that can be manufactured in optimal conditions, is again lack of time to do the manufacturing analysis thoroughly. Luckily, manufacturing constraints and capabilities can be formalised and brought into a design system. The functional design system is then up-graded to a Design for Manufacturing/Assembly system. The contribution to the PSI strategy is clear:

DfMA enables the parallel execution of functional and manufacturability analysis. The barrier between design and manufacturing vanishes. DfMA also enables manufacturing constraints and cost to be considered very early in the design process and so optimise the mechanical performance of the product taking into consideration manufacturing criteria. In relation to DfA, the DfM module enables the simultaneous design of individual components belonging to the same product. Given the limitations of the assembled product, the details of single parts can be modelled in parallel.The integration of the DfA-tool and the Design of Production Facilities tool enables the user of the DfA-tool to relate equipment components like tools, fixtures, parts bins, machines, etc. to part or assemblies of the product. In fact, integration of the Design for Manufacturing, Design for Assembly and Design to Cost is a prerequisite to obtain CSE. The results worked out employing these tools may be different, because each tool alone is aimed at a different goal. On the other hand, these tools use partly the same data. The tools therefore should work closely together and data used from the tools in common should be entered and stored only once. DfMA contributes largely to the standardisation of product design and functionality through the use of standard functional features. The result is in an increased standardisation of manufacturing procedures, methods and tools. This standardisation effort can benefit enormously from the link with a Group Technological Repository, containing in a structured way, existing parts and designs.

Group Technology (GT)

The memory of a company covers 5 years

Re-use of existing designs not only reduces design time but also has a tremendous impact on manufacturing. Due to larger series, procurement can order larger quantities, the number of stock items can decrease, the number of suppliers decreases and there is less chance to make errors, ... Why for then re-invent the wheel again and again? Because you do not know where to find that part you need, that's why GT puts an end to these kinds of discussion. GT collects, structures and allows the re-use of experience from former designs and processes. It allows the optimum exploitation of the company's memory. Using family concepts and group technology previous design effort is recycled in new projects to produce new designs. The GT is perhaps the most PSI oriented technology, the closest to the heart of CSE:

Through the use of GT, an intrinsic standardisation of parts is obtained and hence also of machines, equipment and other resources. The same applies to the processes in a project, so that the project flow can be standardised. By reducing the number of new products, GT is a de facto standardisation generator for resources, projects and products. The same applies to processes that can be standardised and limited in number. As the GT classification not only contains geometrical product data, but also process, cost, ... data, it also contributes to the aspect of company wide integration e.g. If there is a match between the new product and one of the former designs, the matching data can be read again from the RGT tables and transferred to other design tools, like DfMA. GT contributes to parallellisation by avoiding new activities. By retrieving a part, all its other information is available as well, thus avoiding a new work preparation, avoiding a new cost estimation, ... all the information is there ready to be used.

Design for Quality (DfQ)

Smile, tomorrow will be worse

While chasing for reduced lead-times and costs, the demon of Quality is likely to appear and ruin your good intentions. Quality -unfortunately- is one of the strongest supporters of the Law of increasing Entropy. It remains, even after years of brainwashing people, one of the hardest things to grasp. To be able to design quality in a new product one must support the designer by providing him with tools, DfQ-tools. The DfQ tool provides the necessary support for a successful implementation of CSE strategies in different areas. In terms of CSE, a product consists of information about its basic characteristics, like geometry, structure, etc., as well as of a description of the production process, like process and test plans. By the simultaneous treatment of design and production-related aspects, the DfQ tool supports this integrated product view. The DfQ tool supports the detailed planning of the production process at an early stage of the product development. Quality simulation enables the development team to predict the delivery quality of a new product, in order to fulfil quality requirements during the design process. Within the DfQ process, measures for error detection are developed. Their evaluation also leads to possible testing strategies for the product. This means that the development of the production process and of the product itself are integrated by the DfQ tool. Another aspect of integration covers the use of quality data gathered from the current production process. This means that problems detected during manufacturing are immediate input for the development and improvement of the product. Information produced during DfQ provide the basic data (history) for subsequent projects. This also contributes to a standard representation of engineering data.

Design of Production Facilities (DoPF)

A product requires a production system, that is build from products that require...

A lot of attention is given to the development of the product, one should not forget that some of the production equipment it-self has to be developed. In particular, when it concerns specific assembly stations, that depend to a large degree on features of the product, a close integration between product design and design of production facilities has to be envisaged. DoPF-tools provide this functionality. Here again by their nature, one can feel the importance of these type of tools to reach the goals of CSE. The possibilities provided by DoPF-tools to check the design of production facilities and means of production with regard to ergonomic aspects (dimensions, reach ranges, vision ranges, maximum loads allowed, ...) minimises costly use of methods laboratories. Therefore, this shortens the time from planning the production facilities to the realisation.

The close linkage between DoPF and DfA makes it possible for the means of production (tools, parts bins, jigs, fixtures, machines, ...) assigned to the product components, to be faded in directly in the layout of the work system. DoPF supports standardisation by providing the possibility for each user (company) to create its own subset of equipment components. So, executing the design of production facilities only standard components will be used, if sufficient. A further step towards standardisation is the possibility to put the design solution of former projects (or parts of them) into the archives and to access them in later projects.

Intelligent CAD: KnobieCAD

From Computer Aided Drawing, to Computer Aided Design

The CAD systems plays an important role in the total product development process as it is the major source of detailed product information. Design for X tools have to work on the CAD model in order to provide the designer feedback from manufacturing, cost and quality analyses. For these tools to work properly two conditions must be fulfilled: that the information they receive be accurate and conceptually correct, and that the information they need be actually released by previous processing. These requirements go far beyond the capabilities of traditional CAD-systems as it requires a level intelligence currently only available in advanced Intelligent CAD-systems, the KnobieCAD-tools. The KnobieCAD-tools are an important constituent of all intelligent design tools. The information they are able to provide is also important for project-management. They offer the basis for standardisation, integration and parallelisation.

5.3 Product-Oriented Tools

5.3.1 Intelligent Computer Aded Design: KnobieCAD

P. Raiteri

5.3.1.1 Introduction

Integration between CAD and knowledge-based shells

KnobieCAD (KNOwledge-Based Integrated Enhanced CAD) is a system and a development platform implementing communication, and hence integration, between two classes of software objects: CAD systems and knowledge-based shells. The reason for this is the opportunity to augment the functionality offered by CAD systems with the support services that knowledge-based applications can provide, exploiting state-of-the-art knowledge-based programming methods and tools to implement them directly on the CAD data.

KnobieCAD-supported applications could assist the product designer in a virtually unlimited variety of ways, enhancing the effectiveness of the first step in product design: the production of drawings using a CAD. Being based on knowledge-based technology, they are expected to help in capturing the corporate experience in product design, and be able to work in general when problems have a fuzzy nature and data is not well defined.

Since the very beginning of the design process is addressed, benefits in terms of the PSI (Parallelisation, Standardisation, Integration) strategy to Concurrent Simultaneous Engineering, pursued by CONSENS, are expected to be noteworthy.

Applications of KnobieCAD

KnobieCAD applications can span over an enormous range of functions, from the design review to manufacturability review, including review of drawings, various possible kinds of assistance to the designer, and possibly providing parametric capabilities to non-parametric CADs. Moreover, KnobieCAD would ease the task of capturing, preserving, using and distributing to all CAD users the experience about design activities existing in a company. By having access to this experience of different design problems in an on-line fashion, designers would work in a much faster and effective way.

KnobieCAD supports application development firstly by its very existence, which provides an architectural framework, and secondly by providing two APIs (Application Programming Interfaces), established within the CAD system and the knowledge-based shell.

Application Programming Interfaces

The KnobieCAD pilot application in CONSENS demonstrates these concepts and architecture. Indeed, this application allows the CAD user to work with a knowledge-based system that decides the surface protection treatment for parts that touch each other, in the design of an aerospace assembly, without leaving the CAD environment and with connections with the CONSENS Product Information Archive (PIA).

In fact, KnobieCAD is integrated into the CONSENS system by establishing a link between the CAD system employed in the target company and the Product Information Archive (PIA) module. This is important both for the needs of CONSENS, which early in the progress has at its disposal information on the products that are being developed, and for the CAD user and the KnobieCAD KB applications, to access the existing product information.

5.3.1.2 KnobieCAD Support to CSE

CSE (Concurrent Simultaneous Engineering) aims to achieve three main goals:

Goals of CSE

- reduce time-to-market for product development,
- reduce costs,
- improve overall quality.

PSI-strategies of CSE

The PSI-strategies of CSE address these goals by:
1. Parallelisation,
2. Standardisation and
3. Integration

in the product development process.

From a technical perspective, the manufacturing industry may convincingly claim that it has already achieved quite a bit regarding its attempts to parallelise and standardise production processes as much as it is reasonably possible. Treating the manufacturing process as individualisable, modular components, much has been automated to reduce development time, improve quality, and reduce overall costs. A range of deployed CAD, CAE, CAM, CAP, etc. systems document these efforts.

Yet these systems still need to be improved, especially in the framework of today's more and more competitive production and market environment.

As much as Computer Aided (design-related activities) systems (CAX in the following) are a help, they add complexity of their own to the design process. Applications of the same module category maintain separate and redundant object libraries, there may be problems of version control, and not all engineers may be uniformly skilled in using a particular tool. In all cases, engineers must iterate a "hypothesize, evaluate, re-hypothesize" cycle so as to adequately consider all constraints which are imposed on the product design. This cyclic process cannot be avoided, as it is an inherent technique of the design method. But in many situations, it may be automated, accelerated, and condensed, liberating the engineer to concentrate on the creative aspects of product design.

Product design as a cyclic process

Problems with modular CAX systems

There remains one other problem derived from automating the manufacturing process with modular CAX systems. As much as they are modular, hide data and insulate their domain from the distracting influences of their peers, they fail to integrate and communicate well among each other, and inhibit the back propagation of corrective information throughout the entire manufacturing process. Modular CAX systems need a mechanism by which their output may be analyzed and interpreted for "knowledge sharing."

KnobieCAD is motivated to address such problems at an early stage, and in a first-approximation manner, in the framework of CONSENS.

Applications supported by the KnobieCAD system can be developed to match a number of different but interrelated engineering problems. These are very much relevant to the implementation of concurrent engineering, at least in the first phase of product development – design.

Here is a brief overview of the possible KnobieCAD application subjects.

Design Review

Design review

It is one of the most time-consuming activities in design, usually requiring experienced engineers. Often this analysis is essentially qualitative.

The main tools for performing this analysis are engineering and design standards, drafting standards, engineering specifications. Also, project-specific standards and specifications are defined (and managed within CONSENS). These standards can easily be translated in KBS rules, and entered into KnobieCAD to be applied during design.

Drawing Review

Drawing review

Once the drawing has been completed, it must usually be checked for consistency and correctness. This process is not always based on expertise, but it certainly involves a checklist, and consistency rules. These issues can be handled by KnobieCAD applications, either in batch or in interactive mode, the latter to assist the user while the CAD drawing is being developed.

Design Assistance

Design assistance

Product design can often be modularized, to take into account for instance the following areas: material selection, fastener selection, equipment selection, geometric configuration and human factors. To do this, the usual practice is to consult experts inside the company. These types of design problems are ideal KnobieCAD application candidates, and can be effectively faced with a modular, incremental application development.

Parametric Design

Parametric design

Incidentally, particular KnobieCAD applications can be used to transform into "parametric" CAD systems that are not originally so. For instance a KnobieCAD KBS application could examine the parameters for geometry and configuration of the drawing, and call the CAD procedures to directly modify the design model.

What's more, KnobieCAD provides the means for such applications to accomplish their parametrisation task without changing the CAD user interface, having the CAD user continuing to work in the old way, with the old proven procedures. This can help in solving the acceptance problems that parametric CAD systems still have, as they force the user to adapt to a completely new way of producing and modifying drawings.

Manufacturing Review

Manufacturing review

Once a design has been generated, it typically undergoes a manufacturability review. Since drawings do not usually satisfy at first the manufacturing requirement, the result can be a cycle of drawing modifications – reviews, slowing down significantly the design activities. KnobieCAD applications have at hand the design information necessary to implement manufacturability reviews, and machining and manufacturing tolerances checks – at design time.

Advantages of applying KnobieCAD

More general, KnobieCAD applications would provide a distribution of expertise to all CAD users. By having access to the knowledge about different design problems in an online fashion, designers would work in a much faster and effective way.

This is regarded as an important aspect of concurrent engineering, as it saves design iterations, by applying the relevant and exact engineering knowledge at the earliest possible time in the design flow.

Moreover, the possibility to perform a multi-disciplinary review already during the design definition, reduces the time normally taken by the design to be approved once completed. This can also reflect on the time the company needs to develop the manufacturing plans relevant to the specific design. As a result, the design and manufacturing process is more responsive to changes and modifications.

Finally, consistency of design, application of drafting and notation standards, and conformance with specifications, can result in huge savings in a manufacturing company. The use of standard rules for design and design review, as supported by KnobieCAD applications, can improve the ability of the tools of CONSENS following logically in the activity flow in a given scenario to control manufacturing quality and cost, starting from a more uniform basis.

KnobieCAD in the CONSENS enviroment

To implement CSE inside manufacturing companies, CONSENS takes the approach of an integration of a number of engineering tools and services, to provide users with a consistent product data base at the earliest possible time.

This means that, as soon as information is released from a tool, the tool that is logically placed afterwards in the engineering process flow can access and work on that data. This is ensured by the CONSENS framework, the central database system (PIA), and by the functionalities of document management provided by IMS.

The CONSENS tools are in general of the kind of design for X, where X is cost, quality, manufacturing, assembly ... These tools take their input data on products to be manufactured ultimately from the CAD drawings, and a number of different sources (process information, cost information ...) contained in the CONSENS PIA. However, for these tools to work properly two conditions must be fulfilled: that the information they receive be accurate and conceptually correct, and that the information they need be actually released by previous processing. In all cases, no processing can be performed by these tools while the relevant CAD drawings are not released.

In this framework KnobieCAD applications can already implement a sort of pre-design for X, accessing PIA information for what concerns process, quality, and cost and using CAD models well before the official release of the product drawings. This would bring two types of benefits:

- **shortening of cycle time in design acceptance**. Even with the efficient procedures of product information management implemented by CONSENS, one has no way of knowing if a particular product would be accepted on a cost- or assembly-related basis by the relevant design for X tool, before the tool calculations are finished. A specialized design for X tool provides a highly accurate estimate of its relevant product design aspect, but a rough calculation, as the one that KnobieCAD applications could already provide at drafting time, would help to better address the development of particular product features – in a better way – which could cause problems, at a very early stage. A product, represented in CAD drawings preprocessed by this kind of KnobieCAD application, would undergo a very rapid development indeed.
- **accuracy and consistency of product information.** Accessing in parallel PIA and the CAD drawings, KnobieCAD applications check that the manufacturing rules be fulfilled in the drawings, that the suitable notation stan-

dards be applied and that manufacturing specifications be met. This again should result in an improved efficiency during the design for X tools operations, as the design information would be inserted by KnobieCAD applications into PIA "clean and clear", without possible spurious features that would need analysis to be resolved.

5.3.1.3 Notes of Architecture

Parameters with regard to KnobieCAD

From a system-oriented viewpoint, KnobieCAD has fuzzy requirements. Not only the knowledge-based applications inserted in its environment must be able to develop different kinds of reasoning strategies on different sets of data coming from CAD, and thus their domain and number are not pre-defined; also, the number of actors in the KnobieCAD environment is variable and the type of data that can be accessed is totally beyond an initial definition.

Thus, the development of KnobieCAD has followed a few but very firm guidelines, aiming at overcoming these difficulties:

Guidelines for the use of KnobieCAD

- the various processes interacting in the KnobieCAD environment exchange information in the form of messages, with the same format,
- all messages pass through a single communication server, so that checks can be centralized and an effective queue management policy can be implemented in a single process,
- the KnobieCAD user interface exploits the user interface of the CAD system which is inserted into the environment. Thus, the CAD user is not confronted with a different working tool and does not need to learn different conventions,
- external data, wherever it can be found, is accessed by independent processes that open the file or database or interacts with yet different external environments, and communicates with the rest of KnobieCAD only with messages. These processes in KnobieCAD are called drivers. This allows not to change the rest of the system because of a different data format to be accessed. On the other hand, a slightly bigger development effort is necessary to convert the target data format to KnobieCAD standard messages, when such a need arises.

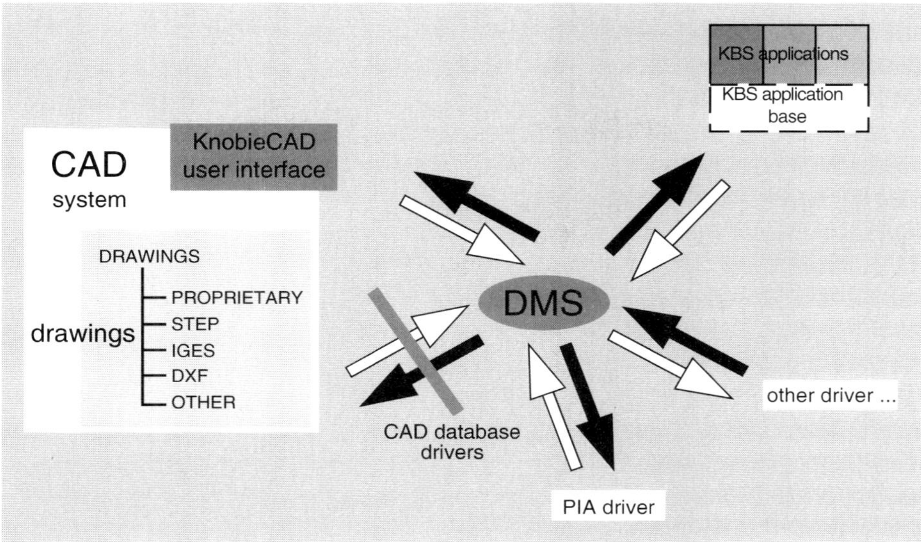

Fig. 5.2: Structure of the client-server configuration

KnobieCAD – Dialogue Manager Server

Thus, we define for KnobieCAD the Dialogue Manager Server (DMS), an independent process, managing dialogue among different KnobieCAD processes. DMS is always running and alert in the background during normal operations, ready to accept requests from its communication clients and pass them to the target process. DMS acts as a message scheduler and queue manager, shortly as a demon, handling communication traffic between its two clients.

The communication clients are the KnobieCAD user interface, one or more knowledge-based applications and any driver that happens to be active in the system.

The overall KnobieCAD system architecture is summarized in the figure above.

Structure and modules of the client-server configuration

The modules appearing in this figure are briefly described here below:

- DMS handles the communication among the other KnobieCAD components, using UNIX sockets. It implements checks on the syntax of messages and manages message queues.
- KnobieCAD-based KBS applications. KnobieCAD is capable of supporting a number of different KBS applica-

tions at the same time. KBS applications also act as servers in the KnobieCAD architecture, in the sense that they are activated by requests coming from the CAD, process them and reply as appropriate.

KBS applications in the KnobieCAD environment are based on an application base which provides them with the communication services they need.

The KnobieCAD prototype employs for the development of its KBS applications a knowledge-based shell called , a product of Inference Corp. (USA). However, in principle there is no reason to limit KnobieCAD to the use of this shell: other shells can be employed as well.

KnobieCAD user interface

- The KnobieCAD user interface, implemented as an application in the environment of the target CAD system. The user interface consists of a series of menus, dialogue boxes and functions the CAD operator can access to control the KnobieCAD services. Within the scope of the CONSENS project, that is, all the KnobieCAD prototypes, the target CAD for KnobieCAD was Intergraph's/Solaris.

Driver processes

- A variable number of drivers, to implement communication with external systems. In general, driver processes are used in KnobieCAD each time a communication, implying a format conversion, is required. Format conversions can be functional (when a need arises to communicate with an environment which already has an internal communication functions suite, different from the KnobieCAD standard) or can identify a data format (for instance to access CAD files).

Support of application developers

To support application developers, KnobieCAD provides two APIs, acting in the points where the effort to develop KnobieCAD applications is more substantial: the CAD interface and the knowledge-based shell. These APIs mainly contain functions that handle sending messages. They are available within CONSENS on the first KnobieCAD target environments: MicroStation V.5.0, and ART-IM V. 2.5 for SUN

5.3.1.4 An Example of Application: STEP into KnobieCAD

Definition of STEP

STEP (ISO 10303) is the ISO standard for product definition, modeling and description, which provides a common environment, a language for product specification and a methodology to allow the sharing of product information. STEP has a complex internal structure of sub-standards, including a file exchange format specification among different CADs, a product definition language (called EXPRESS, ISO 10303-11) and the framework to develop application protocols (ISO 10303-2xx), vertical formalisation of application domains. A possible application family for KnobieCAD would provide a link between CAD drawings and product description in EXPRESS. This is interesting in the framework of CSE because there tools to describe graphically a product structure using EXPRESS-G (the graphical formalism for EXPRESS), but it is impossible using these tools to insert graphical product features, as those that can be extracted from a CAD system, into the EXPRESS model of the product. This would bring a greater effectiveness in the development of products, enhancing the design portability over different platforms.

KnobieCAD and EXPRESS

Currently the KnobieCAD user and communication interface under MicroStation translates the internal drawing model of the CAD into the formalism of the STEP physical exchange file specification (ISO 10303-21). This specification offers the possibility to describe the draft of a product in all its graphical aspects, grouping the graphical entities, to form parts and components, in a very general way. Any kind of graphics can be described, 2- or 3-dimensional, as well as any kind of product, given the generality of STEP.

Mapping and restrictions of EXPRESS

However, the information that can be coded using ISO 10303-21 is not as general as is possible to describe EXPRESS. ISO 10303-21 does provide a mapping of EXPRESS statements and structures into the physical exchange file format, but the mapping is not complete, as the types of data available are reduced and some features of EXPRESS are not implemented in the physical file format. Thus, it is not possible to go back to the EXPRESS data types when a product is specified by only using the ISO 10303-21 formalism.

A specific KnobieCAD application can help to solve this problem. This application would maintain an extended product model with respect to the one which is a standard feature of the KnobieCAD application base (based on ISO

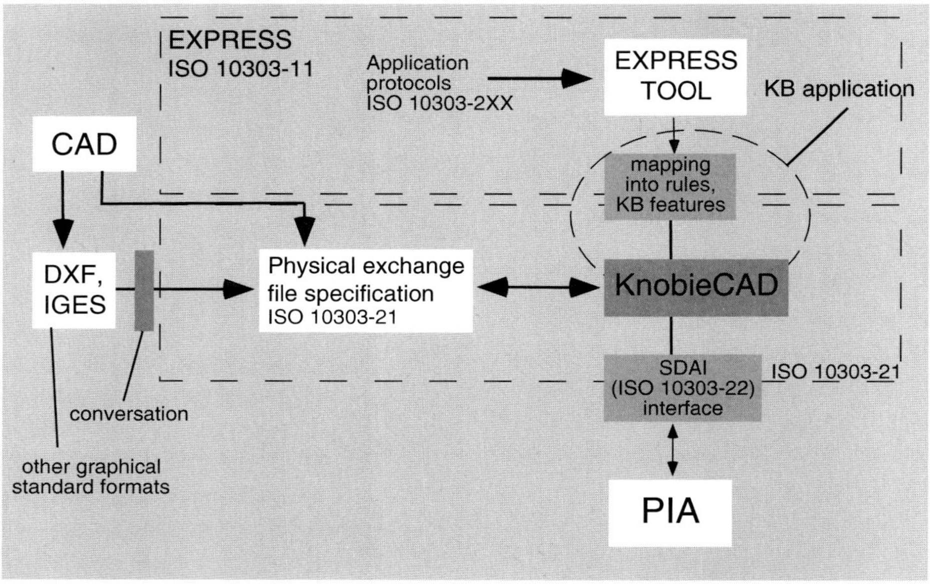

Figure 5.3: STEP-standard KnobieCAD application

10303-21), which would allow to attach the graphical features, taken directly from the CAD interface, to the EXPRESS description of the product under development, also organized into an application protocol. Within CONSENS the product structure can then be exchanged with PIA, using its SDAI tool interface (SDAI is another standard of the STEP family: Standard Database Access Interface, ISO 10303-22). The structure and interfaces of such a KnobieCAD application would be the following:

Exchanging the product structure by means of PIA

Since not many CAD systems of today offer the possibility to save data in the STEP physical exchange file format, an interface to IGES (Initial Graphics Exchange Specification, the ANSI standard for graphics exchange) or DXF (the AutoCAD proprietary standard for file exchange between CAD systems, supported by most graphical packages) could also be required. The two formats offer the possibility to describe drawings at roughly the same level of detail, although in DXF 3-D representation is not completely supported. So, to match the KnobieCAD internal representation, a conversion from one of these formats to ISO 10303-21 is necessary, and, in the framework of the KnobieCAD architecture, should be carried out by a specialized driver.

IGES or DXF formats required to match the KnobieCAD internal representation

5.3.2 Design for Manufacturing

J. L. T. Santos, S. Tilly

5.3.2.1 Design for Performance

The Design for Performance (DfP) tool supports designers, manufacturing, and structural engineers in improving design performance while ensuring manufacturability of the final product. The tool may be used for the design of both mechanical components and built-up structures. Capabilities of this tool are presented in the following sections.

Use for the design of both mechanical components and built-up structures

The DfP Tool Environment

The software is integrated with several commercially available codes such as I/EMS, PATRAN, ANSYS, ABAQUS, ASAS-NL, and ADS with special-purpose modules for topology optimisation and detailed design. These modules share a common User Interface which partially replaces module proprietary interfaces. The backbone of this integrated system is a local relational database (DB). The data is retrieved from application-specific databases or output files, stored into the DB, and then used by downstream modules. Since each module has different views of the data, wrappers have been developed to transform the data into a form which is acceptable to the application program.

Capabilities of the DfP Tool

The DfP tool contains modules to support the definition and optimisation of the topology and of the shape of mechanical components. It guides the designer through the interactive or automated optimisation of dimensions and spatial orientation of structural components. These capabilities are accessed through the four main branches of the menu-driven, Motif-based UI: model, modelling, analysis, and optimisation, see figure 5.4.

Modeling Capabilities

Modeling tools support the definition of the geometric, finite element, loading, material, parameter, performance,

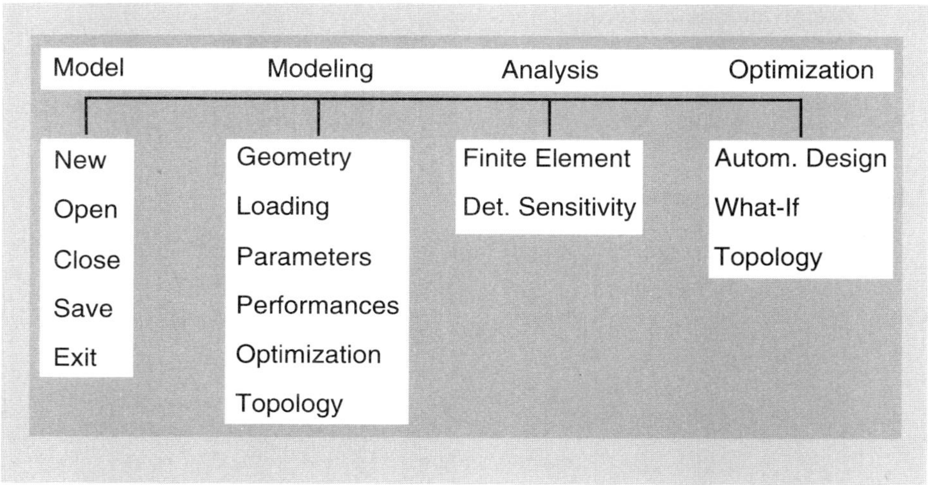

Fig. 5.4: Main menu tree of DfP tool

and optimisation models. The sequence to generate of these models is not arbitrary. Firstly, the geometric model must be built. It constitutes the foundation of all other design and analysis models, see figure 5.4. For the topology and shape component design a CAD modeler such as I/EMS is more suitable. For the size and configuration design of built-up structural systems a geometric modeler such as PATRAN is more convenient. For purposes of analysis the geometric model is used to create a finite element model. Geometric and finite element models together with material and loading models, completely define the structural model. In the DfP tool, the definition of the structural model is supported by I/EMS and PATRAN.

Parametric and performance models

The parametric and performance models must be added to the structural model. Finally, parametric and performance models are used to define the topology and optimisation models. These models together constitute a design model. General dependencies between models as well as the order of their creation are shown in figure 5.5

The definition of the design model is facilitated by the DfP graphical user interface. Figure 5.6 and 5.7 show, respectively, the editor menus for size parametrisation and displacement performance definition. To specify objective and constraint function formulae, a graphical editor is provided to allow for polynomial expressions, see figure 5.8.

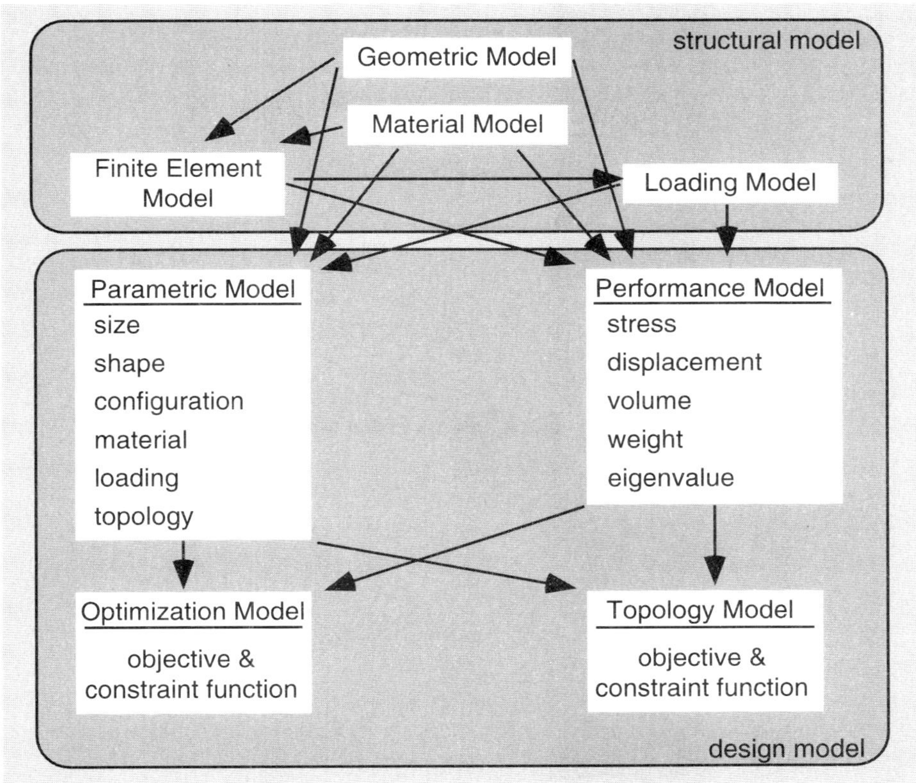

Fig. 5.5: General dependencies between data models

Analysis Capabilities

To support the finite element analysis and design sensitivity analysis of mechanical components and large structural systems, several computational modules have been integrated. For the design sensitivity analysis, the continuum adjoint method is used. It requires only one original analysis and then a restart analysis for each perfomance sensitivity computation. Evaluation of design sensitivity information is performed entirely outside the finite element analysis code allowing to maintain a modular architecture.

The continuum for design sensitivity analysis

Optimisation Capabilities

Within the interactive methodology used in the DfP tool, automated topology, "What-if" studies, and detailed design optimisation may be used to find an improved design. To

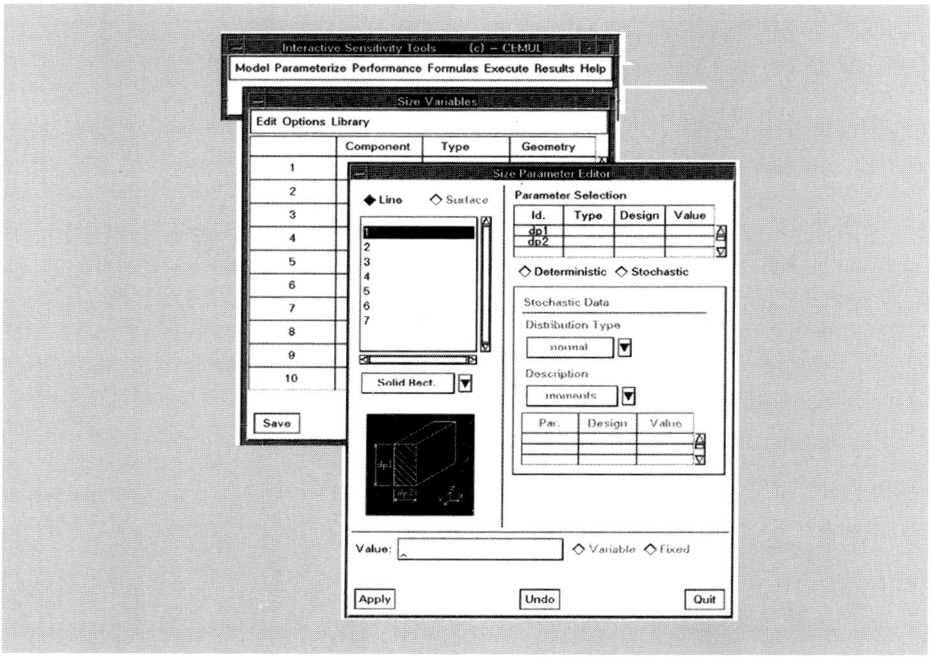

Fig. 5.6: Size parameter editor

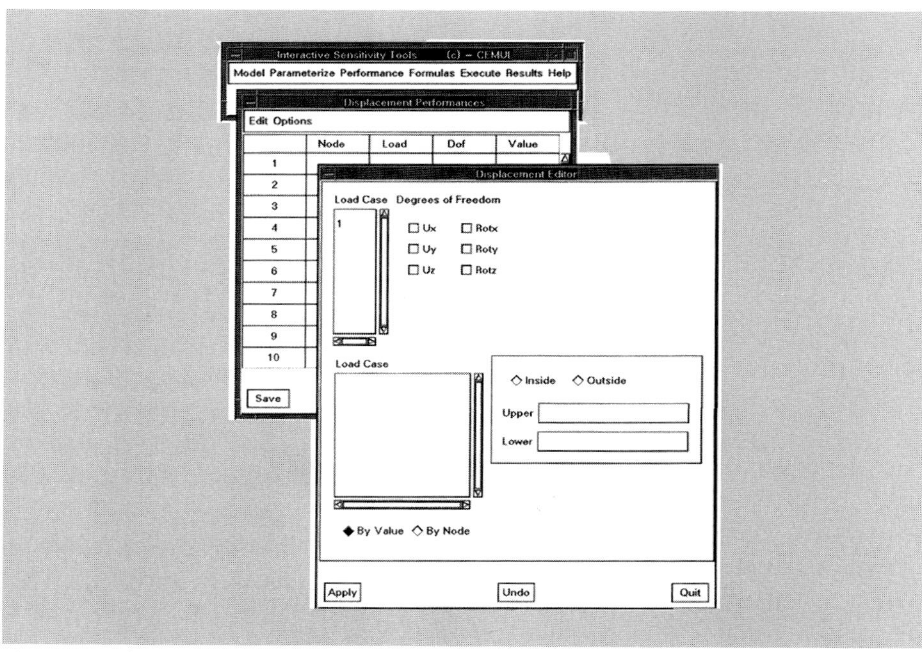

Fig. 5.7: Displacement performance editor

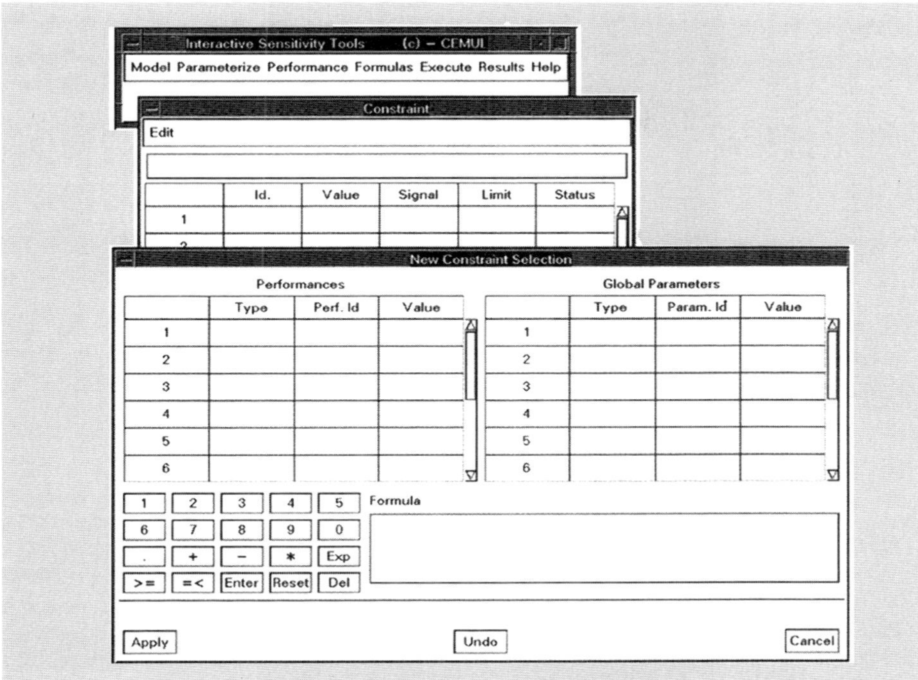

Fig. 5.8: Constraint function editor

carry out topology optimisation, the design engineer applies loads and boundary conditions to a finite element model consisting of cells covering the permissible design space. The optimum material distribution that maximizes structural stiffness for the given applied loads and boundary conditions is calculated to predict the topology of the component.

"What-if" studies

"What-if" studies are used to find an improved design without having to perform new structural analysis. They are based on a first-order Taylor expansion of the current design. During the analysis of design sensitivity information, the user may modify the values of some design parameters and inquire about new values of performances, objective and constraint functions associated with the new design.

Interactive and automated design

Interactive and automated design optimisation may be used to optimise the detailed design of mechanical components or built-up structures. The parametric and variational capabilities available from the I/EMS solid modeler are used for 2-D and 3-D shape design parametrisation. The methodology permits to make use of the capabilities of I/EMS for conceptual design and automatic mesh generation; link analysis, and manufacturing models through a common set of parameters and constraint equations that capture the design intent; and obtain, at any desired design stage, the 2-D drawings required for production.

Interactive and automated design optimisation may also be used to optimise the dimensions and configuration of built-up structures. In this case, the geometric entities created in PATRAN are used to define the size, material, load, and configuration parameters.

Industrial Applications

Torque Arm

Dimensions of the torque arm

Shape optimisation of a torque arm is considered to illustrate the capabilities of the DfP tool for mechanical component shape design. The boundary geometry, represented by lines and arcs, was created in I/EMS. Geometric and dimensional constraints were defined in I/EMS as well, see figure 5.9.

In figure 5.9 a small circle denotes a tangency constraint and a dotted line denotes a vertical/horizontal constraint. The dimensions represented with both, name and value, such as d12 = R54.2, are the design parameters.

For the analysis, a finite element mesh was created, using the automatic mesh generator available in I/EMS. The finite element mesh is made of 481 three-and four-node plane stress elements, with 560 nodes, and 1064 degrees-of-freedom. The finite element model and boundary conditions are shown in figure 5.10

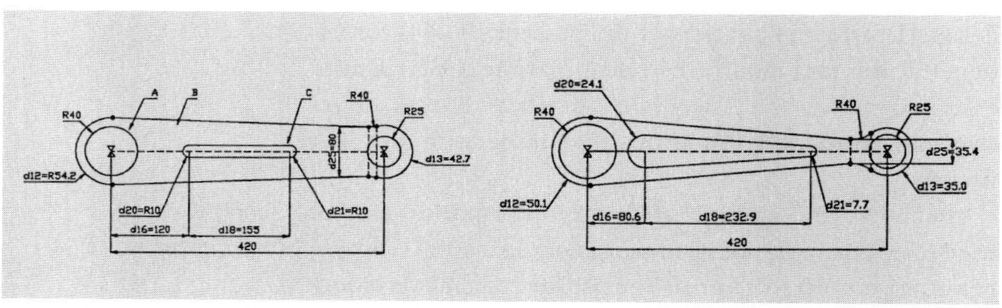

Fig. 5.9: Torque arm initial and optimal dimensions

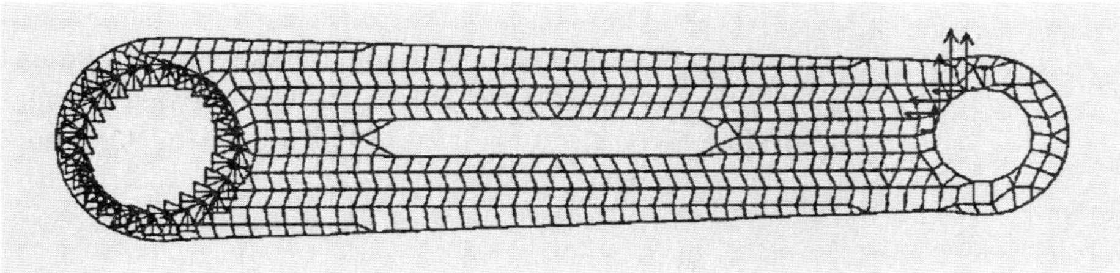

Fig. 5.10 Torque arm finite element model

Optimisation problem

The optimisation problem aimed at minimising the total volume with constraints on the average von Mises stress at three representative elements located at A, B, and C, see figure 5.9 To preserve a minimum distance of 10 mm between the inner and outer boundaries, geometric constraints have been imposed.

Formally , the optimisation problem is formulated as follows:

Minimise volume

Subject to	d12	≥ 50 mm
	d13	≥ 35 mm
	d16 - d20	≥ 50 mm
	d16 + d18 + d21	≤ 385 mm
	d12 - d20	≥ 10 mm
	d25 /2 - d21	≥ 10 mm
	VM stress A	≤ 800 MPa
	VM stress B	≤ 800 MPa
	VM stress C	≤ 800 Mpa

The ADS optimisation package was used for the design optimisation. For the finite analysis and adjoint response reanalysis, ANYSY was used. The computation (small ≤)of design derivatives was carried out inside the DfP tool. The optimal geometry obtained is illustrated in figure 5.9.

Guillotine Shear Machine

Capabilities of the DfP tool for the size and configuration design of built-up structures

Weight optimisation of a guillotine shear machine is used to illustrate the capabilities of the DfP tool for the size and configuration design of built-up structures, see figure 5.11. The machine is designed to cut metal sheets up to 6 mm and 3,0 m wide. The structure consists of two basic parts: a fixed body and a moving ram. The finite element model contains 2000 plates 20 beams, and 10 solid finite elements amounting to 7000 degrees of freedom. PATRAN modelling capabilities were used to create the geometric, load, and finite element models.

The most adequate design parameter linking strategy

The finite element analysis flagged 10 critical displacement constraints and three stress constraints. Using the DfP tool GUI, 31 size parameters were assigned to the plate thickness distribution and one configuration parameter was assigned to the rear wall angle. To decide the most adequate design parameter linking strategy for the side walls, three weight optimisations were carried out considering in alternative 19, 2, and 1 independent size design parameters. The optimised volumes are shown in Table 1. In order to compromise volume decrease and manufacturing cost increase, it was decided to parametrize the side walls with two design parameters. The second parameter corresponds to the thickness of the "C"-shaped reinforcement, whose limits are defined by the stress analysis. To determine the most adequate angle for building the rear wall, for maximum torque resistence during operation conditions, a configuration optimisation machine table was performed. The volume is minimized under displacement and stress contraints. After 5 iterations the optimum configuration was found, see figure 5.12.

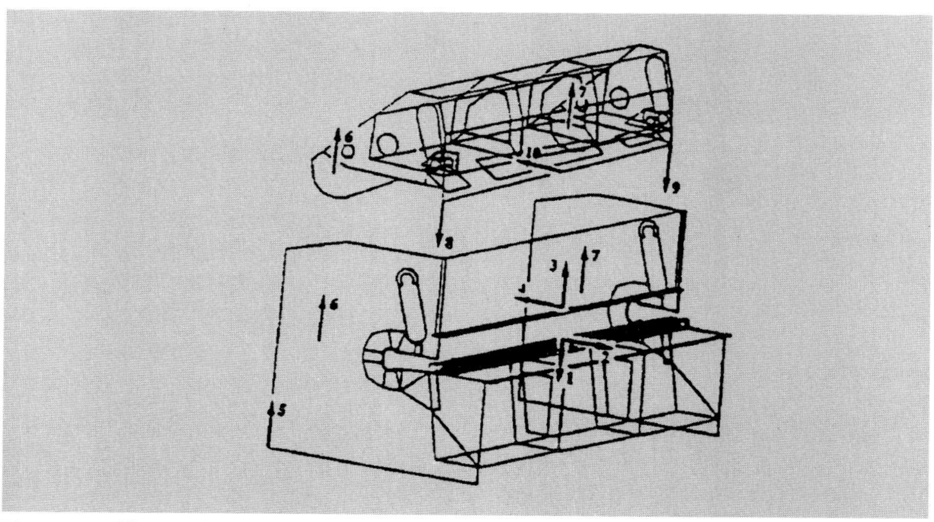

Fig. 5.11: Guillotine shear machine

No. of linked parameters	Machine volume
19	3.3×10^7
2	3.6×10^7
1	11.7×10^7

Table 1. Impact of the side wall parameter linked to the volume.

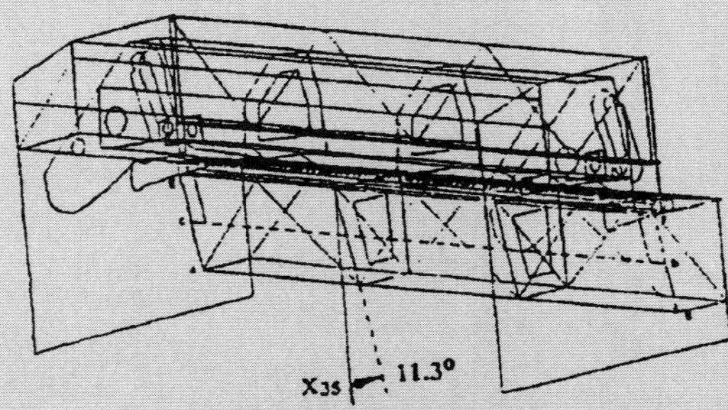

Fig. 5.12: Optimum configuration of the rear wall of the machine table

5.3.2.2 Design for Manufacturability

Design for Manufacturing in General.

Definition

Design for Manufacturing aims at reducing the overall manufacturing cost of a new or updated product. Design for Manufacturing or DfM deals with a wide area of processes and products. In this chapter, the scope of DfM is reduced to the manufacturing of mechanical components.

The first generation of "production-friendly" design systems was developed to solve specific design problems for a specific manufacturing environment. These systems contain as much information as possible, describing the products and processes known at the moment of system development.

Industrial implementation

During practical use of these applications in industry, several problems arise. An industrial environment continuously changes and improves production methods and products. The design system has to add new "experiences" to it's "intelligence". Also during the individual design stages, a large amount of data is generated which are useful for following product life stages. Therefore the design system must be open to share this data with other departments.

The new generation of Design for Manufacturing systems try a less specialised approach allowing a much more flexible use of the software. For this purpose, they focus on easy updating and adding of knowledge and data rather than on using large, highly specialised data sets.

Other design tools

Also, the DfM analysis can not be seen as beeing independent from other analysing techniques such as Finite-Element analysis, Design for Assembly or Design for Recycling. Each of these techniques allows another view of the product functionalities and design requirements.

Design for Manufacturing in the Product Life-Cycle.

Design life-cycle

In many companies, designers have only limited knowledge about the effects of each design decision on production. Most of the time, designers have limited or no access to up-to-date information concerning production capabilities. Also, no simple tools are available for making a rough cost estimation of a new or modified design. The evaluation of a design is restricted to a limited number of evaluation sessions between design and manufacturing teams. Therefore, comparing alternative designs on the basis of manufacturing capability and cost takes a long time. As the Time-to-Market is reduced, insufficient time is available for this type of design checking.

A design tool for manufacturability can be used in several phases of the product life cycle.

- In the conceptual phase, the system provides information for specifying the functional requirements of the product. The designer has access to libraries of standard components and product families. Thus one reduces the number of new designs created. Also the basic shape of the product is defined during this phase. For Concurrent Engineering, these data, typically, are sufficient for starting parallel design sessions.
- During the detailed design of each product, the provided information focuses primarily on the manufacturing information. Based on geometrical and tolerancing information, the system indicates expensive, non-standard operations or impossible specifications. Thus, the design

is improved on-line and specific design decisions can be stored.

- During both phases, the operator can start a cost evaluation of the current state of the product design. Using "Group Technology" techniques, he can compare his design with previous, similar products and with the original product specification. Based on the production cost actions can be taken for changing product specification or detailed product characteristics.

Basic architecture

Linking CAD and CAPP

A Design for Manufacturing system is a kind of merging of both design (CAD) and process planning (CAPP) functionality. In practice a DfM tool must be a kind of assistant process planner who also knows how to design product functionality.

To fulfill this requirement, a DfM tool must use techniques of both environments. The recent developments in Computer Aided Design have lead to a rapid introduction of parametrical or feature-based design systems. On the other hand, features are also used as the basic knowlege and data unit in automated process planning systems. Therefore, the architecture of the DfM module in CONSENS is based on a feature based product model containing both a design (CAD) and manufacturing (CAPP) view of the product. The tool is built using the following techniques:

Modules

- A design model of the product is built using (design)features. These features are used for describing the functionality and geometry of the product.
- A manufacturing model is built, using (manufacturing) features. These features link the geometry of a product with the available manufacturing methods. The process model is a simplified version of the model used by Computer Aided Process Planning systems.
- An analysis module. This DfM module analyses the functionality and required manufacturing operations for the product against the design and manufacturing knowledge of the company. Special user interfaces allow updating of this knowledge.
- An interface between CAD and analysis. This interface allows the embedding of CAD information in the product model and uses the CAD system as the main interface to the analysis module.

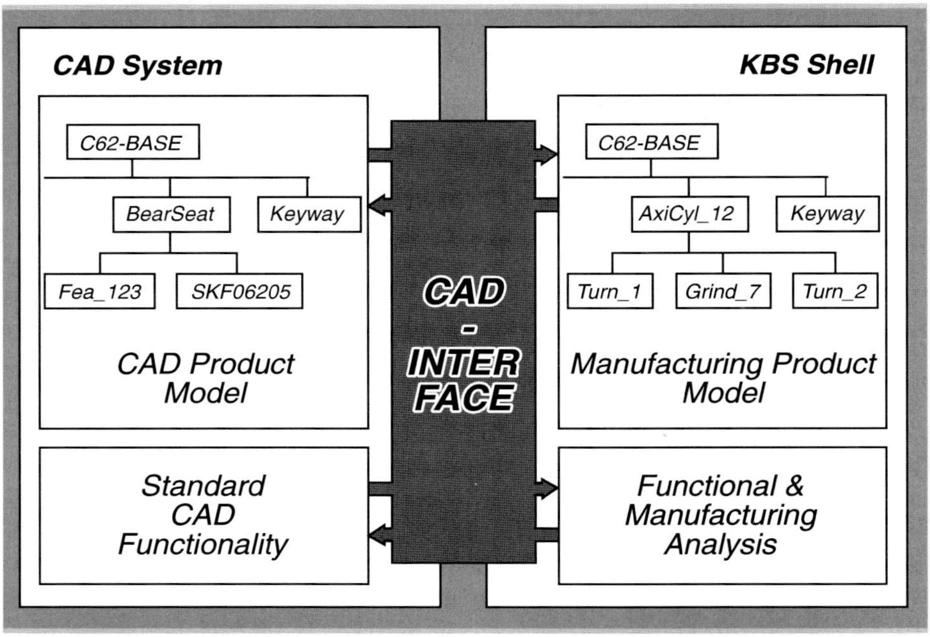

Figure 5.13: Basic DfM architecture with feature based CAD and Manufacturing Product Model.

- Interfaces to the outside world. Both the CAD module and the analysis module provide interfaces to the outside world. Through these interfaces, the system interacts with other tools of the design or CSE environment.

Feature Based Modelling and Feature Extraction.

Feature modelling

The DfM applications rely upon features as their primary source of product data input. A feature is a partial form or a product characteristic that is considered as a unit and that has a semantic meaning in design, process planning, manufacture, cost estimation or other engineering disciplines.

There are many examples of basic differences in the way designers and manufacturers look upon a product. There has been a lot of discussion in the area of CAPP on the need for feature recognition, seen as the opposite of feature based design, to define manufacturing features. In the developed software, both methods are used in a complementary mode.

A parametric feature based solid modeler (Pro/ENGINEER), is used to design the part. Company specific design

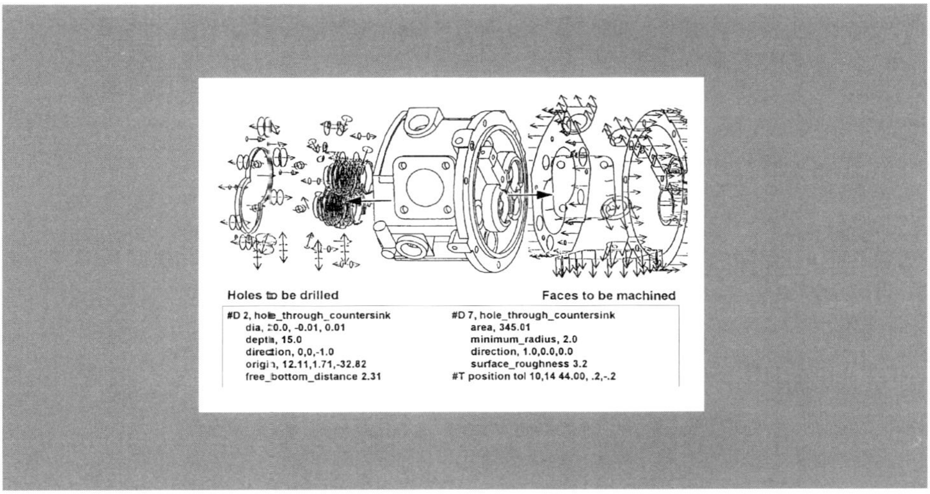

Fig. 5.14: Feature extraction for machining

Feature extraction and filtering

features are used to design and standardise the functional elements of the product, while standard CAD-features are used to design the remaining elements of the product. In addition to the geometry and the topology, technical data such as, surface roughness and dimension, position and shape tolerances are added. From this complete and unambiguous CAD-model, features are recognised for the different DfM/DfA applications.

DfM Manufacturability Analysis

Simplified CAPP model

A generative process planning program (CAPP written in NEXPERT) handles all pre-defined atomic and compound manufacturing features. Priority rules, depending on relative cost, determine the selection of the manufacturing method for each feature. Whenever a high priority method is found not applicable, the exceeded constraint and the corresponding product parameters are recorded. The user receives information concerning the violated constraints and possible actions to eliminate the manufacturing problem.

For each feature the set-up and the required operation sequences are generated. Set-up calculation is determined by the orientation of individual operations and the external tolerances between the features. In a second step an overall optimisation is done, (i.e. minimum number of set-ups, tool changes,etc.) based on the relations and connections between the features .

The set of rules and data is based on the product category assigned to a design. The analysis allows both a selection of alternative manufacturing resources within the chosen product category, or a redefinition of the assigned product category. Thus the effect is checked of a changed production environment on the overall production cost and product manufacturability.

Contribution to CSE.

Parallelisation and Standardisation

The Design for X tools primarily support the CSE action in the area product. In addition, the areas resource and project are supported.

The design for manufacturability focuses on the engineering activity for a single piecepart. The module enables

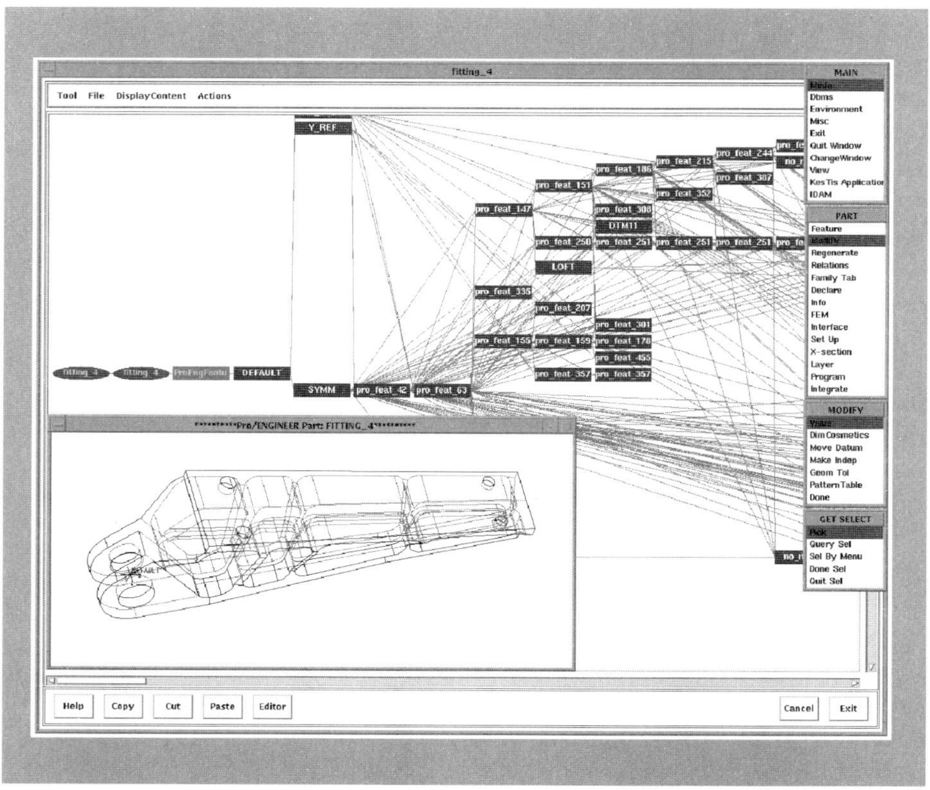

Fig. 5.15: General lay-out of analysis model with the manufacturing product model

the simultaneous design of pieceparts belonging to the same product or sub-assembly. Given the limitations and interfaces of the parentassembly, the co-designers work in parallel on the detailed parts. During the design, the user retrieves features from standard libraries and checks the manufacturing requirements against sets of standard manufacturing resources and capabilities. Through a high degree of standardisation in both the design and the analysis part, the DfM tool allows a further standardisation of the new designs. Through its integration with a powerfull CAD tool, the DfM module has access to a wide selection of design-support tools. Through the integration with the framework, the DfM module links with both project and product-oriented tools.

Integration

Amongst the framework modules, two special relations exist with other modules. The first relation deals with the information on the level of the product (assembly). The relation deals with the overall cost aspect of a product. Together with the DfM tool, the Design for Assembly tool analyses the product for assembly cost. Both tools work on the same product model. As such, the DfM module exports the complete product model (assembly) for use by the DfA tool. The results of the DfA analysis can be applied to the model in the CAD tool and can be run against a check by the DfM tool. Typically, measures saving cost in assembly can lead to higher manufacturing costs for pieceparts. To help decide between these two different "costs", the analysis data are shared with the Design to Cost tools.

The second relation ensures an even higher degree of standardisation of product design. The special link with the RGT tool ensures that a designer can easily reuse previous development effort and store new or update designs for later utilisation.

5.3.2.3 Summary

Design for Manufacturability assist a designer or engineer in optimising product and piecepart design for minimal production effort and cost. DfM provides every designer with his personal assist process planner giving advice about standard components, preferred layouts or tolerance settings.

The key development in this module is the use of parametrical features for both the product and the manufacturing processes model. This approach allows for an easy link between on the one hand, design knowledge and on the other hand process planning knowledge. In addition to the use of features, the module is layed out as an add-on module to an advanced, parametrical CAD system, i.c. Pro/ENGINEER, so that the user can take full advantage of the DfM assistance while working in the standard CAD environment.

The DfM tool does not only support a single user. Within the overall CSE concept, it assists both the team and a team member in providing links to other Design for X systems, e.g. Design for Manufacturing or Design to Cost. Through this link, the product model is shared by different skills and an individual can see the effect of his efforts on the overall performance of the product.

In the final CONSENS demonstration, several types of products are tested. The DfM capability is tested for a number of typical machining and sheet metal manufacturing products and processes. Through the use of features and a Knowledge Based System, the DfM module is easily adaptable to the different environments at the CONSENS testsites.

5.3.3 Design for Assembly

R. Menges, U. Eigenmann

The market situation

In recent years product life-cycles were shortened dramatically. The time span between the market entry of a product and the return of the respective development cost, the payoff period, increased at the same time. This means a cut in the profit margin and, thereby, smaller profits. This is why "Time to Market", the time required to react to the market's demands by supplying adequate products becomes more and more decisive in view of the competition. Hence follows the demand for shorter product development times and lower expenditures. The fact that production costs can be influenced effectively only during the product design phase is well known. The later necessary modifications are recovered the more expenses and problems are caused by these modifications. Most of the production cost do arise when manufacturing the parts and assembling the product. The important decisions concerning production, however, are taken earlier: about 70% of the total production costs are determined by the product design. The designer decides about:

The influence on production costs

- product structure,
- number of parts,
- part geometry,
- joining methods,
- assembly tolerances,
- composition of surfaces and
- materials.

Determination of the assembly method

The interdependence of the parameters listed above also largely determines the assembly method. The possibilities to supply an assembly component are restricted by it's geometry. Part geometry and assembly tolerances limit the level of automation which is possible to achieve within a given cost frame. The structure and design of assembly lines are strongly influenced by the product structure. Jigs and fixtures, as well as needed tools are determined by the assembly method and the shape of the assembly components. Subsequently, designing the product leads to a fairly precise assembly process (see figure 5.16).

Do We Need Design for Assembly?

Does the need for DfA exist?

In order to maintain the ability to compete, enterprises have to reduce costs and shorten "Time to Market". Forty to sixty per cent of the total production period is covered by assembly. In the past an increased productivity in the assembly sector has been achieved by carrying Taylorism to extremes or by increasing the degree of automation. Automation, however is strongly restricted. The productivity gained by automation must justify the corresponding investment. From a financial point of view this is difficult to achieve, considering products with small production rates and products with many variants. Moreover, in many cases machines turned out to be not flexible enough to perform complex assembly tasks. In addition, they could not manage to assemble a wide variety of variants. Therefore, most

Increased production through time and motion studies

of the assembly work is still executed manually. On the other hand, increased productivity in pure manual workplaces has been achieved by consistently applying methods related to time and motion studies, e.g. MTM or Work Factor and by the division of labour. Nowadays, very short cycle times are considered to be inhuman. Frequent repetitive motions are monotonous and can result in injuries like Carpal Tunnel Syndrome (CTS) or Repetitive Strain Injury (RSI). In addition, job dissatisfaction leads to increased fluctuation and absenteeism.

Obviously, to face the demands for a greater quality of work, improved working conditions and higher wages, we have to apply new methods. Regarding the growing competition and decreasing product life-cycles we can no longer afford not to exhaust the potential of an assembly-conform product design. The answer to the question "Do we need design for assembly?", therefore, must be "Yes".

Significance of Design for Assembly

As outlined before, the assembly effort and, thereby, a large part of the product cost are determined by the product design. In spite of the fact that their decisions strongly affect the assembly process, designers far too often take little notice of assembly. They are preoccupied with achieving the product function. With the requirements of the cost-producing sectors manufacture and assembly not being known or not being considered to an adequate extend, today, a large number of products are designed causing too high an assembly effort. The neglect of the assembly process, when designing components and defining the product structure, results from:

The neglection of the assembly process

- a lack of understanding of the assembly problems,
- a lack of knowledge about design oriented assembly,
- the attitude that small inadequacies regarding assembly can easily be compensated in production,
- undervaluation of the importance of assembly within the entire process of fabrication and
- a deficiency in the communication between different departments.

With the assembly cost being determined to the largest extend by the product's design it is obvious that product

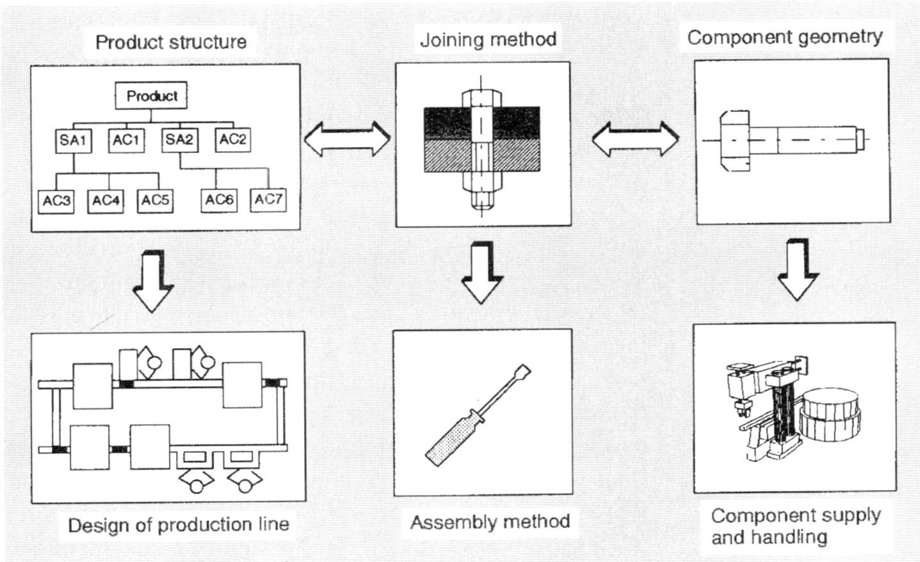

Figure 5.16: Relation between product design and design of the assembly system

design is the first candidate for modifications in order to reduce cost. An assembly-conform product design should be addressed to:

How the assembly cost are being determined by the product's design

- simplification of assembly processes,
- reduction of handling operations,
- avoidance and facilitation of orientation respectively,
- increasing modularisation and standardisation,
- increasing profitability of the assembly equipment and machines,
- reduction of fabrication and assembly cost and
- increasing product quality.

The designer, therefore, has to be able to envisage the consequences of his decisions. He has to know, which design parameters determine the quality of assembly and he must be aware of general design principles to reduce the assembly effort.

Design for Assembly within the development process

The design for assembly must not be regarded separately from other company processes. When struggling to decrease assembly costs, the designer must not lose sight of the total cost. The main aim is to optimise total cost. The reduction of assembly efforts represents an important factor in order to achieve the intended goal, though not the only one. The designer initially aims to guarantee the functioning and the required quality of the product. The product is optimised with regard to the assembly effort, thereby proceeding from those results obtained by the designer.

The superior goal however is to minimise the production costs. Therefore, it is not appropriate to consider the assembly separately from other cost-determining processes, especially manufacture. Parts which are easy to assemble, that also means causing low costs, don't necessarily need to be cheap to manufacture. Regarding the total cost of the product it might even be more expensive to use one part which is easy to assemble but difficult to manufacture than to assemble several standard parts. The designer, therefore, needs a comprehensive knowledge about both processes. He always has to keep in mind the total manufacturing costs.

The product design influences tool design of production facilities

Other important processes to be considered, for example, are tool design and production planning (see figure 5.17). The product design also determines whether fixtures and tools are needed to assemble the product. The best of course, would be to avoid the use of any tool, because fundamentally operating a tool or fixture adds no value to the product. For different reasons it is not always possible to avoid jigs and fixtures as well as the use of tools or machines. The product design, however, should allow the simplest possible fixtures and tools to be used. Parts and subassemblies should not need to be reoriented in the fixture during the assembly process, so that handling operations remain limited to the extend of what is absolutely necessary.

Besides the factors production rate, cycle time and pay scale, the structure and the design of the respective production line is defined by the product structure and by the assembly method, as well as by the tools and fixtures needed to assemble the product (see figure 5.16). Consequently, the costs caused by assembling the production facilities and acquiring machines and tools belong also to the expenditures which are already determined to a great extend by the product design. Necessary modifications which are revealed just before the start of production e. g. when the

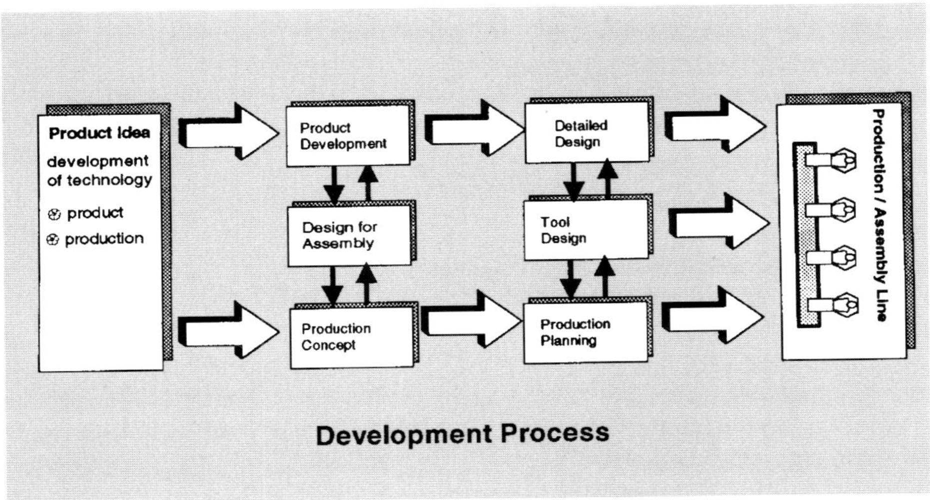

Figure 5.17: Parallel Development of Product and Production System

production line is set up, or even worse, during production cause excessive costs. Therefore, designers, method engineers and production engineers should start to co-operate at a very early development stage. This way, the existing knowledge of the different disciplines can be used. If such a co-operation is not established a lot of cost saving and quality benefits may be missed out on.

DfA-Tool

Computer Aided Engineering

In order to manage the torrent of data and to ensure that results can be transferred in good time from one department to another, there is no doubt that DfA-analyses have to be computer-based. However, such an application first of all has to be easy to handle and easy to understand. A software tool must be capable of giving useful advice and easy to update when new knowledge becomes available. It has to be easily adaptable to company-specific requirements. Using the computer based DfA-tool also must not be very time consuming. A software tool not fulfilling these requirements will not be accepted by the user. Regarding the needs moving along with the parallelisation of development processes in putting data manually and getting a paper-printout is not sufficient. To face the challenge of shorter development cycles, the manual data handling must be reduced and the data flow must be improved. Results have to be easily accessible at any time by anyone interested. This is why a DfA-tool should be integrated into a CSE-framework.

The DfA-tool developed within the ESPRIT Project CONSENS (see figure 5.19) consists of two parts addressed to the assembly of mechanical parts (DfA mechanic) and of printed circuit boards (DfA electronic), respectively. Both parts are described briefly in the following paragraphs. The DfA-tool is employed with the aim to:

Why to employ the DfA tool

- evaluate the assembly effort for different products, depending on product components, production technology and production location,
- identify components causing a high assembly effort,
- increase the usage of standard parts,
- give advice on possible improvements,
- determine draft assembly times,
- assign means of production like tools and bins to assembly components,
- estimate assembly costs.

All these functions support the user in achieving the superior goal: reduction of assembly cost.

DfA Mechanic

The use of the DfA-analysis makes it possible to assess the potential for assembly without having a detailed knowledge of the assembly situation to be found later. Influences like:

- weight,
- part geometry (shape, dimensions),
- material properties (flexible, hard),
- state of surface (smooth, rough, oily),
- particular handling conditions,

are described and quantified.

The method is suitable for both, the evaluation of new and existing products. All the decisions which are necessary to execute the analysis can be made easily and quickly by the user.

The aim of the DfA-analysis is to evaluate the assembly effort for different product variants and to identify those features resulting in high assembly cost. The assembly operation of a particular part is described in a standardised way and judged according to its difficulty. The value assigned to each operation assesses the expenditure (of time) to execute it, related to other assembly operations. Manual assembly operations, thereby, are grouped into:

- manual handling,
- insertion,
- processes.

Manual handling involves the grasping, transportation and orientation of parts or subassemblies before they are inserted into or added to a fixture or partially built-up assembly. A part has to be initially controlled by one or two hands, or by using a tool held in one hand. After this, the part is moved to the place of assembly. While moving the part, correcting and preparatory moves are executed in order to achieve the desired inserting position and orientation. Thereby, the move operation might be interrupted with respect to the part's nature.

Insertion includes inserting a part into another as well as adding a part to another part or partially built-up assembly. The operation either positions a part manually, by means of a tool or with both hands at the desired position (e.g. surface on surface) or inserts it (pin into bore).

All operations which only indirectly aim at positioning parts on or into one another are called processes. According to this definition, processes may be manually executed like screwing in by hand, but may also be automated operations like pressing-in operations.

Firstly, the manual assembly process, executed according to the product structure, is described. The aim is to evaluate each assembly operation with regard to its degree of difficulty due to the properties of the part handled. Part in this sense means a separate part as well as a subassembly. Influences on the degree of difficulty are the characteristics of the part to be assembled (e.g. dimensions, shape, weight) and the location of assembly (e.g. accessibility, fit). These characteristics are determined by the product design.

Two fundamental principles have to be taken into consideration when performing the DfA-analysis.

- Separate parts are assembled one after the other. The assembly of each part is described and evaluated separately even if it would be a simultaneous action or if two parts would be handled at the same time in reality.

- The improved design of the work system and use of fixtures is not taken into account.

The analysis shall emphasise the weak points of the product design. Considering jigs and fixtures would lead to a false result because these are suitable to compensate faults of the product design.

Evaluation of the product design

The DfA-analysis tool is actually represented by evaluation tables. Depending on the influences, a value selected from the evaluation tables is assigned to each assembly operation. All values indicated in the table are time units, expressed in TMU (Time Measurement Unit, 100 000 TMU = 1 hour) as applied in the MTM-method. The total assembly effort of the product is quantified by the sum of all values assigned to single operations (see figure 5.18). The first result of the DfA-analysis, therefore, is a figure of time measurement units (TMU). The comparison of the results of the DfA-analysis from before and after the product design improvement shows the approximate percentage-cut of assembly time.

The computer aided DfA-analysis is not able to make creative suggestions to improve the potential for assembly of a product. Restrictions resulting from security requirements, desired quality, maintenance and possibilities for recycling are unknown to the system. The designer's creativity may not be replaced by the DfA-analysis. It is a tool demanding the designer to systematically reflect the design of the product with respect to the potential for assembly. The application of the DfA-analysis, above all, leads to a systematic approach. Potential improvements are furthermore visualised.

The DfA-analysis sheet gives a complete breakdown of the assembly effort required for each product component. Assembly operations showing high assembly times are highlighted. The evaluation of the analysis sheet is based on two indicators. The time for each handling and inserting operation, respectively, is considered and compared with an ideal handling and inserting time. The highest values are marked, indicating that these are the assembly operations with the highest improvement potential.

Some potential for improvement is indicated by the codes used to describe the operation. Codes beginning with a "P", for example, represent processes which generally hinder the progress at work. During the assembly, processes basically should have very few time shares. Further hints are given by codes indicating obstructed access and/or restricted vision (ensure adequate access for part insertion) or the time required for holding something down (design parts to be self locating and aligning).

Product improvement

Basically, improvement measures can be divideed into four groups:

1. • measures concerning separate parts,
   - reference surfaces and stop faces should be used, if possible,
   - separate parts should not consist of shock sensitive materials,
   - low specific weights should be striven for etc.

DFA-analysis

assembly process		handling		insertion		fre-quency	total time
No	description	code	time	code	time		
	Product before improvement						
1	cover	E10	48	B47	76	1	124
2	screw	D11	50	A42	17	4	268
3	process			P51	150	4	600
							992
	Product after the first step of improvement						
1	cover	E10	64	A41	31	1	95
2	screw	D11	60	A42	17	1	67
3	process			P51	150	1	150
							312
	Product after the second step of improvement						
1	cover	E10	48	B45	65	1	113
							113

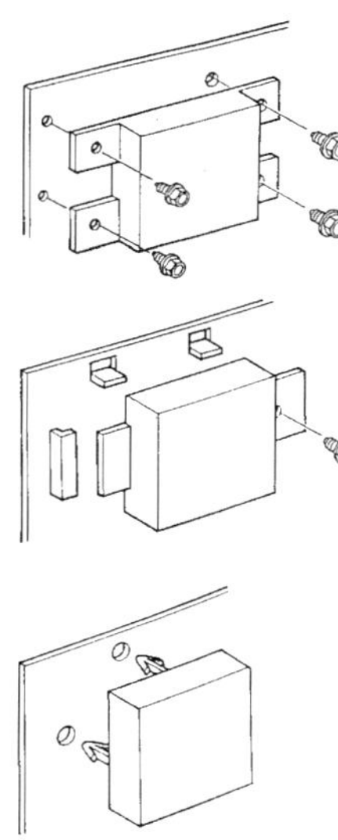

Fig. 5.18: DfA-analysis

2. • measures concerning subassemblies,
   - parts which don't have to consist of different materials should be combined standard parts should be used, if possible etc.
3. • measures concerning the product structure,
   - all function units of a product should be analysed with regard to their necessity,
   - the function structure of a product should be as simple as possible,
   - function units creating variants should be combined in separate subassemblies etc.
4. • measures concerning connecting techniques,
   - separate fasteners should be avoided,
   - the tasks of inserting and fixing a part should be combined, if possible,
   - only few different connecting techniques should be used etc.

To improve the design of a part or subassembly the designer is supported by the technology information data base. The DfA-tool itself is not able to make sound suggestions on design improvement, because some information (security requirements, manufacturing aspects, etc..) is unknown to the tool but obvious to the design team. However, the technology information database provides the user with general knowledge and with company-specific examples for alternative solutions, the scope of these solutions, as well as the advantages and disadvantages.

Structural analysis

The assembly analysis is completed by a structural analysis aimed at the reduction of parts. Each part is checked and examined by a few questions. The result is a classification of the parts into A-components (essential) and B-components (avoidable). Based on the number of essential and avoidable parts and an ideal assembly time per part, a figure is calculated, characterising the quality of the product design. Depending on the company-characteristic products it is possible to determine a benchmark value for the average assembly effort needed for those products. Comparing the average assembly effort with the actual assembly effort analysed allows the evaluation of the quality of the product design.

DfA Electronic

DfA for Printed Circuit Boards

The printed circuit board (PCB) assembly primarily involves the insertion and soldering of electronic components onto PCBs. The insertion of electronic components is carried out manually or automatically. Additionally, robots are employed mainly for the insertion of those components which otherwise would have to be inserted manually due to a non-standardised shape of the components. A combination of both, automatic and manual insertion, is often used because of oddly-shaped or non-standardised components which cannot be handled by automatic insertion machines or which cannot withstand wave or reflow soldering temperatures.

Unlike the DfA for mechanical Parts (DfA mechanic), the DfA (electronic) is not addressed to the improvement of the shape or the integration of functions of assembly components(see figure 5.19). Printed circuit boards are equipped with standard components. The task of the DfA (electronic), therefore, is different from the purpose of the DfA applied to mechanical products.

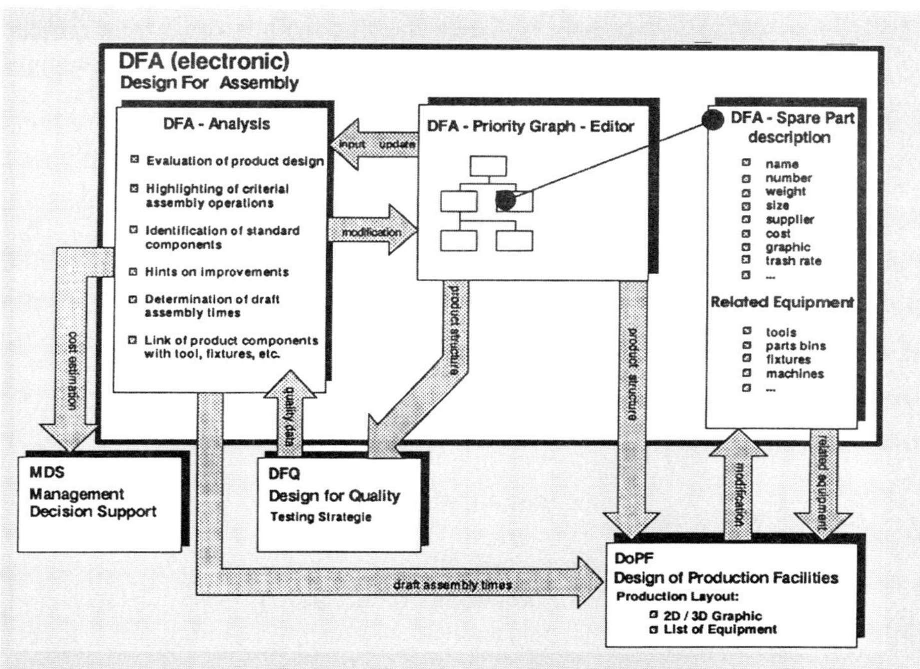

Figure 5.19: DfA within CONSENS (TEMIC scenario)

The DfA-analysis for printed circuit boards is used to indicate and emphasise components which causes the highest assembly costs. This assembly cost calculation is based on the insertion times, the time consumed by rework and the cost of the equipment. The aim is to achieve low assembly costs. The visualisation of the costs caused by each component is important to show the user the impact of employing certain components or techniques. The effect of modifications becomes obvious. High assembly costs indicate the most promising starting-point for possible improvements. By employing the DfA-tool it is possible to compare alternative solutions based on insertion times, machine rates per hour, wages, different production locations, the equipment used, depreciation, etc. The result of this comparison is a break-even analysis showing the number of products which have to be assembled by a certain technique to become cheaper than employing another technique.

The aim of the DfA-analysis (electronic) is to support the user in the selection of the appropriate technique and assembly location. Surface mounted devices, for example, are more expensive than through-hole components. On the other hand the cost caused by the assembly of SMD-components, as a rule, are lower than those caused by through-hole components. The gap between the product and the design of production facilities is bridged by DfA (electronic), relating product components and means of production. The DfA analysis results not only in an abstract figure but has actually a concrete impact on the design of production facilities. By assigning machines and equipment to assembly components, these can be displayed in the graphical layout of the worksystem. Furthermore, this direct relation is the basis for a sound assembly-cost calculation.

Summary

Competition increases when a market becomes more global and more open for foreign goods. At the same time, cycles of product innovation become shorter. A growing demand for more individual products results in smaller batch sizes. All that puts tremendous pressure on enterprises to rationalise. They are looking for effective measures to reduce cost and to increase productivity. Several studies in recent years have shown that about 70% of all the costs for fabricating a product are determined by the product's design. That means, all succeeding production areas, e. g. manufacture, tool design, assembly etc. are strongly influenced by the product to be fabricated. Therefore, it is reasonable to start all efforts to reduce production costs by analysing the product. Since assembly covers a big part of the total production time and represents the main part of the costs, it is obvious that possible improvements should first be addressed to the assembly sector. Above all Design for Assembly using the DfA-tool supports Concurrent/Simultaneous Engineering:

How does the DfA-tool support CSE?

Avoidance of costly design modifications and keeping within given cost frames:
- by visualisation of the assembly effort,
- by unbiased evaluation of different design alternatives,
- by early inclusion of cost caused by the assembly process and the required means of production and
- by not purely focusing on assembly cost, but also on related equipment and means of production.

Reduction of the time consumed by production planning:
- by parallel consideration and development of the product and related means of production and
- by assigning required tools, machines, fixtures and means of production to assemble the components.

Avoidance of failures caused by working on obsolete or wrong data:
- by integration of the DfA-tool into a CSE-framework,
- by ensuring that only valid and up-to-date product data is worked on and
- by avoiding the manual transfer of product data.

With regard to the facts described above, it becomes clearly visible that assembly can first and foremost be rationalised by modifying the product, so that the assembly becomes simplified or is completely avoided. Consequently, applying Design for Assembly is the most promising method for rationalising assembly.

5.3.4 DESIGN TO COST

E.Engelborghs, J. Frech

About 70–80% of a product's cost is determined during its preliminary design phase. The designer influences every decision regarding shape, size, manufacturing and assembly features and all other details like surface roughness and tolerances. Many experts agree that performing more analyses and involving manufacturing personnel earlier in the product development cycle, may cut manufacturing cost while raising product quality. The problem with this, is that gaining access to the analysis and manufacturing resources needed to design for manufacturability and low cost early on, is difficult without the medium all engineering specialists understand namely a fairly refined engineering drawing. However, the existence of such a document generally marks the end of conceptual design, and by then most of the opportunities for cutting cost have passed. (Deitz 1988)

Advantages in early design

The consideration of cost in early design stages implies some advantages:

- Modifications of the product structure are the most fundamental and farreaching actions to save money in design, since the product structure defines most of the subsequent design, manufacturing and assembly operations necessary to produce the product.
- In early phases modifications of the product structure are done easily, in later phases the designer has to modify his already created drawings, which even in CAD requires a substantial effort and will not be done often to analyse other ideas.
- Only in the early stages of design is the whole concept of the design still open to do structural changes and analyse the trade-offs, e.g. between a multitude of simple parts with considerable assembly effort and complex parts with very little assembly.

However, there are also some disadvantages and problems involved in trying to optimise design cost-wise in early design stages:

- All the available information in early design stages is very general and may change easily. Therefore, any calculation software has to deal with these insecurities.
- The result of any cost-calculation is in its accuracy linear to the amount of input. Therefore, early cost-estimations can only provide rough calculations with limited accuracy.

  This leads to a compromise in terms of cost estimation: As soon as possible and as late as necessary. The actual compromise depends on the type of product, the variables to influence by the designer and the scope of the system, which can include manufacturing, assembly and in years to come, some recycling aspects.

Compromise as soon as possible and as late as necessary

One solution to the problem of early cost estimations is multidisciplinary design teams that draw on the expertise of manufacturing, assembly, marketing and other areas. Others try to use sophisticated software solutions to support the engineer in his complex tasks. Since design teams are not always possible and the efficiency is sometimes insufficient, Design to-Cost software is becoming more and more utilised in industry. As a result, this task was included in CONSENS as the start of any design process.

Organisation on Software

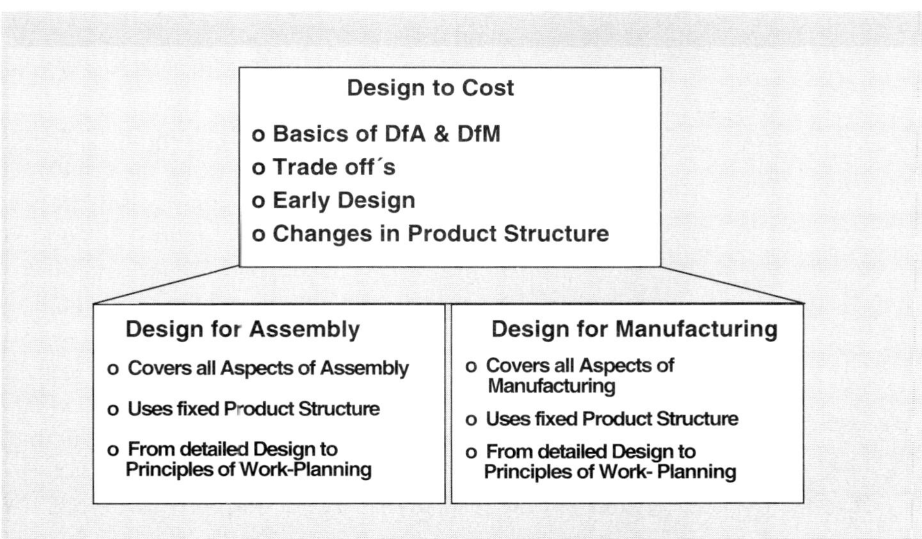

Fig. 5.20: DtC, the trade-off between DfM and DfA.

DtC = DfA + DfM

Design to Cost includes the two basic parts: Design for Manufacturing and Design for Assembly. These last two areas concentrate on either the manufacturing of the parts or their assembly. The trade-off, which is necessary for substantial improvements, is not covered by either one of them. This is the domain of Design to Cost Tools, which include both aspects in appropriate detail.

Trade-off is necessary for substantial improvements

Since the trade-offs from an assembly requiring design and a design with complex (and expensive) single parts are taking place quite early in the design-process, a Design to Cost tool is used. This necessitates the change of the product structure due to cost considerations. In the specialised methods and tools, these changes are not possible. Therefore, the implications of more detailed aspects further down the design process, such as the layout of a specific workplace, are consequently included in these systems.

The CONSENS Approach to Cost-Effective Design

Integration of different tools

Current Design for X tools are not flexible enough to cover all these aspects simultaneously as they cannot provide a suitable surrounding for the rough and detailed design at once. But, within the CONSENS system, these different tools are integrated to provide working conditions similar to those of the most complex tools without the effort required to build these and with the advantage of flexible insertion of tools, according to the specific needs of a company.

For the initial cost estimation usually a simplified view of products and processes is used. Therefore, the calculated cost is only a rough estimation of the actual costs of the new product. The general rule is that the amount of data used for a calculation, consists for one part of new data for the designed object and for the other part of known data, like machine-hour-rates, standard times for processes or similarities to existing objects. For completely new products or processes the later data is only very partially known or inaccurate and the resulting estimation has a low probability. Input into any calculation process, therefore, has to be gained by the available experience in similar matters, either by teamwork or by using appropriate databases. The type of cost-calculation depends on the type of product and the

Experience by teamwork or by using databases

involved processes. For complex assemblies of parts, the emphasis is put on the assembly operations, as they take up the larger share of the cost. For complex parts, the manufacturing processes determine the larger part of the cost. To reflect this duality in cost calculation, two separate components are built into the Design to Cost module. A first module, the Assembly Cost Estimation Tool (ASCET), focuses on the assembly related costs of a product. A second module, the Flexible Cost Calculation System (FCCS), focuses on the manufacturing cost of a complex product.

Different methods for assembly and manufacturing

Assembly Cost Estimation Tool (ASCET)

To support the designer in the scope of assembly and cost efficiency, the IAO has developed the engineering tool ASCET. This enables the designer to estimate the assembly time of a product or an assembly-group as well as to estimate roughly involved parts cost, for different variants of a product design during conception. In this way an evaluation of assembly efficiency will be given by the comparison of the values of assembly times from different variants.

Designer estimates assambly times

ASCET uses the object-oriented data base HyperWork, for storage of all data involved in the calculation, as well as for the actual objects (equivalent to program code) of the calculations. For a clear structure of the generated data base, the parts, assembly-groups and products can be allocated to suitable classes. Objects needed at once can be found by using a convenient similar parts search. The designer can save generated parts, assembly-groups and products, and can define their assembly processes and estimate assembly times using two possible methods. The generated assembly protocols are also available for further analysis, which are, for example, based on some modifications of parts or assembly-groups.

Object-oriented data base HyperWork

The internal descriptions of parts are limited to the real necessities for assembly, which means here, the three dimensions and the weight of an object. This limited description allows a fast definition of parts, assembly-groups and products, a crucial condition for encouraging the designer‘s creativity.

Only minimum data

ASCET is usable not only with existing but also with virtual objects. So, based on the existing data base, virtual objects such as parts and assembly-groups can be created, modified, and finally, also integrated in real products by substitution of real components. In this way, the designer can replace parts and/or assembly-groups for getting better solutions of assembly efficiency in early stages of designing.

Easy modification

These characteristics make ASCET a tool for early phases of design. The designer works on the current product or assembly-group and improves it in as many loops as necessary. This is in principle the same as conventional procedure, but it is much faster and cheaper than the conventional way, which includes other departments for second opinions and is therefore time consuming and expensive.

To use ASCET the user has to define the database first, which involves inserting the assembly-relevant data (object specification, assembly features, dimensions and weight). By setting up their configuration, products and assembly-groups are defined as objects consisting of several parts and, optionally, of other assembly-groups.

A configuration of assembly-groups or products defined in this way can now be evaluated in relation to assembly time. Therefore, the user has to trace the assembly procedure of an object by fixing the assembly steps. This precondition for estimating assembly times can be performed by two different ways of assembly definition.

Feature-based method

The feature-based method uses pre-defined assembly features for the determination of assembly steps. Based on this definition, the assembly time will be calculated by evaluating the feature specific assembly linkage.

Pre-defined assembly operations

The other possibility of getting assembly times in ASCET is using pre-defined assembly operations for the determination of assembly linkage. Every assembly operation

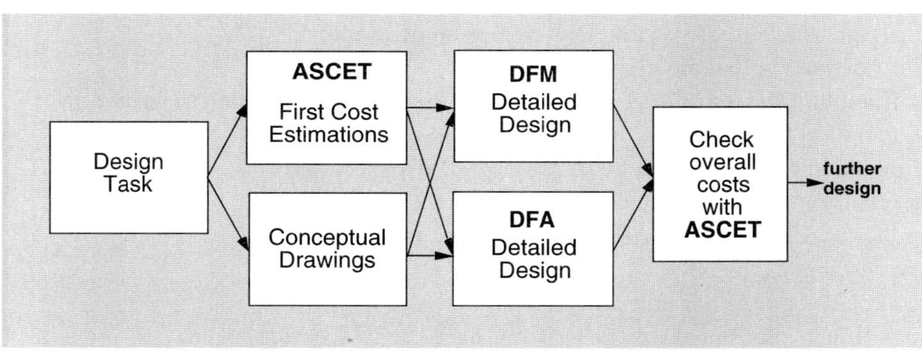

Fig. 21: Process flow with ASCET

reproduces a list of MTM-base-movements, which will be evaluated in the operation-based method of estimating assembly times by including the assembly object's specific data (MTM = Methods Times Measurement, a system of predefined times).

Both ways have their own specific advantages and problems. The most important problem is the additional information necessary to work with the assembly features, although this method may be very helpful if such a system is closely connected with a CAD-System that already uses features. Further development of this system may go in this direction.

Growing data base

Once defined, assembly objects (parts, assembly-groups and products) are available for further analysis. So a growing data base offers a numerous range of variations, which can at least point out the capacity to reduce assembly times in early stages of designing. The whole of ASCET is developed with this use of data and the very early use of the tool in mind. This leads to the following process flow with ASCET.

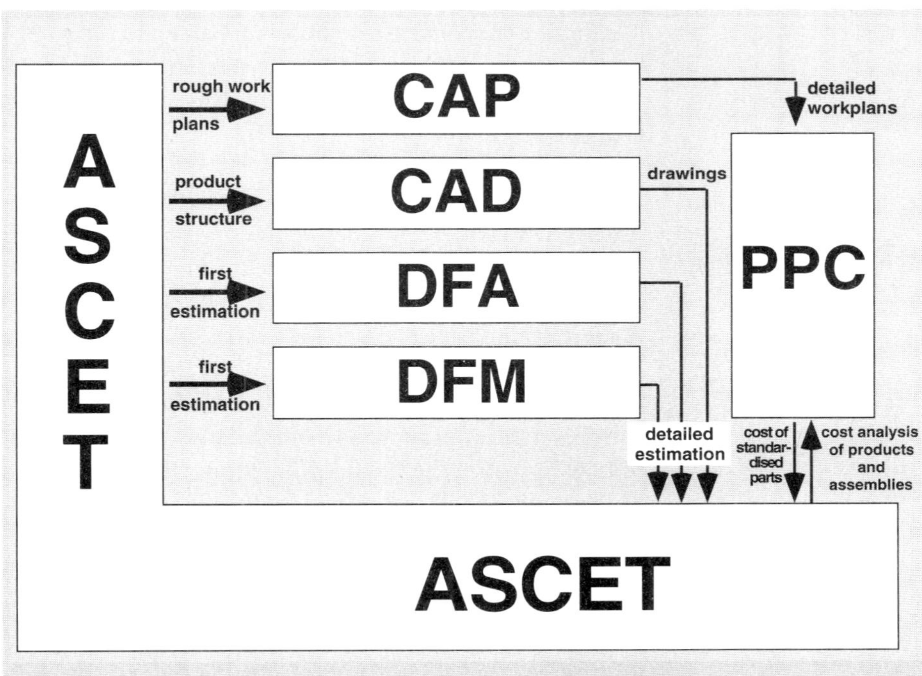

Fig. 5.22: The major data flow to and from ASCET.

ASCET's Support to CSE

ASCET is a tool which supports CSE very well even on ist own and is even more powerful if used in a CSE-environment. ASCET effects all three basic principles of CSE very positively:

Supports CSE

- **Parallelisation**
  After the initial use of ASCET, the module allows cost-calculations at any time with the actual data to be made. Standard procedure in most companies is to make a calculation when design is completed only. With ASCET the calculation can be updated anytime and done in parallel to other tasks.
- **Standardisation**
  The information flow regarding cost data is highly structured in an environment where ASCET is used. In CONSENS, ASCET is used in conjunction with the common database PIA, which makes the processing, storage, retrieval and demonstration of cost data easier.
- **Integration**
  ASCET as an integrated tool in the CONSENS framework, is the central module to access and process cost data. Cost information from various other tools is collected in PIA and summarised by ASCET. This gives the user a consistently accurate picture of the cost situation of the product under development.

Data flow

The picture 5.21 describes the major data flow to and from ASCET with other modules. The data flows with other sources are not included to ease the reading of the picture.

Example for use

A short example may help to understand the workflow:
The designer gets the requirements of the new product from the marketing department. He has a lot of experience with this type of product and immediately has some ideas of how to realise the required functions. He makes some rough sketches by hand on paper of the proposed product. He then uses ASCET to test two or three different possibilities with different product structures. Within ASCET he finds many standardised parts and assembly groups from other products. He uses these whenever possible and inserts new ones whenever necessary. For the existing ones, the cost data is very accurate, for the new ones ASCET, pro-

vides a first estimation. The result of this session is a cost-optimised product structure.

Now the detailed process of design starts. The product structure is the guideline for the designer of how to set up the product. A first rough assembly plan later helps to understand the product structure and eases the preparation of a detailed work-plan. In the course of the design process, more cost data is created, gradually improving the estimation. This information is collected and processed within ASCET, which allows for an overall cost picture all the time; if some parts or assemblies turn out to be more expensive than expected the effects on the overall product are clear; they may be compensated by less expensive parts or assemblies. This gives the project manager – in contrast to common procedure, where the first reliable calculation is done when design is finished – the possibility to change or stop the design-process at any time, saving time and money. CAD provides the data necessary for visualisation of the objects within ASCET. PPS originally stores some of the data used or generated in ASCET.

Overall cost picture

The Future of Design to Cost Systems

Even the broad understanding of Design to Cost in CONSENS will not be sufficient to fulfil the necessities of the future. The cost responsibility of the designer does not end when the product leaves the factory. Tightening legislation, combined with environmentally conscious customers, has lead to a return of products at the end of use. Therefore, recycling costs have to be calculated as a company expense and must be minimised.

Recycling costs have to be calculated

Again, this is mostly the responsibility of thedesigner, who is able to reduce disassembly and landfill cost and to increase profit from materials or parts sold by changing assembly operations and material specification. Again, this domain can be sensibly covered only if all relevant aspects (manufacturing and assembly and recycling only to name the most important) are covered at once and tradeoffs between these various aspects are possible. This will lead to a further increase in functionality of Design to Cost systems.

The other major change in this area will the increasing integration of Design to Cost systems in general engineering software, most likely CAD. The use of 3-D-Systems, a feature oriented work process and general integration of all CIM-components with new interfaces like STEP or solutions like Engineering Data Management Systems will give further moment to the use of Design to Cost systems.

Solutions like CONSENS will soon become standard procedure in industry. The benefit of such systems will lead to broad applications within the near future.

Flexible Cost Calculation System (FCCS)

To support the designer with the cost efficiency of manufacturing, WTCM has developed the engineering tool FCCS, which enables the designer to estimate the manufacturing time of a product or component. The Flexible Cost Calculation System (FCCS) module of the Design to Cost (DtC) tool is based on two things: group technological retrieval of similarities between parts or assemblies and cost functions allowing parametric description of cost relations. These underlying principles represent the designer's behaviour when estimating or calculating the cost of the product or assembly when designing it: trying to find similar items designed and/or manufactured before and building and using cost functions based on rules-of-thumb or experiences.

Group technological retrieval

The developed DtC system allows fast cost calculation and estimation by using as much as possible the power tools offered by modern database technology. An example is the fast retrieval of data by relational queries and evaluation of functions and formulas in a much similar way to spreadsheets. Those two specifications are very essential for a generic system with powerful cost calculation possibilities. The FCCS system on PC is designed as a stand-alone unit which the designer can consult as a kind of "pocket calculator" while making his concept and design. However, this PC is linked with the CONSENS system. Making accurate cost calculations is usually a problem for companies producing in small batches or making piecework. At the moment, cost calculations are usually handmade on the basis of global rules of thumb and experiences from the past. Accurate results are not always evident in this way.

However, very often it is possible to formalise these rules of thumb and experiences and put them into a computer. By entering a few parameters concerning the global characteristics of a product, the cost can be estimated.

Advantages

The advantages of a systematic cost calculation are obvious:
- accurate cost calculations can be made very quickly,
- the experience can be recorded.

The specification for the FCCS can be divided into three major areas:
- **cost calculation**
  Cost calculation
  e.g. The possible cost calculation methods must be group technological as well as parametric, the user definable reference pieces or families must be organised in a tree structure and user definable methods must be linked to these families, multidimensional look-up tables for experience values, rates standard costs etc., decision tables allowing logic to be entered in an easy way.
- **quotation management**
  Quotation management
  e.g. Customer files with user definable attributes, making and printing quotations: user definable format by means of a powerful report generator, classifying and retrieving previous quotations by any combination of criteria.
- **general topics**
  General topics
  e.g. Import and export in ASCII and DBF for linkage with accounting, process planning and the CONSENS system. Userfriendly interface using windows, pull-down windows, analogy with spreadsheets, maximum flexibility due to user definable screens, methods, variables and formulas.

When using the CONSENS system, the optimal way to use the FCCS system is to consult it as a background system active while designing parts or assemblies.

In order to integrate the FCCS's functioning more closely in the CONSENS system, WTCM developed a more integrated version of the FCCS module. This additional version of the FCCS module is necessary for a closer integration between the CONSENS DfM and DtC module and eventually a DfA user tool (see figure 5.23). This version of the FCCS calculates the manufacturing costs based on the operations as they are defined during the DfM module. If some information is missing for the calculation, the user is asked to complete the information via the OSF/Motif based user-interface. The formulas, which are used to calculate the costs for every type of operation, are stored in a separate

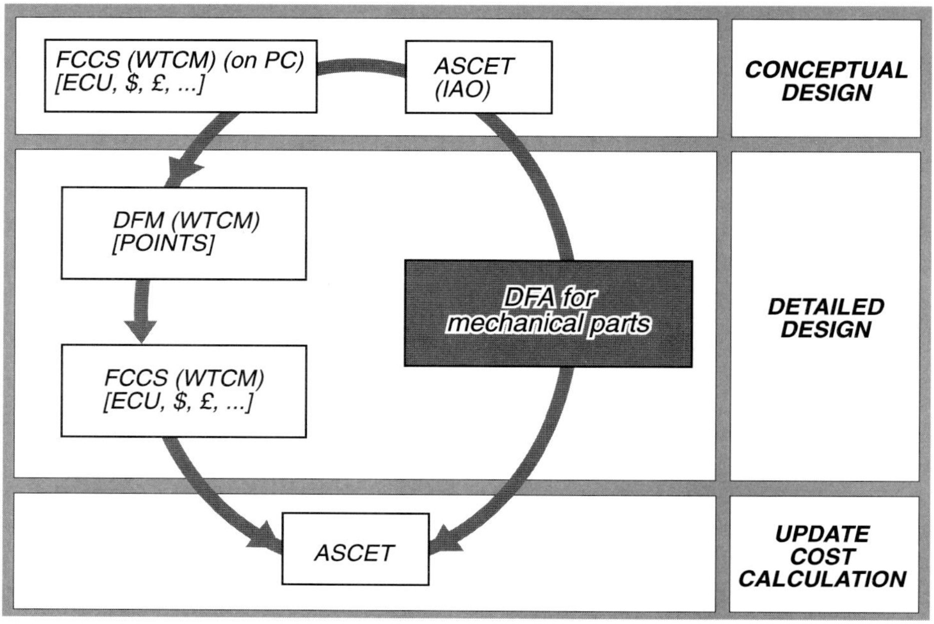

Figure 5.23: Integration of the DtC modules in CONSENS

ASCII file which is easily adaptable. The cost calculation functions are then parsed in the FCCS.

The original FCCS module (on PC) will be able to make cost estimations (before Detailed Design), whereas the integrated version will only be a cost calculator (after Detailed Design).

The FCCS modules contribute to the PSI approach since FCCS tends to parallelisation taking into account assembly and manufacturing costs. In addition, it helps in standardising products and is integrated with the DfM module to allow easy information exchange. The system is designed in such a way that it is able to implement other functionals for modified cost estimation (eventually adapted to company's requirements and specifications). The implementation of other cost calculation models is made very easy and user friendly. At the present time, cost calculation models for machining of mechanical workpieces is implemented in the system on UNIX and for sheet metal in the PC version.

5.3.5 Design of Production Facilities

R. Menges, U. Eigenmann

The relation between product and production facilities

In order to fulfil the superior objective of an enterprise, to achieve an established position and to make profit, the enterprise's products have to meet the requirements of the market regarding the function, design, quality, individuality and terms of delivery. At the same time the product has to be offered -and sold- at a price which yields sufficient profit. If it is not a proprietory product, enterprises have little influence on the market's pricing mechanism. The goal, therefore, must be to reduce expenditure concerning the development and fabrication of products.

A significant reduction of the time consumed by the development and planning of products and production facilities, however, can only be achieved by executing the product design and the design of production facilities in parallel. One of the objectives of concurrent and simultaneous engineering, therefore, is to synchronise product development and production planning. The early consideration of assembly requirements when designing the product, avoids costly modifications of the product in advanced planning phases, as well as changes in the production method.

The need to start the design of production facilities in the early product design stage becomes obvious if we consider that

- the assembly method,
- required tools and fixtures,
- form and complexity of jigs and fixtures,
- design of workplaces and
- cycle times

are largely determined by the product's design. The problems and costs which may arise from different design alternatives and how they affect manufacture and assembly, should be discussed early in the planning process. Therefore, a close co-operation between all departments which share the development and production planning process is necessary.

The question why production systems should be planned and designed can easily be answered: to save money.

The need for a Design of Production Facilities

Markets are becomeing more global. Products of the same company have to fulfil different national regulations and standards. Also a different mentality and special preferences of potential customers have to be taken into consideration. This results in an increased number of variants. The fact that people tend to buy more individual products intensifies this effect. An increasing number of variants and upward tendencies to extend the functioning of products lead to production systems which are more complex with high demands on the planning team. In addition, the extra expenditure caused by the growing number of variants and additional product functions cannot fully be compensated by higher prices. Competition, therefore, forces enterprises to reduce development and production costs. Since the financial latitude is getting smaller, production systems which do not work well for a longer time can rapidly cause a loss in profit.

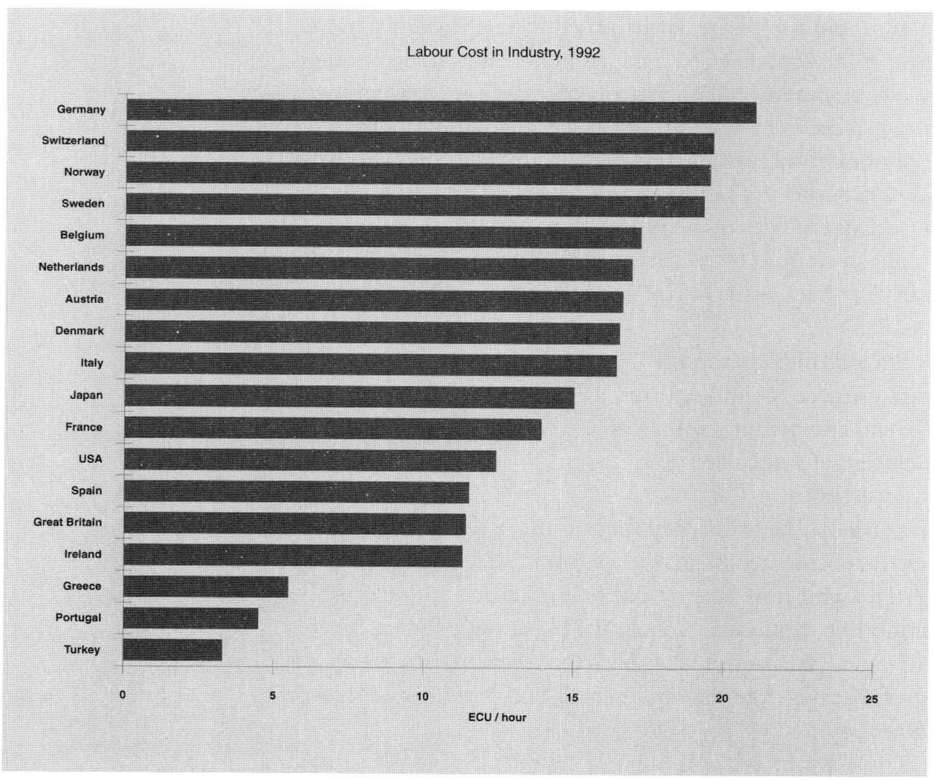

Figure 5.24: Labour cost in industry, 1992 (Source: Fischer Weltalmanach 1995)

Workers and their work stations – the most important resources of a company – often receive too little attention. This leads to poorly designed work stations where neither work satisfaction nor full performance can be achieved. Failures in the design of work stations result in low performance, stress and unnecessary fatigue. Considering increasing labour costs in production (see figure 5.24), productivity is more of a decisive factor than ever before. The significance of designing production facilities becomes obvious when we consider that the later failures and problems are revealed, the higher is the expenditure to solve the problems. Failures and problems can only be avoided if the product design and the design of the production system are well-suited to each other. The production systems, therefore, have to be designed simultaneously with the product.

Design of Production Facilities with ERGOMAS

The task to design production facilities is manifold. Material flows have to be optimised, costs must be calculated, production lines have to be designed with regard to optimised cycle times and to ergonomic requirements. Shorter development cycles, the reduction of operating costs, the optimal use of existing capacities and the optimisation of material and information flows are especially vital with regard to growing competition. These task can only be managed quickly and efficiently if computer aided tools are employed.

ERGOMAS (Ergonomic Design and Optimisation of Manufacturing and Assembly Systems) is a mature Computer-Aided Planning system for the preliminary and detailed design of manual and partly automated production systems. The main modules are based on a comprehensive database management system (DBMS) and a newly-developed graphics system for three-dimensional representation and modification of layouts on screen. ERGOMAS however does not only focus on the pure visualisation of production lines and manual workstations, but supports the user in all relevant planning processes concerning the production system. The scope of ERGOMAS is:

The scope of ERGOMAS

- Layout of production systems,
- Balancing production lines,
- Organising material flow,

- Design of work stations,
- Evaluation of work station design with regard to ergonomics,
- Cycle time analysis according to MTM,
- Calculation of assembly cost,
- Archiving and document planning results.

Benefits

The most important benefits from using ERGOMAS are:

- Support of Concurrent Simultaneous Engineeringresulting in a reduction of planning time and planning costs,
- Fast and efficient layout of production systems,
- Prevention of costly mistakes in production planning,
- Assistance in developing workstation layouts designed to improve the worker's comfort and safety by using ergonomic principles,
- Reliable and fast cycle time analysis using recognised MTM systems,
- Reduction in workstation set-up time.

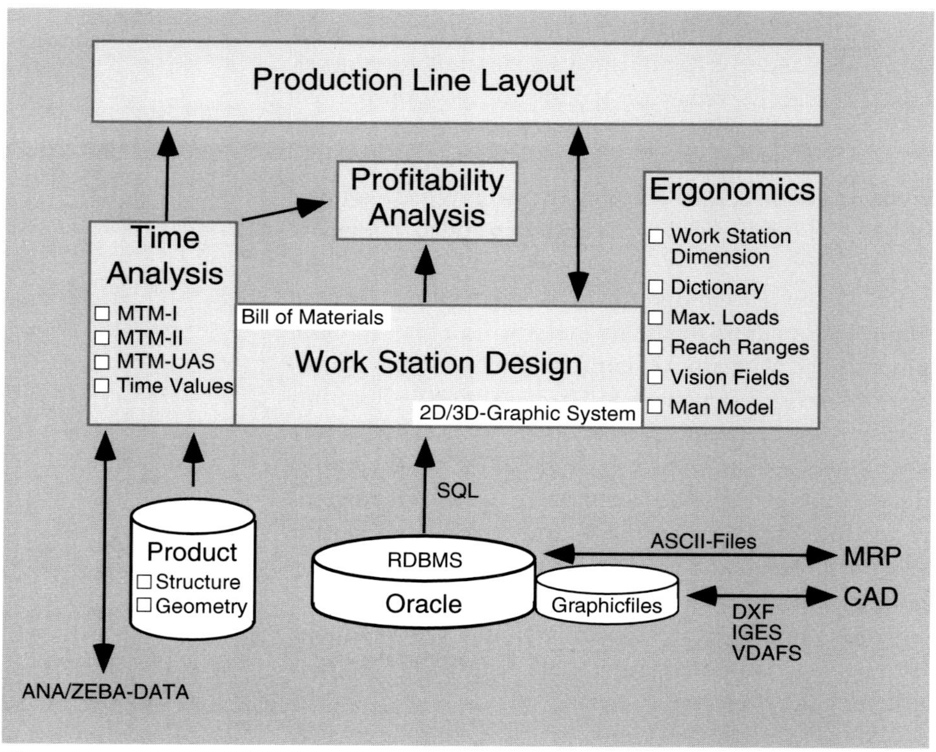

Fig. 5.25: ERGOMAS-Architecture

ERGOMAS consists of a multitude of single functions which are permanently updated (see figure 5.25). These functions are explained on the following pages and classified by section headings. Its open interfaces and modular architecture allow the integration of ERGOMAS as a CAP (Computer Aided Planing) module into an CSE-environment.

Architecture

ERGOMAS Component Data Base

The database contains a multitude of commercially available workstation components like tables, chairs, parts bins, tools, shelves, conveyor belts, etc.. All work station components are supplied with a 3D graphic representation. The windows-based software allows the user to search for components according to their individual characteristics, like supplier, order number, price, length, width or special properties. The ERGOMAS component database is constantly extended and regularly updated in co-operation with different suppliers. Moreover, the user can easily add company specific components to the database.

Component Database

Definition of the Assembly Structure

The assembly structure, i.e. the configuration of the final product, consisting of components, assemblies, subassemblies and parts, is defined by the product editor, either directly on the screen or via an interface to product archives or CAD systems. Handling operations, processes and workstation components can be assigned to subassemblies and parts. The allocation of operations to various workstations is made by reference to the product structure. This allows the creation of assembly sequences in which the existing relationships between the product structure, parts bins, tools and fixtures are taken into account. The product structure can be extended or modified at any time to accommodate product alterations or variations.

Definition of the Assembly Structure

Design of Work Systems and Work Stations

In order to set up a layout, the necessary components have to be retrieved from the data base. All components selected from the database are automatically entered into a list of equipment and graphically represented in the layout at the same time. During the design process, any number of dif-

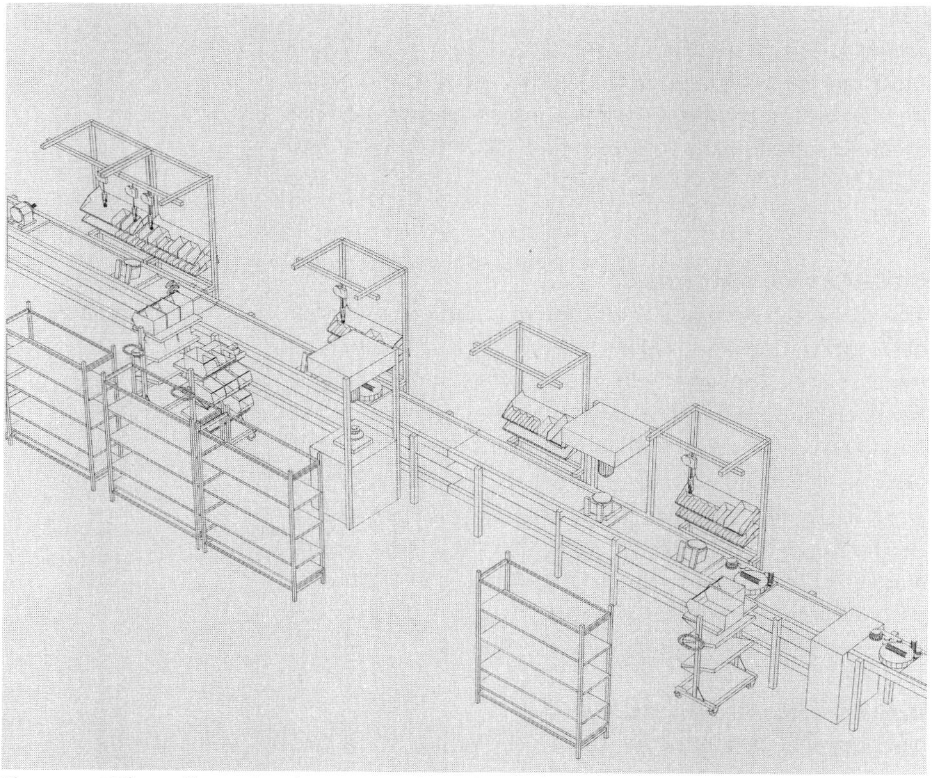

Figure 5.26 Three-dimensional representation of a production line

ferent views, i.e. top views, side views, front views and 3D views, can be displayed. ERGOMAS offers a full range of tools for the development of three-dimensional representation of work systems. Inserting, placing and modifying components are the main functions. The movement of components occurs simultaneously in all views. The user can select and place graphic objects in any view.

3D graphics

There is a special group of objects with simple graphics available for the purpose of defining rough work system layouts. This group contains the most important components needed for a rough planning of the production line layout. It includes, e.g. simple graphics that represent manual work stations, automatic stations, roller track sections, work piece carrier circulating systems, shelves, etc. It is, of course, also possible to use all the other components stored on the database.

Groups of components of such a rough line layout can be assigned to existing or not yet created work stations as substitute graphics. The use of substitute graphics has a positive effect on the resource usage and often increase the clarity of the production line layout as well. Anyhow, ERGOMAS provides the possibility to represent the work stations included in the production system as detailed or rough graphics.

The three-dimensional representation (see figure 5.26) makes it easy to immediately detect errors and weaknesses in the layout. Logistical and material flow dependencies and problems can thus be easily recognised and resolved. The agreement on the planning result with other departments, managers and manufacturing or the outfitter is made much easier, since all those involved in the planning process are able to understand the plan quickly and correctly. Misunderstandings and wasted time are essentially eliminated. The time consumed for setting up the production line is reduced.

Ergonomics

Ergonomic guidlines

The ergonomic analyses check the compliance with accepted ergonomic principles at each stage of the design process. All inputs are made interactively. Immediately after the modification of a parameter, the new results are displayed accompanied by a detailed documentation. The following functions are available:

- Determination of recommended work station dimensions for production (DIN 33406)
- Determination of recommended height dimensions of assembly work stations (TÜV Rheinland)
- Handling loads, both sitting and standing (VDI / REFA)
- NIOSH-Guideline (OSHA, USA)
- Determination of recommended limit values for transferring loads by hand (SIEMENS / Burandt)
- Determination of maximum muscle efforts for hand-finger and hand-arm systems (Schultetus / VDI)
- ABA-requirements and load capacity analysis
- Health risks to the musculosceletal system

Normally, the consequences of failures in the design of manual work stations become apparent during the implementation phase. Ergonomically insufficient work conditions do not only have a negative effect on the work performance but also increase absentteeism. Such conditions usually are the reason for increasingly frequent production failures, dissatisfaction and reluctance to work. Work stations which are not designed according to the rules of ergonomic design, therefore, result in low productivity.

Reach Rangers and Vision Rangers

ERGOMAS supports quick and reliable design and evaluation of work stations by the use of anatomical elements such as the 3D model of a person, reach ranges (see figure 5.27) and fields of vision. Anatomical elements are suitable for simulating the situation at the workplace from the view of the worker. Fields of vision show the work station seen with the eyes of the operator. The entire figure shows the field in which it is possible to view objects in focus without having to move the head but with moving the eyes, while the usable field of vision (white section) is the field within which both eyes can see clearly with no head or eye movements. Reach ranges depend on the person operating the work station. According to sex and percentile of the operator ERGOMAS shows the selected reach range (see figure 5.27). Available reach ranges are:

- ideal reach range,
- physiological maximum reach range,
- geometrical maximum reach range and
- reach range with upper body movements.

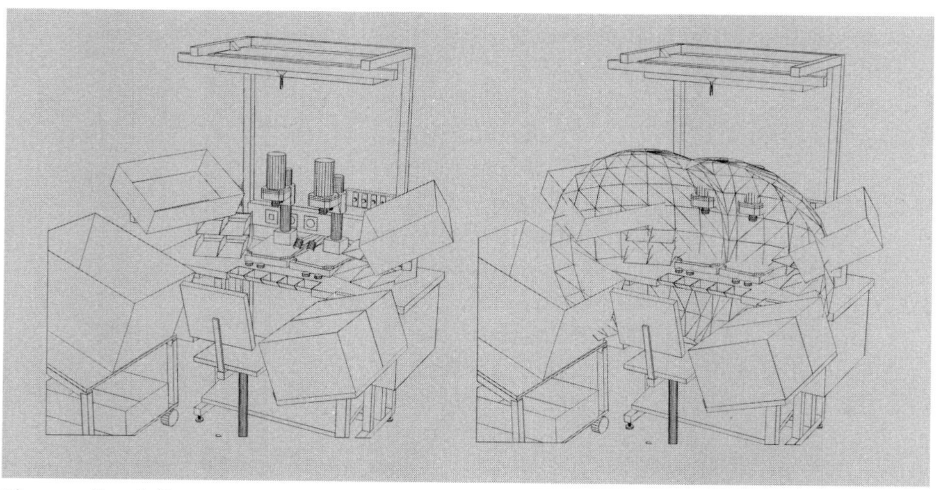

Fig. 5.27: Reach Range

Restricted vision or access to parts bins, tools, fixtures, machines or to the product to be assembled can easily be detected evaluating the work station using anatomical elements. This way, a work station design leading to postures and movements which are not ergonomic, is avoided.

Human movement simulation

ERGOMan, an optional software module, provides the ability to simulate the operation at the work station. A biomechanical 3D volume man-model, according to chosen gender and percentiles is placed in the work station. The various human motions defined by the MTM analysis are animated. The man-model may be seated or standing. Loads on joints which occur during the motions are calculated and graphically documented. Possible collisions with parts of the work station are indicated.

Hypertext dictionary

In addition, the user is assisted by a hypertext dictionary. The dictionary supplies important information, via keywords, on environmental issues, ergonomic standards and solutions to specific problems, such as sound absorption or safety regulations. The included dictionary can be modified and extended by the user to incorporate company guidelines.

Standard Time Analysis

MTM (Methods-Time Measurement) is used for standard-time analyses. It permits a quick and efficient determination of cycle times with high precision. There is automatic access to the defined assembly components and the operations on which they are used, as well as to existing work station layouts, for the determination of the starting and finishing points of motions. The computer calculates the correct distances and, if essential locations are moved in the work-station layout, updates them. The cycle time and the methods analysis can be displayed on the screen at any time.Recurring sequences can be saved as macros. They can be recalled and modified in future analysis. Fixed sequences are given an identifying code, defining them as new time elements, and are then handled like all other MTM codes.

Relation between layout and standard time analysis

The program reports errors made in the MTM analysis. Keywords can be attached to each analysis. Printouts are available of the MTM analysis sheet, the sequence of operations card and the time calculation, with a drawing of the relevant workstation layout.

MTM Analysis coversheet

DELTA Industrie Informatik GmbH	work system **91-06-05/01 DELTA Assembly System**	created: **23-MAY-91** changed.**10-JAN-95**
	work station **05.06.91/01-01 Lamp Assembly**	created: **23-MAY-91** changed.**12-JAN-95**

description :	code :	E-TMU :	TT :	A :	C-TMU :
Assembly of lamp	**ASS-033-0001**	**0.0**			**560.0**
operation :		operation number :			

output
cycle time (100% MTM)g :0.00
set-up time : 0.0000
product :
part or group :
work protective methods :
quality notie :
garbage disposal :
remark :

machine :
equipment :
device :
workshop :
conveyance :

no	description	tg(min)	tg(TMU)
1	lamp socket in mounting	0.0420	70.00
2	device, spring on socket and	0.0330	55.00
3	in lampholder		
4	lampholder inl.mounting dev.	0.0420	70.00
5	pick up socket	0.0540	90.00
6	pick up spring	0.0660	110.00
7	screw on socket	0.0240	40.00
8	insert socket with spring	0.0360	60.00
9			
10	press in lamp sockets		
11	move pressleaver down	0.0270	45.00
12	adjust socket with spring	0.0120	20.00
	sum	0.3360	560.00

std time with allowance		
TE :	0.336	min/unit
	0.006	hrs/unit
PART	2.98	pcs/min
	178.57	pcs/min
	1428.57	pcs/8.00
TR :	0.000	min
	0.000	hrs

department	date	created	checked	validity	
	12-JAN-95				ERGOMAS

Fig. 5.28: MTM-Analysis

The main advantages in performing MTM-analysis using ERGOMAS are:

- quick, accurate and efficient determination of cycle times,

- full access to the work station layout and the pre-defined assembly components with their related operations,
- easy access to existing MTM analyses,
- automatic checking of MTM analyses for correctness and completeness and
- full integration into the existing MTM software.

Profitability Analysis

Cost transparency

The assembly cost per unit of a production provide an important basis for judging the profitability of a production system. The assembly cost function determines the total production unit costs at both the level of work stations and the level of production systems. This is a static cost calculation that is best suited for comparing and evaluating different variations. The calculation is based on tooling costs, cycle times, company labour rates and company costing practises. In order to be able to compare two execution alternatives, break-even analyses and amortisation calculations can be performed. The break even analysis calculates the number of pieces that have to be produced for an alternative method to be either cheaper or more expensive. The amortisation calculation compares investment to requirements and determines the time it takes for an investment to pay off.

Benefits Gained from Designing Production Facilities Employing ERGOMAS

Increased productivity by computer based work system design

The costs produced by manufacture and assembly are determined to the largest extend by the product design. Therefore, it is too late to start to plan production facilities when the product design is finished. The later necessary modifications of the product are recovered, the more expensive are these modifications. Often deficiencies come to light only when setting up the workplaces and assembly lines. This leads to inefficient and insufficient ergonomically designed workplaces. At this stage it is usually too late and would be too expensive to modify the product's design.

To avoid costly mistakes it is most important to simulate the circumstances at the workplace (virtual prototyping). Three dimensional graphics of the workplace make it easy for everyone to imagine how the later workplace will look, even if he is not used to working with drawings. This way, each member of a heterogeneous team is able to contribute to the discussion. For example, normally it is difficult to involve the workers themselves in the design of new production facilities or the modification of existing ones. Nevertheless they have important and valuable knowledge about the difficulties and inadequacies of the actual assembly process. Using realistic and understandable 3D-graphics to show how future workplaces and assembly lines will look like, even people who are usually reserved, may drop some useful hints.

Evaluation of work place design

Evaluating the designed workplaces by reach ranges and vision ranges, problems and mistakes can be recognised and removed in time, before the design is finished or the workplaces are set up in method laboratories or, even worse in production. Ergonomic aspects are taken into consideration from the very beginning of the planning process. Ideal working heights, depending on the working posture, sex and the task to be performed and suitable chairs are considered as well as loads limits. This protects people against sever injuries and consequences concerning their health.

Cycle times are determined by using the MTM method. Thereby, full access to the work station design is available. Starting and finishing points of motion can be taken directly from the layout. The co-ordinates are assigned to the respective work station components. This way, the cycle time analysis is updated automatically when the components are moved and the work station design is modified. This procedure ensures that the analysed cycle time goes with the corresponding layout and at time.

In addition the profitability analysis enables the user to evaluate different design alternatives from a financial point of view. Thus, decisions can be taken on the basis of objective cost data. Moreover, cost awareness is strengthened.

Why to use ERGOMAS?

This is why ERGOMAS

- increases productivity,
- cuts planning time,
- increases planning quality,
- avoids planning errors and
- supports Concurrent Simultaneous Engineering.

Summary

Growing competition and the tendencies towards shorter product life cycles forces enterprises to accelerate their development processes. New products have to be put on the market in shorter cycles. Only new products with unique features and characteristics guarantee the enterprise a monopoly for a short time. Successful products are quickly copied by competitors (Me-too-products). Thus, "Time to Market" becomes more and more important.

Concurrent Simultaneous Engineering

Concurrent Simultaneous Engineering is the appropriate approach to achieve objectives like an earlier market entry of new products, an increased product quality and less modifications due to planning failures. The polymorphic requirements of production planning, however, necessitates the use of computer aided tools. Lean production systems demand continual improvement of delivery periods, production costs, capacities, material utilisation and information flow. Integrated software applications for the design of production facilities, therefore, are imperative on the way towards the factory of the future.

5.3.6 Design for Quality

K. Jeschke

5.3.6.1 Requirements for Electronic Producers

Today, the requirements placed on electronics producers supplying automobile components are very high. The trend towards just-in-time production has already led to complex requirements. Apart from a faultless product quality, a flawless logistics quality was and still is part of many specifications and contracts. This is expressed by the requirement to deliver faultless products directly to the assembly line at a time agreed upon. In the future, this will mean for a system supplier, that not only the production but also the product development will be bound to the customer's time schedules. In order for prototypes to be made available for qualifications in time, internal time schedules have to be adapted to the customer's schedules. The fairness of price and the quality of a product may not be forgotten as regards these new challenges. The relocation of production plants to cheap labour countries or the trend towards worldwide resourcing require a lasting reduction of costs, and the quality parameters of a product do not only concern its function but also its optical properties.

Quality management plans

To fulfil the quality requirements, enterprises increasingly put their trust in quality management systems comprising the overall business. It is expected from these that errors and malfunctions are avoided and thus costs and throughput times are reduced. Quality management systems mean that quality-assuring measures are planned already in the product definition phase to be fixed as quality management plans then [2]. These include:

- the quality targets to be achieved,
- the processes, methods and job instructions to be applied and
- suitable programs for quality inspections, examinations or audits.

Measures of quality management

Some examples of process-oriented measures of quality management are shown in figure 5.29. In order to identify design errors or unsuitable components, digital or analog simulations are used. Design errors can also be checked by means of error simulations, design rule checks or the simulation of electromagnetic compatibilities (EMC). Error simulations consider the behaviour of a circuit when undesired events are occurring, e.g. the failure of a component. Design rule checks include the observance of design rules, such as distances between conducting paths, space required for test pads or boring distances. EMC simulations compare the parasitic with the ideal circuit behaviour in order to examine the effects of electromagnetic influences.

Lacking reliabilities can be primarily explained by temperature overloads. Temperature simulations are therefore used for reliability checks.

Manufacturing errors are detected by means of different testing methods. Here, the spectrum of testing methods ranges from visual inspections, in-circuit and functional tests to system tests. Run-in tests and burn-in tests complete the above spectrum by tests under stress conditions. The run-in test aims to recognize early failures of a componentry, the burn-in test identifies early failures of individual components. By applying the above measures, it is possible to ensure the quality and reliability of a product or a

Method	Description	Design error	Component error	Lacking reliability	Manufacturing error
Digital simulation	Simulation of the logic time behaviour with min., max., worst case values of the components and unfavourable operating conditions	●	●		
Analog simulation	Simulation of the electric behaviour of the circuit with min., max., worst case values and unfavourable operating conditions	●	●		
Error simulation	Simulation of the circuit behaviour when errors are occuring. Testability check.	●			
EMC simulation	Simulation of ideal and parasitic circuit behaviour	●			
Temperature simulation	Simulation of the temperature distribution on the printed circuit board	●		●	
Design rule check	Layout check of observance of design rules	●			
Visual check	Check on identity and position of components and on the visible conditions of joint connections				●
In-circuit test	Check on identity, position and function of individual components under static loads				●
Functional test	Check on the function of a componentry under static or dynamic loads	●			●
System test	Check on the system function under static and dynamic load	●			●
Run-In-Test	Functional check on a componentry under stress conditions to detect early failures			●	
Burn-In-Test	Functional check on a componentry under stress conditions to detect early failures			●	

© iw 210-03-01 ○ detects special error

Fig. 5.29: Process-oriented Measures of Quality Management

procedure. Testing methods to detect manufacturing errors and early failures are traditionally used. The cost of the tests is considerable. For consumer electronics the investment costs amount to approx. DM 10.000.–. The adaption costs reach approx. DM 100.000.– or even DM 1.000.000.– and depend on a multitude of factors. In addition to the technical complexity and the test-friendly product design, an important part is played by demands depending on the product, i.e. the demands for testing depth and diagnosis capability as well as criteria of testing and validating test programs including the documentation [10]. In the field of the automobile industry, the costs may be ten times higher for safety-relevant products.

Cost cutting by preventive quality assurance

A special potential of cost cutting for quality assurance is included in the measures of the early detection of design errors, compare [8,9]. The simulation methods mentioned above shall be profitable, as the cost to eliminate an error which was caused in the design phase, may rise ten times the amount at the time of the manufacturing planning and to a hundred times the amount at the start of production. By detecting and eliminating the errors early, these costs can be avoided. Like the cost of tests, the cost of simulations is dependent on many factors. The expenditure of the temperature simulation is for example determined by the complexity of the package or the availability of a current component library. Therefore, it is to be considered from case to case which measures for quality assurance shall be selected.

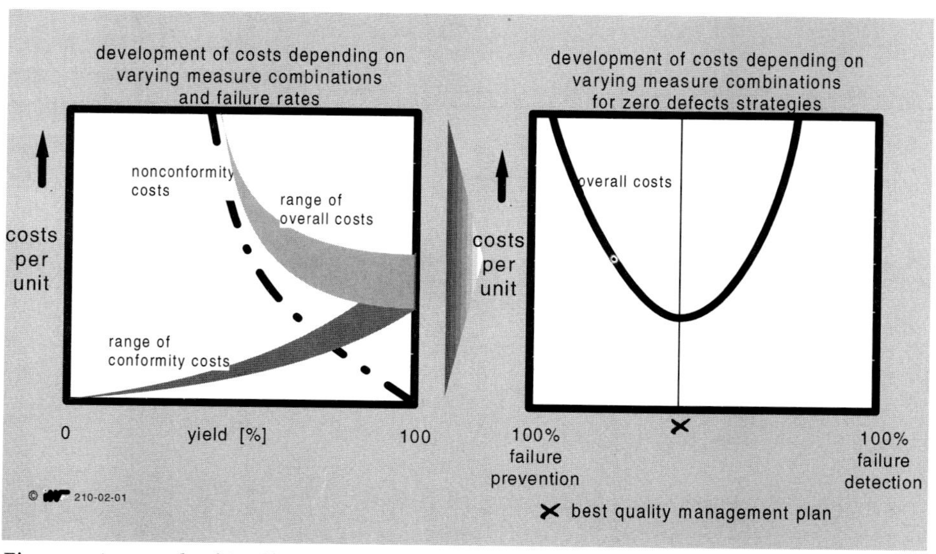

Fig. 5.30: Approach of Optimising Quality Management Plans

For this consideration, the total cost of a quality management plan is to be examined. In this connection, a difference is made between the cost of measures which leads to the conformity with the customer's requirements and the cost of measures which is necessary to eliminate nonconformities'[11], figure 5.30. The cost of conformity comprises the already mentioned expenses of tests and simulations. The cost of nonconformity is determined by the expenses in case of rejects, rework, complaints and warranty cases. According to the principles of modern quality philosophies – to perform all activities in an enterprise without faults – the total costs are the lowest in case of error-free products [1]. In particular cases, however, it is left open which measures – whether error-avoiding simulations or error-detecting tests – are the most favourable ones for ensuring the conformity of the product properties with the customer's requirements.

5.3.6.2 Aims and Focuces of the DfQ Tool

This problem is taken up by the Design for Quality Tool (DfQ), figure 5.31. The system has the aim to plan and optimise quality management plans at the start in the early design phase until the end of production of a product. For this purpose, testing strategies are optimised and the application of simulation methods is assessed.

Cost parameters

The criteria to assess quality management plans are the parameters cost, quality and time limit, the focus lying on the factors cost and quality. For a quality management plan, the system calculates the cost of:

- Error-avoiding measures (related to the order and piece),
- measures limiting the effects (related to the order and piece),
- creation of test programs,
- check of test program,
- design of adapter,
- adapter manufacturing,
- set-up of test equipment,
- set-up of the componentry or the device
- testing of the componentry,
- error diagnosis,
- dismantling of the test equipment,

- rework,
- rejects,
- complaint.

Quality oriented parameters

For a quality-oriented assessment of a QM plan, the delivery quality, the reliability and a risk priority number is calculated. The delivery quality, here, means the number of products that are free from defects before going into operation [4]. The number of operative products in the field is described by the reliability. The risk priority number is calculated by multiplying the assessment quantities for the probability of occurrence and detection of an error type and cause by the error severity, as with the failure mode and effects analysis. New developments of products are often closely combined with the customers' time schedules. For assessing a quality management plan in terms of time, the longest realisation period of all quality-assuring measures is used as a basis.

Promotion of Simultaneous Engineering

Apart from the quality and reliability planning, the system aims to promote the philosophy of Simultaneous Engineering. For this purpose, a team-oriented way of proceeding has been conceived in which employees from the

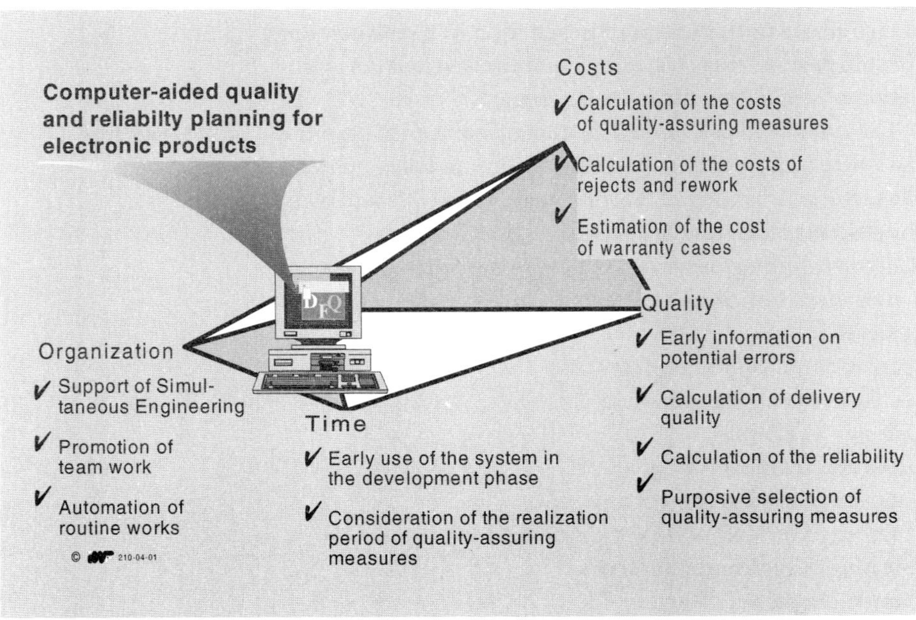

Fig. 5.31: Aims and Focuses of the Design for Quality System

design departments as well as from the test and manufacturing planning are commonly planning the spectrum of quality-assuring measures. Detailed analyses are made based on division of labour. Rough analyses and the coordination of the detailed results are prepared in the team. Routine works such as the calculation of the assessment criteria or the variation of the possible quality management plans are performed by the system. As already mentioned at the beginning, many factors influence the actual costs of quality-assuring measures. The quantities influencing the quality and the time limit are equally complex, figure 5.32. Therefore, the calulation model of the DfQ tool is limited to parameters which are either existing in the enterprise or can be obtained from field or laboratory examinations. The influences coming from human beings or the environment are not considered by the system, as they can be determined at great expense only.

Component categories

Great importance has been attached to a detailed consideration of the product parameters. The system uses the classifications according to DIN IEC 56 [3]. They consider the function and design of electronic components. In addition, characteristics of the reliability calculation and data of error types and causes as well as their probabilities of occurrence have been assigned to the categories. Each category can also be described in more detail with respect to the outside supplier and the technology (wired or surface-mounted). To determine the expense of rework and the cost of rejects, the cost of materials and rejects rates are considered too.

Process types

Just like products, processes are also considered in detail. They are classified according to process types and machine types. Typical process-related error types are managed together with their probabilities of occurrence in connection with the respective defective component. The error causes are treated in the same way. This connection is necessary as process errors often arise only if specific product features are existing.

Quality-assuring measures are the heart of quality management plans. The assessment of the single activities, therefore, requires some detailed data. The data model of the DfQ tool distinguishes between measures for detecting errors, for avoiding errors and for limiting the effects. The

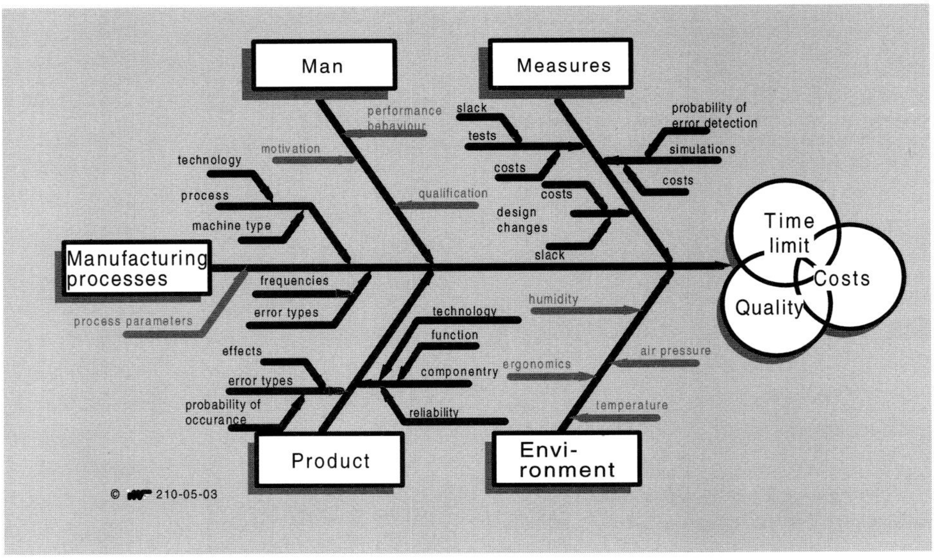

Fig. 5.32: Quantities Influencing Cost, Quality and Time Limit

performance capability of a testing method is marked by the probability of detection. Test equipments often have error slacks which are determined both by the component and the error type. Therefore, error probabilities are acquired and managed depending on the error type and the component type. The range of measures for avoiding errors and limiting the effects is very large and reaches from simulation methods and prototype trials to design changes. These measures can be flexibly assessed when the cost related to the order and the cost related to the unit is indicated.

5.3.6.3 Way of Proceeding

The systematics and way of proceeding of the Design for Quality tool is based on teamwork performed by the employees from the departments design, manufacturing and inspection planning and inspection, figure 5.33. The systematics is passed through several times until the start of the series production.

In the fundamental phase the planning results are not very expressive. With formulation of the product and production conception as well as the gaining of experiences from simulations and trials, the expressiveness is increased

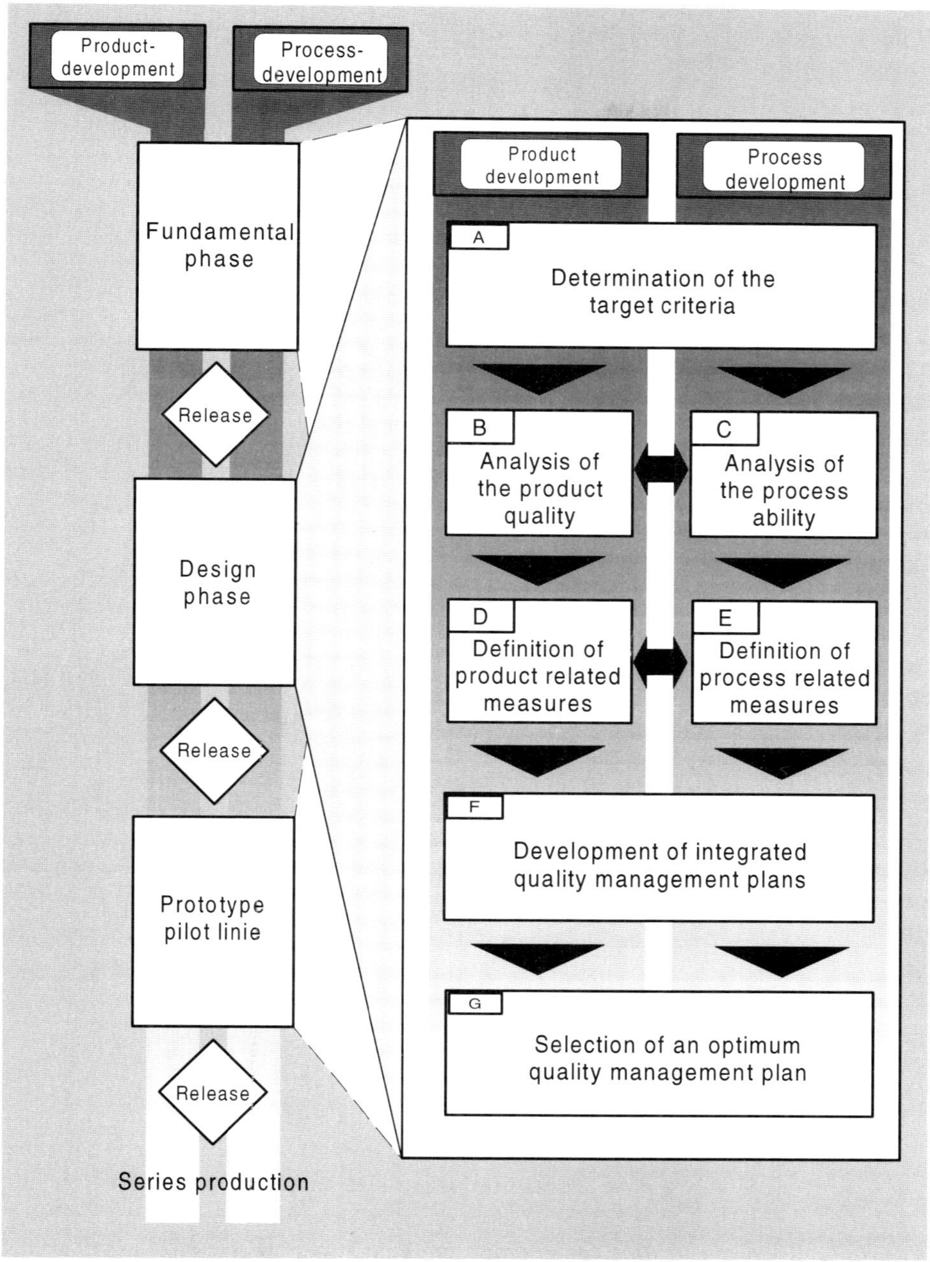

Fig. 5.33: Way of Proceeding for Quality and Reliability Planning

in the design phase and trial phase as the knowledge gained is supplemented step by step.

Determination of target criteria

A DfQ project starts with determining the target criteria for products and processes (step A). On the basis of information from the customer service and marketing department as well as knowledge of competitors' products and sales markets, the DfQ team finds out customers' and market requirements. New product and process technologies are considered in parallel to that in order to determine technology potentials. This forms the basis of specifications regarding products and processes. The product-related requirements comprise product functions and structures as well as delivery qualities, reliability, target costs and target date. As to the processes, potential technologies, process capabilities and availabilities are specified.

Analysis of product and process quality

Rough drafts of the product and production conception should already be in hand for the next DfQ step, i.e. the analysis of the product quality and process safety (steps B and C). The analysis is made following the methods of the risk and reliability analysis [6]. On the product side, the product functions are determined and their loads are estimated. According to the Pareto principle, product functions and components that are subject to risk are identified for further analysis. A function is estimated to be subject to risk if the fulfilling of the specifications seems to be critical. For the analysis of a process chain the procedure is similar. The performance utilisation of the processes is estimated in order to identify critical processes. The performance utilisation can be rated as an indication of the process safety, as precision and performance often behave contrary to each other.

Furthermore, the team prepares a reliability and risk analysis for critical product components and production processes, figure 5.34. On the basis of similarity analyses for product components and manufacturing processes, the DfQ tool supplies experience data on potential errors as well as on their consequences and causes. The experience data are founded on quality data from the production and the field which have been determined within the scope of problem-solving methods. Here, reference is made to the documentation of the problem-solving process, e.g. 8D-Report, for more detailed hints for avoiding errors. Each error and each error cause is assessed with respect to its probability of occurrence and detection. In case of failures

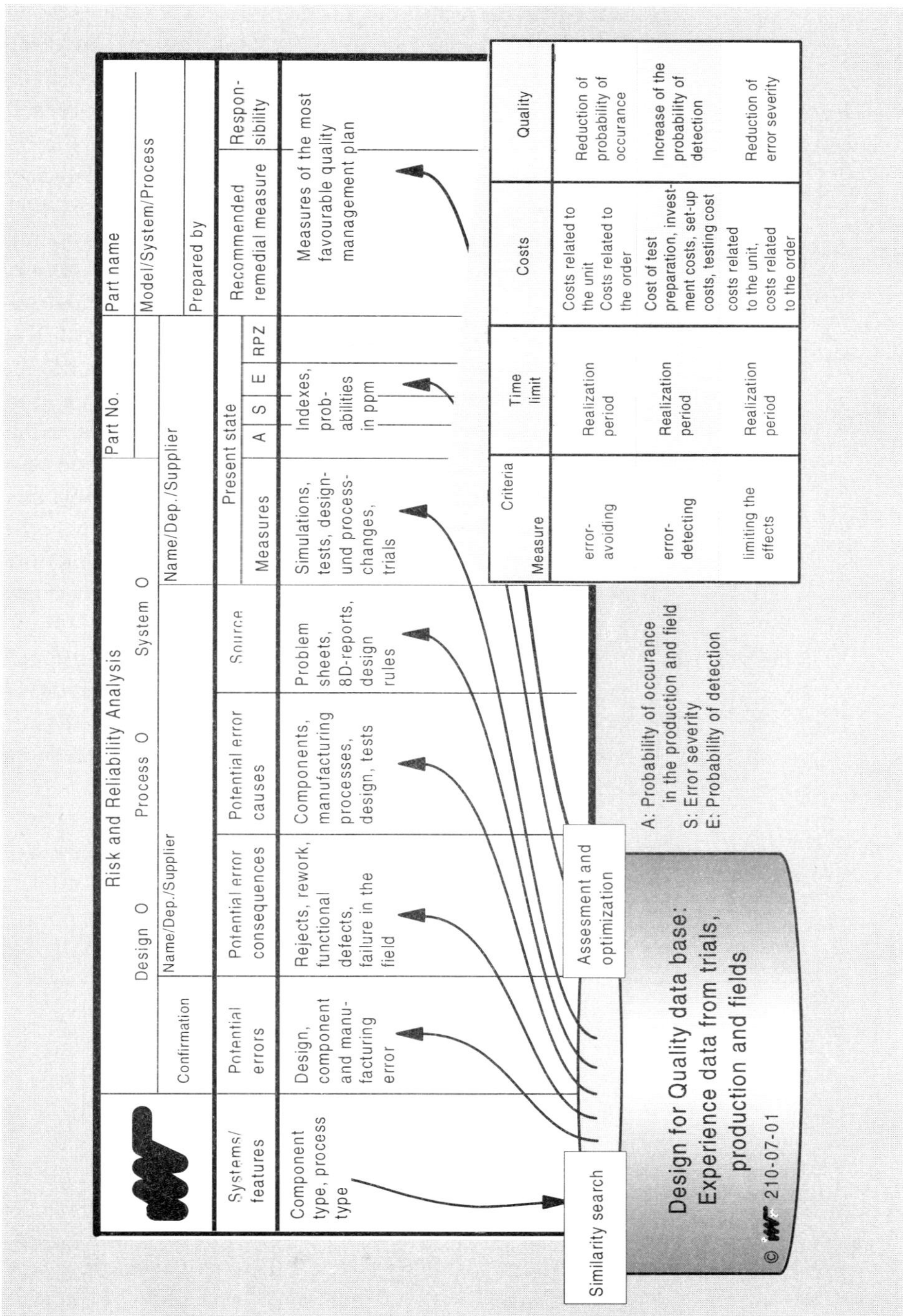

Fig. 5.34: Information on Potential Errors and their Causes, Collection of Potential Measures

in the field the probability of occurrence corresponds to the failure rate. To evaluate the error consequences, the error severity is determined.

Definition of product and process related measures

Then the team fixes product-related and process-related measures to avoid and detect errors or to limit the effects (steps D and E). These subsequently become part of quality management plans. Each measure belongs to a product component, a process and an error type and is evaluated with respect to its benefit and expense, figure 5.35. To assess the benefit of a measure, the new probability of occurrence and detection of an error is rated in ppm, or, if this is impossible, with the usual scale of assessment from 1 to 10. The error severity is assessed with a value between 1 and 10 in any case. To estimate the expense, the costs related to the order and the unit as well as the realisation period of a measure is considered. In error-detecting measures, this expense is determined by the product component in some cases. For the in-circuit test, for example, the running time of the test program is determined with the help of the component type.

Tree structure

This structurised and assessed collection of different measures forms a basis for the development of quality management plans (step F). These can be derived by means of a tree structure, figure 5.36. The example of the testing of a

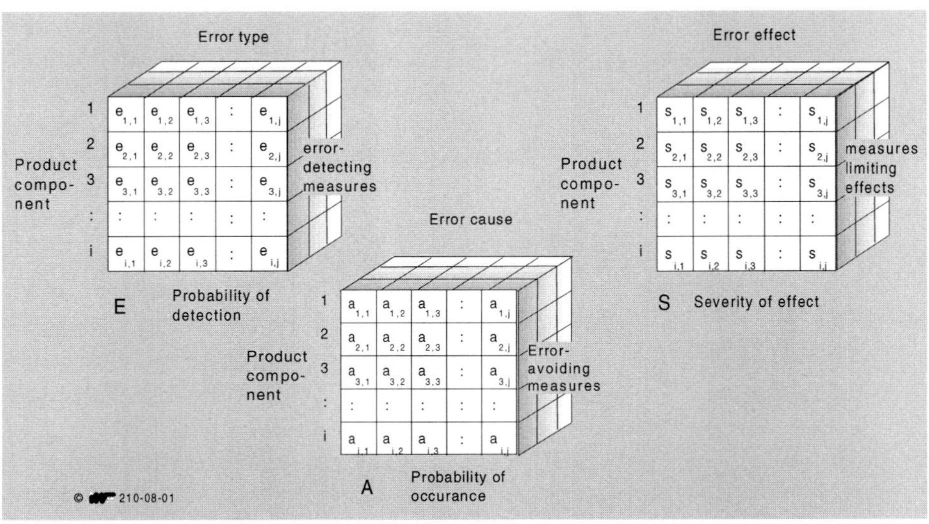

Fig. 5.35: Data Management to Assess Quality Management Plans

product component shows that the methods 1/1 and 1/2 and the omission of the test 1/0 are available within the first testing level. At the second testing level, the testing method 2/1 or the omission of the test 2/0 are possible. In case of further testing levels, the tree of measures can be extended to any size. For the generation of quality management plans, specific branches of the tree of measures are activated and assessed with respect to the parameters delivery quality, reliability, risk priority number, costs and realisation period [5].

Calculation of delivery quality

The delivery quality, in this connection, is determined by the delivery qualities of the product components. These, in turn, are calculated from the probability of occurrence and detection of an error type. Both quantities depend on the measures activated. The probability of occurrence of an error type, for example, can be reduced by changes in the layout, and tests contribute to an increase of the probability of detection.

Figure 5.36 illustrates the calculation of a probability of detection when using different test equipments. For a product component, the probabilities of detection E without quality-assuring measures are determined from the data base. These values are multiplied by the probabilities of

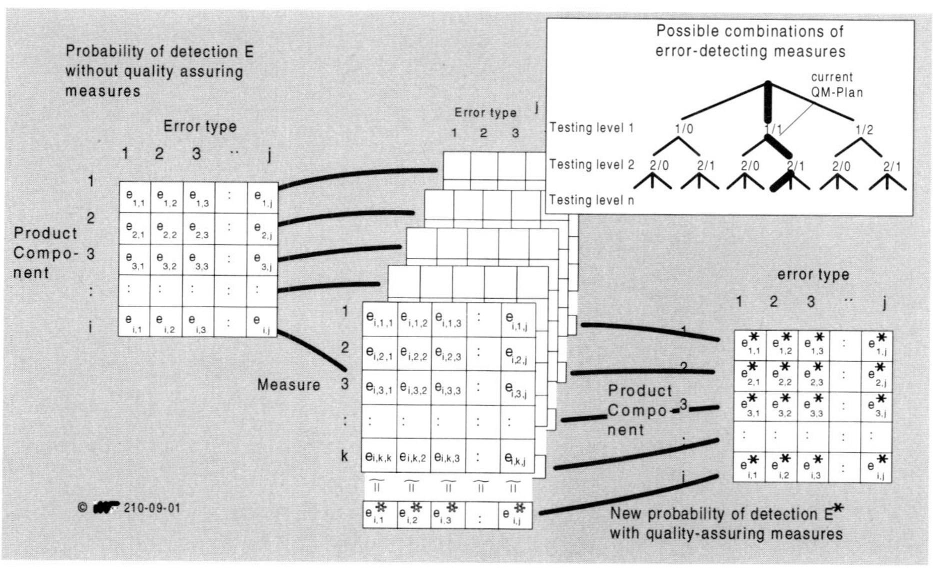

Fig. 5.36: Creation and Assessment of Quality Management Plans

detection of the measures respectively activated. This calculation results in a matrix which includes the probabilities of detection of each error type and product component when all activated, error-detecting measures are applied. The same procedure is used for the calculation of the probability of occurrence. Instead of the probability of detection, here, the probabilities of occurrence have to be inserted in the matrix only. By multiplying the parameters of all product components for each error type, the parameters for the total system are obtained. In this connection, a simplified product structure is being started out from consisting of product components connected in series only. Multiplying once more the parameters related to the error type leads to a compression of the parameters without differentiation of the error type for the total system.

Delivery quality and reliability of systems

The separate consideration of single error types allow a comfortable calculation of the delivery quality and the reliability at the system level. The delivery quality of the system is determined by the multiplication of the probabilities of design and manufacturing errors. The reliability results from the error type "failure in the field". The risk priority numbers per product component and error type can be determined in the same way. The highest of all risk priority numbers then is used for the assessment of the activated quality management plan. To characterize the date, the longest realisation period of all activated measures is considered.

Cost calculation

The calculation of the quality-relevant costs is more complicated. The individual cost types have been introduced at the beginning. The cost of conformity is determined by the activated measures. In case of error-avoiding measures and measures limiting the effects, the cost related to the unit is multiplied by the estimated sales volume and added to the order-related costs. The cost of error-detecting measures is determined depending on the product components and the complexity of the system. Nonconformity costs result from the probabilities of occurrence of the error types as well as the expenses of diagnosis and rework of the defective product component. For the consideration of the cost of rejects, the DfQ tool includes the rejects rates per components. Together with the cost of materials, the cost of rejects can be determined from this.

After the quality management plans have been generated and assessed, the most favourable ones are selected (step G). Here, unclear, i.e. fuzzy, and changing expectations have to brought in line with each other. The conventional theory of sets and logic with its restricted range of values "true" or "false" does not sufficiently describe human decision making. Thus, the risk priority number of a final product which has the value of 120 on a scale between 0 and 1000, may be "just good" or "quite good" with the value of 60.

Fuzzy Logic

This can be remedied with the help of the fuzzy logic, which also permits other values than "true" and "false". Statements may be fulfilled in part only, thus having truth values between 0 and 1. A risk priority number of 60 may be "good" with the degree 0.7, and a risk priority number of 120 may be "good" with the degree 0.1. Overlappings of adjacent fuzzy sets are possible, too. The risk priority number of 120 may be "good" with the degree 0.1 and "bad" with the degree 0.2. In fuzzy sets, the membership values indicate the degree of membership of elements to a set. The function mapping this is called membership function. Such a membership function is assigned to each fuzzy set as a mapping rule.

By means of the fuzzy logic, the 5 assessment quantities of a quality management plan are processed to form one output quantity using processing regulations in the form of rules. The sharp input quantities become sharp output quantities by using fuzzy, qualitative knowledge. The single components of a rule-based fuzzy implementation are shown in figure 5.37. In the fuzzyfication part, sharp input

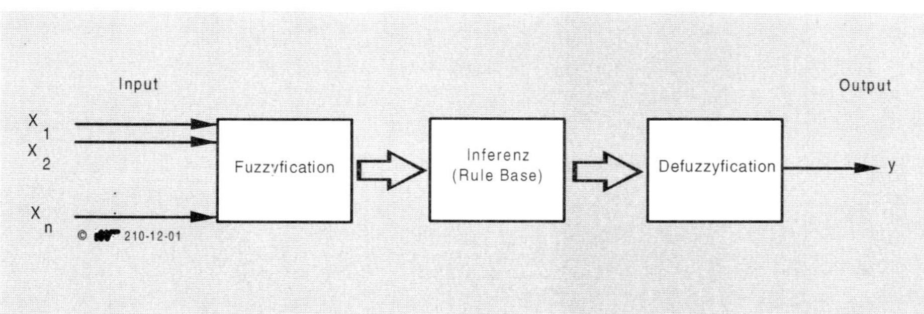

Fig. 5.37: Components of a Rule-based Fuzzy System

quantities become fuzzy values between 0 and 1 by the mapping through membership functions. The interference scheme is a basis of rules containing knowledge in the form of rules expressing if/then situations and a regulation how these rules are to be interpreted. On the basis of these rules the fuzzyficated input values are mapped into an output fuzzy set. The resulting output fuzzy values, in turn, are within an interval (0,1) and are converted to sharp output values in the defuzzyfication part. For a more detailed consideration of the fuzzy theory, see the bibliographic references in the appendix, especially [7] and [12].

Fuzzy Implementation

Fig. 5.38 illustrates an example of assessing quality management plans. In the fuzzyfication part, step 1, the membership values of the sharp input values are determined. For the parameter costs, K = DM 20.000.–, a membership value of 0.8 to the attribute "low" is the result, and the delivery quality, Q=300 ppm, is assigned at 60% to the attribute "good". The risk priority number, RPZ=800 units, has a membership value of 0.85 to the attribute "high", and the realisation period, T=55 h, is assessed with 30% to belong to the attribute "long". Then, the interference schemes are used, step no. 2. The DfQ tool uses the most widespread MIN-MAX operators. For more detailed information on the properties of interference schemes reference is made to [7] and [12]. The first rule defines an AND relation. Therefore,

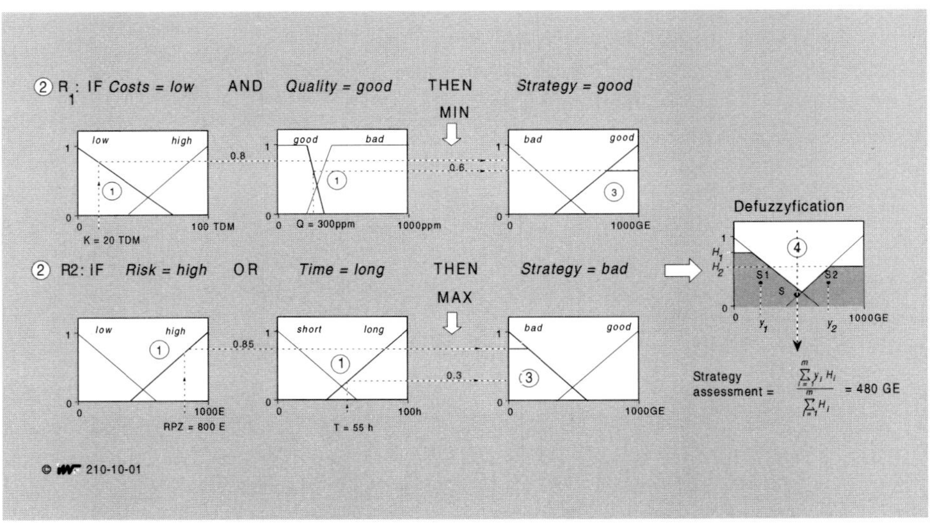

Fig. 5.38: Assessment of Quality Management Plans by means of Fuzzy Logic

the result is the smaller membership value of the parameters costs and delivery quality. The second rule contains an OR relation and results in the higher membership value of the rsik priority number and the realisation period. For the defuzzyfication part, the surfaces under the membership functions of the strategy quality, step no.3. Then, the focus of both surfaces is calculated and the strategy quality is determined from this, step no. 4.

The method of the fuzzy logic allows the setting up of an order of ranks of the generated quality management plans from the five assessment criteria of the DfQ tool. With the membership functions for the delivery quality, reliability, quality-relevant cost, realisation period and risk priority number it is possible to define the target values even in case of unclear limits. The rules and the interference schemes fix which criteria lead to the preference given to a quality management plan. Within the scope of de-fuzzyfication, the strategy quality is finally determined.

Results

The results of the DfQ tool are represented as a portfolio. The diagram clearly shows the top ten quality management

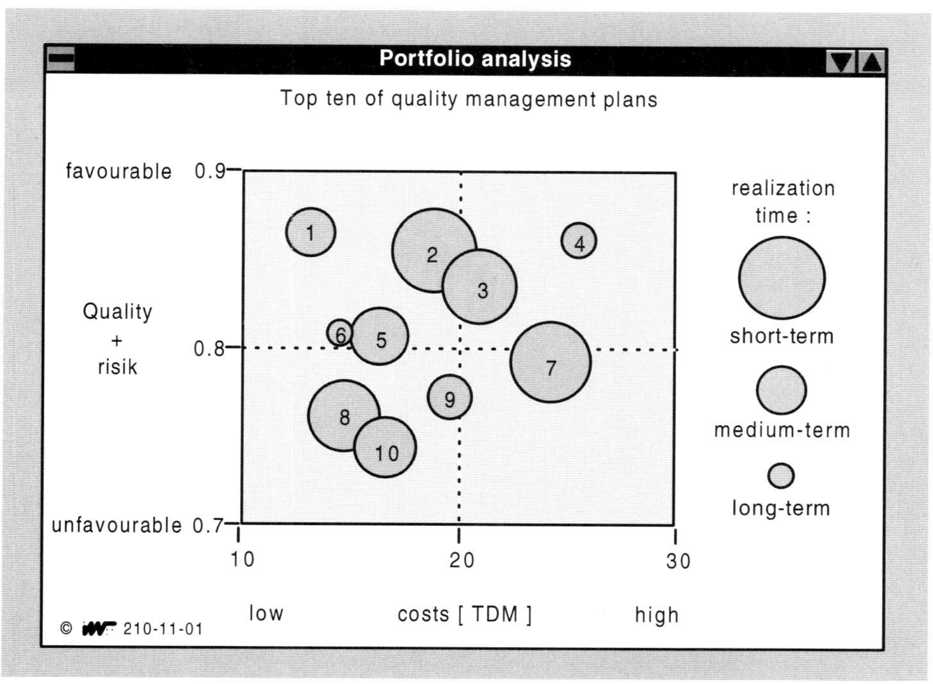

Fig. 5.39: Result Portfolio

plans, figure 5.39. Detailed information is given in a further window. Here, the cost-benefit potentials of each process-oriented measure are shown.

Summary

The DfQ tool offers comprehensive possibilities of quality and reliability planning to the user. The accurancy of the planning results, however, will always depend on the quality of the input data. To achieve good planning results, the system uses real quality data from the production and the field as well as results from laboratory examinations by component manufacturers. The sum of the influencing parameters is relatively large. For the acceptance of the system not to be endangered by an excessive amount of pre-work, all master data related to the process and the product are managed in a hierarchy of three steps. Thus, it is possible to start with rough data and where necessary, the data can be particularized. The system will have the greatest advantage in a team-oriented application, including developers as well as production and inspection planners. If the prerequisite for team work is not existing in the enterprise, it is also possible – due to the universally valid systematics – to perform single tasks such as the optimisation of testing strategies or the reliability calculation.

References

1 Andreasen, M. Myrup; Kähler, S.; Lund, T.(1988) Design for Assembly. Second Edition. Springer Verlag, Berlin Heidelberg New York Tokyo.

2 Baratta, Mario von (1994) Der Fischer Weltalmanach 1995. Fischer Taschenbuch Verlag, Frankfurt/Main.

3 Boothroyd, Geoffrey; Dewhurst, Peter (1991) Product Design For Assembly. Boothroyd Dewhurst, Inc. Wakefield.

4 Böcker, Franz (1990) Marketing. Third Edition. Gustav Fischer Verlag, Stuttgart.

5 Crosby, P. (1979) Quality is Free, McGraw-Hill Book Company, New York, NY, USA

6 Deitz, Daniel (1988) Tools for Total Quality. In: Computers in Mechanical Engineering 8 July/August S. 8 – 13.

7 Dory, R. (1992) An Introduction to STEP, DIGITAL Press, February, 1-34

8 Faria, Luciano (1994) CONSENS Esprit Project no. 6896. In: Knudsen, J.K.H et al (Eds.): Advances in Design and Manufacturing, Volume 5: Sharing CIM Solutions. IOS Press, Amsterdam Oxford Washington DC.

9 Feldmann, Klaus (Eds.)(1992) Montageplanung in CIM. Springer Verlag, Berlin Heidelberg New York, London Paris Tokyo.

10 Frech, Joachim (1993) Kostengerechte Produktgestaltung in einer objektorientierten Engineering-Umgebung. In: Objektorientierte Informationssysteme: 3. IAO-Forum 8. Juni 1993/-hrsg. von Hans Jörg Bullinger. Berlin [u.a.]: Springer, 1993, S. 178-193. (IPA-IAO Forschung und Praxis; T 39)

11 Jeschke, Klaus, Linz, Alfred (1993) Die beste Teststrategie herausfinden, Elektronik, Heft 23, S.52 – 55

12 Jeschke, Klaus, Westkämper, Engelbert (1993) Optimierte Teststrategien zum Test von Elektronikprodukten,Qualität und Zuverlässigkeit 38, Heft 11, S. 641-644

13 Jeschke, K. (1994) Methoden und Hilfsmittel, in: Was der Produktionsingenieur von der Qualitätssicherung wissen muß, 2. Auflage, VDI-ADB-Verlag, Düsseldorf

14 Kahlert, Jörg; Frank, Hubert (1993) Fuzzy-Logik und Fuzzy-Control: Eine anwendungsorientierte Einführung mit Begleitsoftware,Vieweg, Braunschweig, Wiesbaden

15 King, B. J. (1992) A STEP in the right direction. Professional ENGINEERING, 11, 12-15.

16 Lay, Klaus; Menges, Raimund (1992) ERGOMAS – Ergonomic Design and Optimisation of Manual Work Systems. In: IMD Newsletter Issue No. 4. Hawksworth Graphics & Print Ltd.

17 n.n. (1994) DIN ISO 9004, Leitfaden zum Aufbau von Qualitätsmanagement-Systemen, Beuth, Berlin

18 n.n (1992) Zuverlässigkeitsvorhersage für Bauelemente der Elektronik, Beuth, Berlin

19 Westkämper, Engelbert (Hrsg.) (1991) Integrationspfad Qualität, erschienen in der Buchreihe der CIM-Fachmann Springer, TÜV Rheinland, Berlin; Heidelberg; Köln
20 Westkämper, E: Lean (1992) Production oder CIM – Müssen wir unsere Fertigungskonzepte ändern? In: VDI-Berichte 970: Produktionslogistik. VDI Verlag GmbH Düsseldorf.
21 Westkämper, Engelbert, Jeschke, Klaus (1994) Moderne Methoden des Qualitätsmanagements, in: Dale, Jammernegg, Lang, Steiner (Hrsg.), Integrative Strategien des Qualitätsmanagements, Proceedings zum internationalen Hochschulkurs im Rahmen von Comett II, Wien, Service Fachbuchverlag
22 Weyerer, Manfred (1988) Prüfbarkeit elektronischer Schaltungen Hanser, München, Wien
23 Wildemann, Horst (1992) Kosten- und Leistungsbeurteilung von Qualitätssicherungssystemen, Zeitschrift für Betriebswirtschaft 62, Heft 7, S. 761 – 782
24 Zimmermann, Hans- Jürgen (1993) Fuzzy Technologien: Prinzipien, Werkzeuge, Potentiale, VDI Verlag, Düsseldorf

5.4 Support of CSE by Project-Oriented Tools

E. Sleeckx

The objective, being to reduce lead-times with respect to cost and quality targets, requires a stringent control of the total product development process. It requires project-management with special attention for flexibility and alertness to reach the level of pro-active control where possible problems are countered in advance.

Engineering Process Manager (EPM)

Management support by EPM

As the load on project-management is extremely heavy in a complex and fast moving CSE organisation, one must exploit information technology as much as possible to support the decision making in this area. EPM-tools combine the power of computing with the creativity of man. As EPM has to control the PSI area's, they have a major impact on them: EPM supports standardisation by the possibility of classifying projects. The project may be classified by size, task, product or what ever is feasible for the company. The classification contains:

- lists of work packages with result structures,
- lists of milestones,
- lists of reports and checklists for different users of EPM, etc.

The project standards help the project manager and the team to plan the schedule and control the project. To be able to manage, a consistent and standardised information platform is necessary. This is done by EPM. A very important aspect is to exchange the necessary information between the members of the project. Planning, scheduling and controlling of necessary result exchanges is organised by EPM. The edition of project work package structures is the basis for parallelisation. The project members put in the work package structure their information on what they will do and what they expect to get from the others. EPM serves this process by handling and distributing that information. Decision-making is supported by either compressing data for the project-management which has to interpret it, or, it has to be read by a Management Decision Support System (MDS).

The Interface to an Information

Management System (IMS) gives the required document structure which has to be filled in the course of the project. Parallel work structures are now to be planned. The known interrelations between different work packages allow the optimal project schedule to be determnined. Together with information on capacities and resources in the project teams, the schedule can be tuned and co-ordinated with parallel ongoing projects.

Beneath planning and controlling the messaging and early warning system supports parallel work during project execution. Uncertainty and systematic information deficiencies become visible during the run time of the project. EPM supports detection of problems on the project level and helps to generate task forces to solve the problems by giving access to the transparently stored information on the project.

Management Decision Support (MDS)

MDS: A good decision to start with.

CSE requires strategic decisions to be taken far more often than in the traditional organisation which has a settled culture of decision making. Decisions are taken because they have been taken that way for many years and they are very predictable. In fact they are not really decisions but a consolidation of the natural flow. This way of decision-making it not at all suitable for CSE where perpetual stability is a long way off.

In CSE, a decision can no longer be left to natural evolution, neither can a decision be taken in isolation. Decisions happen to effect the CSE company as a whole. Therefore, again support has to be given to those responsible for the difficult task of taking the right strategic decisions: MDS-tools are being developed for this task. The MDS-tools should support the PSI-strategies by providing the right advice to the decision makers by

- providing project overlapping synchronisation of the product developing process and by providing assistance in the scheduling of the product developing processes,
- providing a simulation model to analyse the effects of decisions to be taken during the development process on the costs resulting and on the capacities involved on the technologies to be mastered and on the scheduling.

5.5 Project-Oriented Tools

5.5.1 EPM – Engineering Process Manager

S. Berndes, A. Stanke

Abstract

EPM is a generic IT-tool that supports the CSE process. It has an underlying co-ordination method which reflects the requirements stated by companies which perform complex product development processes. EPM helps to collect, disseminate and interpret data in order to plan, schedule and control this process. After it has been finished, EPM supports the evaluation of stored data to improve and update the standards for project-management in the company.

In order to demonstrate the EPM approach to the management of complex product development processes, the subject of investigation is clarified, requirements to the organisation deduced and the problem more exactly stated. The following section views the state of the art and shows in which areas currently used methods and tools should be improved by EPM. After stating the thesis, we introduce the elements of EPM and demonstrate the co-ordination method. In the final section, results of the use of EPM are stated, discussed and verified.

EPM helps to collect, disseminate and interpret data to plan, schedule and control the Product Development Process (PEP)

Introduction

Classification of planning and co-ordination problems – subject of the investigation

Planning problems occur in three different types. The complicated co-ordination of many jobs in production lines belongs to the first type. The processes which have to be performed to fulfil each job are known as well as resources. The difficulty is to find an optimal relation between job and resources (see [3] p. 15 ff.). For this type, central PPS and their underlying strategies are useful. The second type is the complicated co-ordination of one strongly interrelated program which occurs for example in civil engineering and big aerospace projects. This program consists of many subjobs which are strongly interrelated and the interrelations are known. Here is the field to use the net-planning technique (see [3] p. 17; for a detailed introduction of the net-planning technique [8]). The type we are interested in is the complex co-ordination of a few strongly interrelated jobs. It is typical in product development. The processes for each product development are partly known only (see [6] p. 99). The job is inductive, product development processes need creativity and in a development phase challenging problems occur whose solutions may change the course of the project. Here, project-management is used for the development of complex products. To make matters worse, as we know from consultancy projects, the resources – mostly specialised engineers – necessary cannot be exchanged for different tasks. This is difficult because many companies work on a few developments in parallel: one new product, a few change developments and there is daily business (such as working on urgent change requests) for each engineer too.

Proposed project organisation – an assumption for the following

In order to cope with the conditions of complex co-ordination of a few strongly interrelated jobs, the projects of these complex products should be performed by specialised teams which are coordinated by one project manager for each project. Each team is responsible for a set of tasks which are repeated in each product development. Teams will be manned often interdisciplinarily. The teams have also the responsibility to improve their knowledge about their area of work and their processes.

Decentral, interdisciplinary and partly autonomous teams: The "dip-teams"

The co-ordination task is divided between Project Manager (PM) and team speakers. In this de-central co-ordination, the PM does the frame-planning. He is responsible for the discussion and decision processes inside his project and he stands for the project. Project-managers and the top management of the company, have the task of solving conflicts between different development projects and daily business. The departments do a de-central detailed co-ordination under constraints of uncertainty due to the process and product complexity, time pressure and planning efforts. The teams also prepare project decisions or decision alternatives and meet together with the PM to come to final decisions. The teams do the (personnel planning) de-centrally also. Therefore the teams can be called de-central, interdisciplinary and partly autonoumos. We will abbreviate them as "dip-teams"!

The problem which has to be tackled is the development of a method and a tool to support the project-management of de-central, interdisciplinary and partly autonomous teams (dip teams) which work on complex product development processes.

State of the Art

Projects are divided into three main phases (see figure 5.40). A definition phase in which project planning is done on the basis of information on project goals, experiences from other projects and the current situation in other projects in run time. During project execution, the actual state is measured, compared with the target or planned state and in case of major deviations project steering will be done. It can be either by changing the targets or by influencing the ongoing process by adding resources, solving problems and so on. After finishing the project, the termination is used to collect all important information which can be used to improve the processes and the project-management.

Today, IT-support in product development is mostly given by project-management systems (PMS) including net-planning techniques (NPT) (see [8] and [7] p. 122). But the

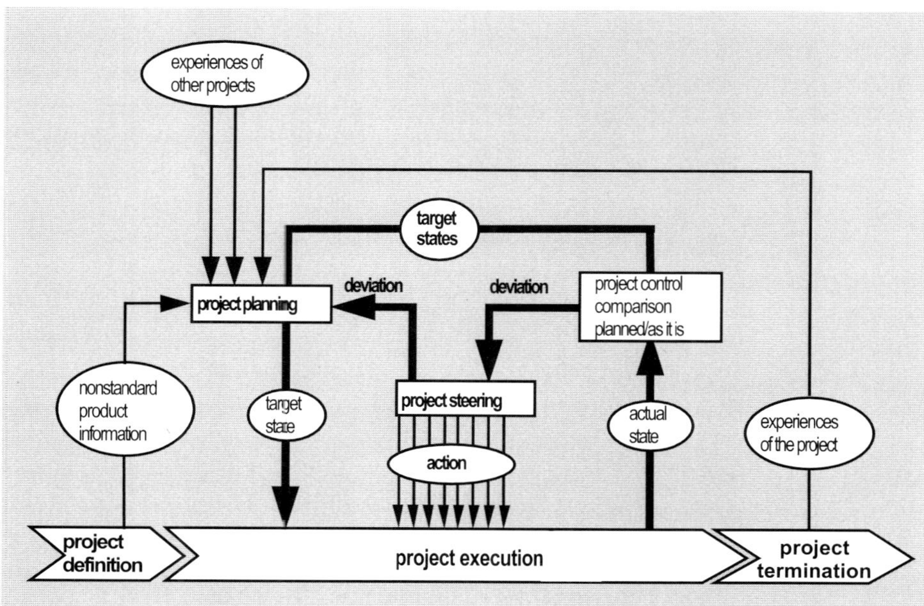

Fig. 5.40: Project Control Loop

PMS tools lack attributes in supporting the management of complex product development processes

PMS-tools available on market lack a few attributes supporting the management of complex product development processes as we know from our experiences in industry (another line of critique is given in [6] p. 113). During project planning, the work-package structures are defined centrally by the project manager not by the "dip teams" who have the knowledge of their processes. All current PMS require a central resource management but companies with an organisation with dip teams need a de-central ressource management.

PMS do not handle inconsistencies

PMS based on NPT do not handle inconsistencies. We assume that NPT is sufficiently known to the reader. NPT needs complete and consistent information about all work-packages, with times, efforts resources and so on. Open relations, cycles and schedule inconsistencies lead to errors. But complete and consistent information is often non-existent or hard to achieve. Typically, a rough plan can be constructed and the detailed co-ordination of work is done for the near future only. The Project manager who wants to use net planning technique has two alternatives. Firstly, he only concentrates on a rough plan and he starts to set up a set of further plans for detailed co-ordination. The effect is a huge amount of work to update and coordinate. This is not inter-linked and distributed by information pools. Secondly, he has to elaborate a detailed plan. Here project managers often end up frustrated. They would have to collect a huge amount of information to update the detailed global plan during the project runtime. Here we expect an information overflow. So they work with old plans and start to edit detailed plans too. In both cases, the unity of information and a universal database has been given up in favour of an eclecticistic management method. Therefore, a method and supporting tool is necessary which is capable of accepting that information about the co-ordination of a project is neither complete nor consistent.

Central data input for efforts, cost and time is necessary

Central data input for efforts, cost and time is necessary for most PMS during project controlling but the task is to support de-centrally acting teams. The reason is a missing multi-user ability of the PMS-software. Along with this goes the problem that PMS support multi-project-management insufficiently because information presentation crosswise to the projects is difficult. Another critique on PMS relates to the huge amount of effort needed to hold the plans actual. The reason lies in the NPT once more with its need for consistent and complete plans which have to be maintained centrally. Resulting from this the project plans become older because there is not the time for a central agent and the project manager mostly adapts them to the actual situation.

For project controlling it is necessary to have a handy concept for the measurement of goal attainment (see [7] p. 122 and [1]). This is missing as long as the percentage of the whole effort spent for each work-package, the opinion of the person who is in charge of finishing a work-package, or something similar, is seen as insufficient by practitioners. Another point is, these figures depend on the estimation of those who elaborate the results of the work-package only. Furthermore, most PMS lack a reporting module which helps to generate and deliver reports or other important project information. Last but not least, PMS do not support the evaluation of project experiences which should be used to improve project-management and the development processes.

No handy concept for goal attainment

PMS support the evaluation of project experiences marginally only

Thesis

In the area of the development of complex products, EPM is an IT-tool with concepts which support the specific attributes concerning co-ordination of de-central, interdiscplinary, partly autonomous teams better than currently used tools. Project-management efforts will be reduced, information will be at all places in a current, true and complete manner.

EPM – the System and how the Demonstrated Method Supports Management of ComplexProduct Development Processes

EPM is a computer program which is implemented in C/C++. The user interface is constructed with TCL-TK and the database bases upon the Object Manipulation System (OMS) which is part of SIFRAME. EPM is white box integrated in CONSENS and provides multi-user-abiity of the system. figure 5.41 shows the principal architecture.

The system EPM

The data model is part of the OMS (Object management system) which contains the project specific. A generic version is implemented which can easily be customised. The data model handles central project data and data adminis-

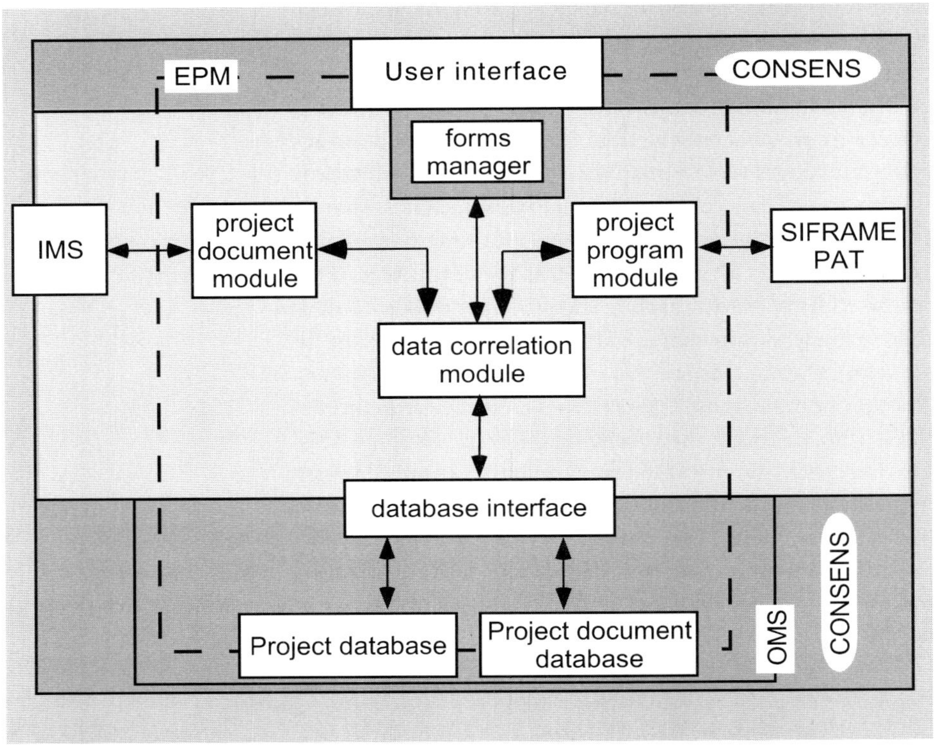

Fig. 5.41: EPM-architecture

tered by the dip teams. The forms model describes the output on the screen. The forms contain information about data represented there and how they are presented. User interaction is done through the forms. The forms manager is the function which leads the user through EPM. Users have different roles in the project and different phases of the project have their specific tasks. So the forms and the sequence of forms presented to the user by the forms manager depends on the user role, project phase and current events. After invoking a form the forms manager starts a database query and data is read from the database. Then data is returned to the form after preparation in the data correlation module. It contains functions to derive information to be presented in the forms from data stored in databases. The database interface interface handles all operations on the data stored in different databases. It is part, of SIFRAME and enables the whitebox-integration of

EPM into SIFRAME. In the object-oriented project database, all information handled by EPM is stored and modified. In the project document database specific project planning and controlling data are stored. This is done in distinct time sequences or at special events like milestones. Also mails, reports and project audit information get documented.

The project documenting module is used to establish the interface to IMS. IMS focuses on documents, EPM focusses on co-ordination of the dip teams. EPM sends IMS information on the completion of results which are the contents of documents. EPM uses the structure of expected documents to plan the project on a global level. The project progress manager is the functional interface to SIFRAME. SIFRAME handles processes performed in every dip team, whereas EPM coordinates the exchange of results between the dip teams. The project progress manager gets the status information of the processes from PAT. If there are modifications in the cell and process structures of SIFRAME, a message is sent to the dip teams which are directly affected. So an immediate response to changed processes is possible.

The method of EPM

Of course work-package structures, netplans, Gantt-Charts and others are powerful means to structure, plan and control a project and the presentation forms are rather expressive. Mostly, the project manager uses as we know from consultancy, if he has a PMS, the result charts only. If project managers use the PMS as a scheduler and presentation form editor without using the variety of concepts inherent to NPT, we should keep to the presentation forms and rethink the constraints NPT has. And as suggested above we have to support the whole project-management, which is, in a dip team environment, a task of the project manager and the dip teams.

Terminology – work-package (wp)

We stick to the terminology of the known project-management concepts. Work-packages help to structure the whole project and form a hierarchy. Each work-package (wp) has a start and an end date. For planning process reasons, these dates have two different values, a bottom up and a top down value. For each work-package, an effort is definable. For the effort values again, a top down and bottom up value is stored. The bottom up value equals the sum of the

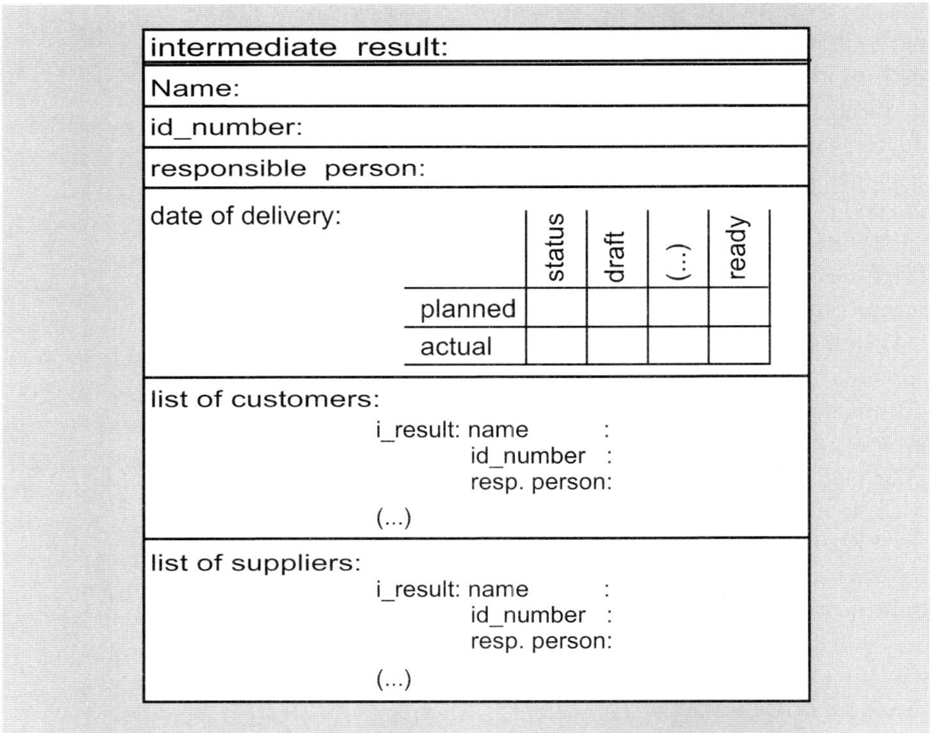

Fig. 5.42: Intermediate Result

efforts of the sub-work-packages below the work-package in question (see figure 5.42).

Terminology: intermediateresult (i_r)

Each work-package has one intermediate result (i_r) which defines the goal of the task. For each of them, a responsible person must be named. Each intermediate result may have a couple of status such as first draft and second draft before the final status "ready" is reached (see figure 5.42). For every status, a date of delivery is defined. The date of delivery for the ready status is typically the bottom up end date of the work-package. Intermediate results serve as mean to control the project progress. As far as we know, the best thing to control the state of work without the problems mentioned above is to build up customer supplier chains in which the customer accepts or rejects an intermediate result on a planned date in a defined status. In order to hand over the intermediate result to the group of persons who rely on it from the point of view of the deliverer, a list of customers of this intermediate result for every status is stored. Also, for producing each intermediate result,, a list of at least one incoming intermediate result is necessary. They are listed in the list of suppliers. The planned dates here are expressed relative to the end date of delivery of the intermediate result's ready status. In conducting this procedure by all persons in charge of an intermediate result, a dually interlinked intermediate result network is defined which is a schedule, a clear responsibility chart and a presentation which expresses information about the big customer-supplier chain or network which is the project.

The actual end date of a work-package equals the actual date of delivery where the intermediate result is delivered and accepted.

Terminology: milestone (ms)

In order to control a project regarding global questions such as

- do we already finish in time?
- are their global problems?
- has the market changed entirely and has the resulting product a good market position?

the concept of milestones (ms) is used (see figure 5.43).

Each milestone defines the end of major development phase (specifiation, design, prototype testing, series start). A milestone can include a date, budgeted efforts or costs which can be expended in the period in question. These costs or efforts equal the sum of the efforts of the work-packages which belong to the phase. All those work-packages belong to the phase in question which lie between the preceeding milestone and the milestone of this phase. The milestone is defined also by results which should allow the evaluation of the project. Each result can be related to a list of at least one intermediate result. The milestone is reached if all results are there, but in many practical cases not all results are there in time. In order to go on with the project, it is possible to postpone i.e. to relate some results to later milestones.

The project planning and control methods are described along the functions of the project control loop: planning, re-planning, control and audit.

Idea of planning with EPM

The idea of the EPM planning method is as follows. We are looking for the best or optimal project plan but we know only some parts. It isn't feasible to carry together all the

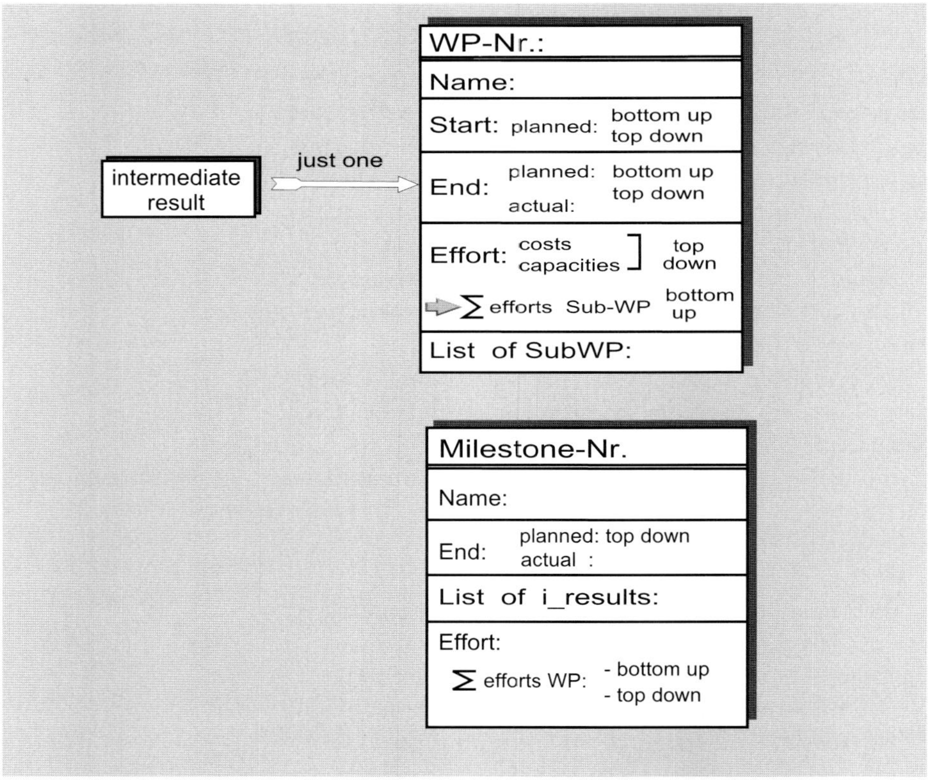

Fig. 5.43: Workpackage and Milestone

information which is stored de-centrally. Therefore, we distribute the planning tasks onto the dip teams, arrange discussions between them, classify the topics of discussion and accept inconsistencies which occur from the complexity of the task, not knowing of vacation, illness, ressource shortages etc. This is a multi agent approach (a similar approach is given by [6] p. 113). It is just like the following picture, symbolised by a kind of loom (see. figure 5.44). In the near future, the threads must be in a clear order, but nobody cares about the position of the threads in the distant future.

Multi agent approach

Top down draft of rough project plan by project manager

The project manager defines the project with the major goals: quality, time and cost and inputs them in EPM. He defines the project organisation with the involved dip teams. He builds up a hierarchical work-package-structure with maximum efforts to every wp. These are the top down efforts. EPM supports this with the help of a work-package-

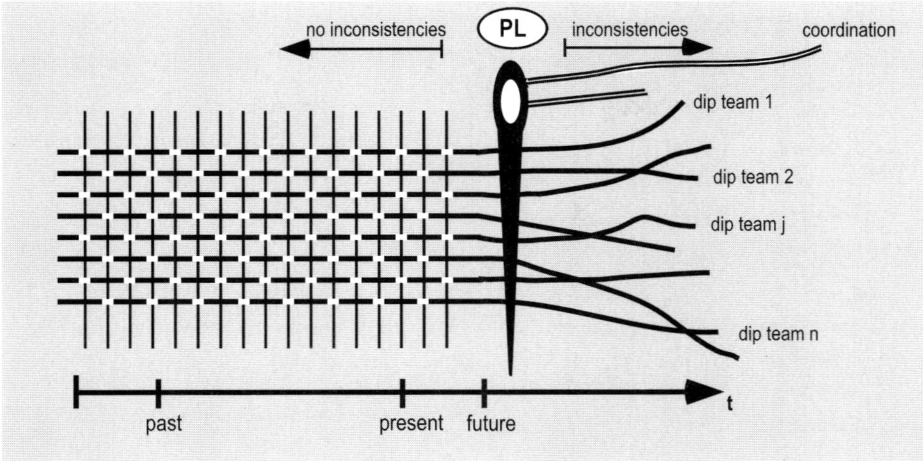

Fig. 5.44: The Loom – a planning principle

editor form. A list of milestones with an explanation is selected from standards and distributed on the time schedule. The PM delegates the planning tasks to all involved dip teams. EPM sends a message to all dip teams.

Bottom up elaboration of rough project plan by the dip teams

The dip teams refine and elaborate the project goals interactivly with EPM. They know about resources and technical or other risks. They make a statement to the main schedule and the cost frame. They consider who will be assigned to the project. Each dip team builds up its wp-structure, using de-centrally stored wp-structures. In order to use experiences from former product developments each dip team stores information, concerning wps, typical dates, efforts and some explanations concerning resources. They define their planned efforts, the bottom up efforts.

Planning the effort distribution by the dip teams

The dip teams perform a rough planning of the schedule for their wps. Without any relation to other wps the dip teams distribute their wps on their capacity chart. The sequence of wp is fixed by their position on the capacity chart. To plan in EPM the wp-dates, there is a pick´n drop mechanism to distribute the wps and by pulling and pushing the wp the start/end date can be selected.

The information about the projects and the efforts each dip team has to deliver, is stored in the capacity chart of EPM. The capacity chart shows the available resources over a period of time. To come up with the capacity chart, each

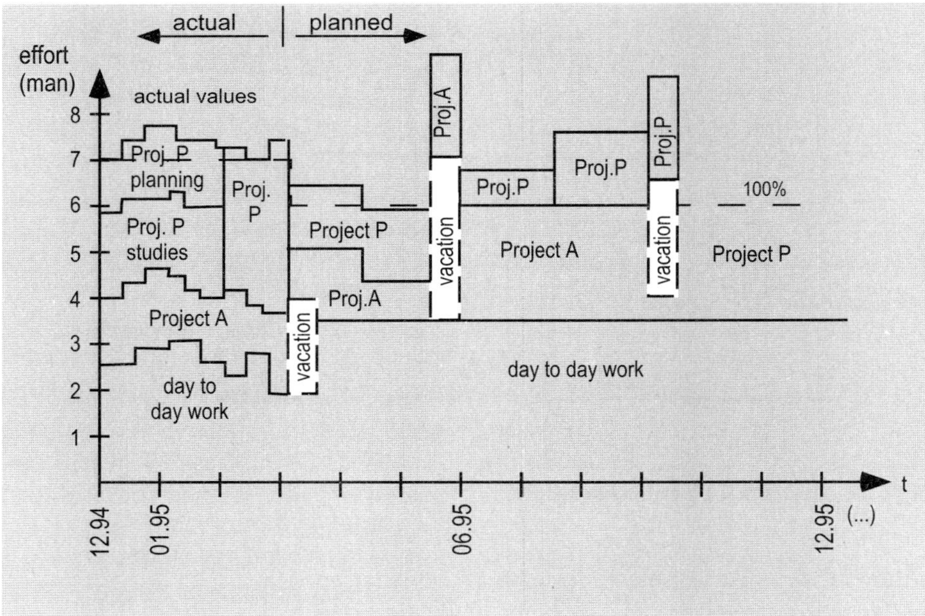

Fig. 5.45: Effort Charts – Planning and Controlling

dip team defines a 100%-line and a daily work line. On the basis of the planning and actual values for efforts an effort line for each project is computed (see figure 5.45). All efforts of wps of one project phase for each project are added and this leads to a phase effort of each dip team. This effort will be distributed equally among the phase time.

Effort controlling

How can you get actual dates concerning effort? Every week engineers have to input into EPM which project they are working on or they were on vacation etc. This information is put into the capacity chart. The planned efforts are defined as those which have been decided on – in terms of the wp-definition it is the bottom up-value. Efforts for wps, which have not been ended during the phase runtime are put in the next phase. Due to the fact that project employees do not input effort values on the level of work-packages, the equal distribution of efforts over a phase is justified.

Negotiate and decide on the rough plan

The project managers and the dip team speakers evaluate the resulting plan. They look at project run time, project efforts and resource bottlenecks. Meetings are held between the dip team speakers to discuss and to solve emerging problems and meetings are held with the top management to prioritize different projects. Here, EPM supports the generation of info charts like a milestone schedule, a wp-plan with efforts, dates and effort/costs distribution for the dip teams and the project. Minutes and decisions can be stored in EPM to document the decision process to enable organisational learning.

This procedure can be repeated until a suitable rough plan is found. The aim of this planning phase is: generation of a stable project plan, definition of milestones (MS) and schedule, check of general ability to realise the project

Modelling of the process structure using intermediate results

Then the dip teams define their intermediate results and relate them to the ms-results. Each work-package has a bottom up end date and this date equals in the first iteration the i_r final delivery date in the status "ready". If the intermediate result bottom up end date is later than the MS-date, there is a confict. This is computed by EPM automatically and solved in talks between the project manager and the dip team speakers.

Next, the dip teams build up the intermediate result network. Each dip team asks which intermediate results are necessary to elaborate their i_r and relate them to it and ask who relies on their intermediate result. EPM supports this by storing the relations to intermediate results from former similar projects and printing lists of intermediate results of all other dip teams.

Evaluation of the intermediate result network and handling of occuring inconsistencies

Of course, in the interlinked intermediate result network, a "few" inconsistencies may occur. EPM helps to discover them. The following classes of inconsistencies are found: Open relations make the dip teams and the project manager aware of co-ordination problems. Information exchanges are not clear. The solution of such problems is rather simple:

Open relations

- What does the relation mean?
- Why do (not) you expect an information flow?
- Which relations should be added or resolved?

Storing the relations in the i_r lists makes the results of these discussions available for future product developments.

Cycles

Cycles are often found in engineering. One dip team relies on the result another dip team elaborates on the basis of the result of the first dip team. A method to resolve them is to establish a sequence of smaller steps, by exchanging draft-information.

Schedule inconsistencies

Schedule inconsistencies are easily found if the i_r planned delivery date of the intermediate result to be delivered is later than the planned date of supply of the intermediate result. The first thing one can do is to check whether the necessary i_r must be delivered on that date. Next, one may try to break down the to be delivered i_r into smaller ones and in doing so define a few further sub-work-packages. Or just try to exchange draft-information . A last solution is to postpone the intermediate result and the work-package.

After discovering the inconsistencies with EPM the task of project team meetings is solving them. Project manager and dip team speakers work out alternative solutions on EPM by input and storing them separately and in evaluation of alternative scenarios, to discuss them and input the result in EPM. The idea is to solve the appearing inconsistencies one after another. Main inconsistencies are solved first and the others are solved at the time when it is possible to foresee the near future. The result network must be consistent for the near future only – consistency just expresses good co-ordination between the dip teams. Of course, for each kind of product development and company the time frame of consistent co-ordination is different. Trying to have a whole consistent intermediate result network is very expensive and needs much effort to adapt to daily changing conditions but it is required from NPT.

Net-planning technique and the presented planning method

The main difference between net-planning technique and our proposed planning method is the introduction of intermediate results. Net-planning technique assumes that resource planning and scheduling of the project progress can be done with the help of work-packages only. We differ here. In creative and inductive processes, the connection between results and effort is not known exactly. Therefore, we introduced intermediate results. The schedule is controlled by the intermediate result network and the effort spent by the work-package hierarchy and this hierarchy is used to get a basis for a sufficient estimation of efforts.

Re-planning

The re-planning procedure will be done at each milestone or any other moment wherever it seems to be necessary. For example, if the planned and actual effort numbers differ significantly from each other. In order to be able to control the process, both the new and the elder planned numbers are stored. Re-planning means just using the same method as in planning. Smaller inconsistencies in the schedule and the intermediate result network may be handled by the dip teams on their own, only a short message is given to the PM. But the decisions are documented in EPM.

Project control

Project control contains data acquisition, distribution and evaluation of data. The time-dimension is measured by comparing planned bottom up and actual data. The cost-dimension is measured by comparing planned bottom up and actual data for efforts and other project costs. And the quality-dimensions are measured with the help of the established customer-supplier chains through the i_r network. The quality of i_r is checked by the customers of the specific i_r who accept them. This is process control on a detailed level. The global project progress is controlled at the milestones by checks and reviews. At each ms, a global check is necessary. Between each milestone the intermediate result exchange, effort spent and schedule progress can be measured at arbitrary points of time.

EPM presents data for different user classes

EPM as an information system presents data for different user classes (similar [6] p. 113). The PM needs information concerning schedule, effort, cash flow and quality, because he is responsible for the project. The dip team speakers must know about effort tables and schedule situation. The project employees have to know where they are involved and what they have to do. At least the top management wants to have information for setting project priorities.

In doing project control, problems can be found and will automatically be presented. EPM supports also agenda setting and the management of project change requests by presenting, mailing and documenting of specific forms to the involved persons.

Audit – EPM supports learning

In doing so for each project, a tremendous amount of data is processed and stored which serves as an evaluation base. The evaluation is done in audits which are used to reflect the problems. Changes in ms, wp-structures and the i_r network can be proposed. So EPM will become the database of an learning project-management organisation. In this prototype, audit support is given by evaluation forms.

Support of multi-project-management

Multi-project-management is resource distribution between different "competing" product development processes and is the establishment of a preference list of projects and jobs. EPM supports a de-central resource management approach and supports multi-project-management. Each team has the task of administering its own resources. The results of the co-ordination are documented in the schedules.

The task of deciding on preference lists is done by the top management. EPM here supports multi-project-management by giving the top management access to comparable, true and full information on the project and its progress.

Results

EPM bases on experiences with project-management in industry

The generic project-management method is based upon experiences and analysis of enterprises ranging from 500 to 10000 employees and in project size from designing an ABS controllers to the design of aircraft. The enterprises analysed were TEMIC and DASA (see case studies). Moreover, user requirements to EPM have been written by DASA and TEMIC itself. They were met as it was mentioned in the test documents of DASA and TEMIC. User friendliness was mentioned as well as the simple and comprehensible underlying project-management method. Because there was not yet enough time to test EPM in a normal product development process at DASA and TEMIC, experiences with a basic version of it should be mentioned. This version is in operational use at LOEWE, a medium sized company in the electronics branch which has already implemented dip teams since 1992. This test proved the major direction of the concepts to information collection and evaluation for planning and controlling, as well as evalution, to be feasible. Prototypes of EPM are already installed at DASA, TEMIC, Hidrosorefame and ASF, Berlin. Further results will be available in the near future.

Prototypes of EPM are installed

Conclusion

The problem we tackled was the development of a method and a tool to support the project-management of de-central, interdisciplinary, partly autonomous teams which work on complex product development processes. In the state of the art part, we presented several problems which a tool should overcome. If EPM overcomes these problems, the suggested thesis would be true. Now, we repeat the main critique points once more. The tool and its underlying method should be capable of accepting incomplete and inconsistent information on the co-ordination. EPM solves this by using the dually interlinked intermediate result network and by doing rough plans and schedules with the help of effort considerations only. EPM also supports the de-central resource management which PMS lacked on the marketplace. EPM´s multi-user ability overcomes the problem of central data input and output. De-central resource management, the permission of inconsistencies and incompleteness, together with the multi-user ability, reduce the amount of the effort to service the project data to a minimum and distribute the rest of effort to the whole project team. And as shown, EPM supports by using the de-central resource management approach intrinsically multi-project-management. Very importantly, EPM has a very strong measure for project goal attainment which directly relates to the construction of customer-supplier chains. This is also supported by establishing via EPM a reporting and documenting system. Last but not least, EPM supports the evaluation of project experiences in the audit phase.

In the area of the development of complex products, EPM is an IT-tool, which contains the specific attributes concerning co-ordination of de-central, interdiscplinary, partly autonomous teams better than currently used tools. Project-management efforts will be reduced, information will be at all places in a current, true and complete manner.

Further Developments and Ideas

EPM is now available in a prototype but some expanding ideas are already there. In the context of CONSENS the connection between intermediate results and the product structure which is stored in PIA should be more thoroughly investigated. Maybe one can directly relate parts of the tasks and result structures from the product structure. In order to improve the EPM as a learning system, further components should be added as a module for text and protocol evaluation, a component for communication evaluation and its simulation.

Finally, we are very interested in the question of how the dip teams in different enterprises will make use of the EPM. Will they have in the end very stable intermediate result networks? Will they tend to work with defined milestones?

How will they organise their conflict and inconsistency management and relate all these questions to different types of complex product development processes?

References

1 Bürgel, H.D. /Kunkowsky, H.R. (1989) EDV-unterstütztes FuE-Controlling. In: Technologie & Management. 38/3. p. 25 – 32.
2 Bullinger, H.J. / Warschat, J. / Berndes, S. / Stanke, A. (1995) Simultaneous Engineering. In: Zahn, E. (ed.): Handbuch Technologiemanagement. Stuttgart. Schäffer-Poeschel. p. 377 – 394.
3 Domschke, W. / Scholl, A. / Voß, S. (1993) Produktionsplanung. Ablauforganisatorische Aspekte. Springer. Berlin Heidelberg New York.
4 Eversheim, W. (1991) Simultaneous Engineering – eine organisatorische Chance. In: VDI-Gesellschaft Produktionstechnik (ed.): Jahrbuch 90 / 91. Düsseldorf. VDI-Verlag. p. 189 – 216.
5 Kusiak, A. / Yang, H.H. (1993) Modeling the design process with Petri nets. In: Parsaei, H.R. / Sullivan, W. (eds.): Concurrent Engineering. Contemporary issues and modern design tools. London Glasgow New York. Chapman & Hall. p. 447 – 464.
6 McGrath, M.E. / Anthony, M.T. / Shapiro, M.R. (1992) Product development success through product and cycle time excellence. Butterworth Heinemann. Stoneham, MA.
7 Möhrle, M.G. (1991) Informationssysteme in der betrieblichen Forschung und Entwicklung. Verlag H. Schäfer. Bad Homburg.
8 Schwarze, J. (1990) Netzplantechnik. Eine Einführung in das Projektmanagenent. Herne Berlin. Neue Wirtschafts-Briefe. 6. Aufl.

5.5.2 MDS: A System for Decision Support in the Economic Efficiency Analysis and Controlling of the Product Developing Process

G. Korn

Improvements by using the MDS – System

During the definition phase of a new product it is necessary to evaluate the single product conceptions with respect to their economic feasibility. This means in particular that the material and manufacturing costs have to be calculated and the product developing processes have to be assessed in terms of finances, capacities and timing. The single conceptions, however, can only be evaluated within a short time and with the necessary accuracy and safety when data from previous development projects with similar requirements with similar requirements is available.

This line of reasoning has defined the scope for requirements regarding the functions of the decision support system MDS (Management Decision Support) [10], [11]. The MDS System is a tool for the manager, supporting him during the early phases of the product developing process. The system has a modular structure. The application-oriented modules "economic feasibility", "project planning" and "controlling" directly support the respective user in performing his tasks. The system-oriented module similarity search allows the classification, storage and search of projects and processes as well as of products, components and parts. By means of the interactive concurrence of all modules, data of completed projects can be made available for analyzing the feasibility of new product conceptions.

In the constantly intensifying competition, enterprises are increasingly forced to decide quickly whether they shall react on new market demands – especially on definite inquiries by potential customers – with the development of new products [4]. Such a decision can only be made with the required safety when it is possible to elaborate a decision basis which indicates the economic feasibility of the development project [2].

Questions such as the following ones have to be answered precisely and quickly:

- What material and manufacturing costs are to be estimated for the product?
- What is the amount of the prospective non-recurring product developing cost?
- What will be the prospective behaviour of the inflows and outflows in respect of time?
- When will the project pay off?
- What is the amount given to the contribution margin and the percentage of the profit margin?
- Does the project contribute to the utilisation of production facilities according to the operative planning standards?

The MDS – System supports the whole product development process

The MDS System supports the manager when these questions are under consideration (see figure 5.46).

It offers the possibility of simulating the economic feasibility of different project scenarios and thus considering different products. The system, upon request, is able to produce characteristics of already concluded, similar projects available as a basis for the simulation. Based on the results of the economic feasibility analysis, the MDS System supports the definition of the project plan for the product and production development. In this context, the development project is planned in terms of time limits, capacities and costs.

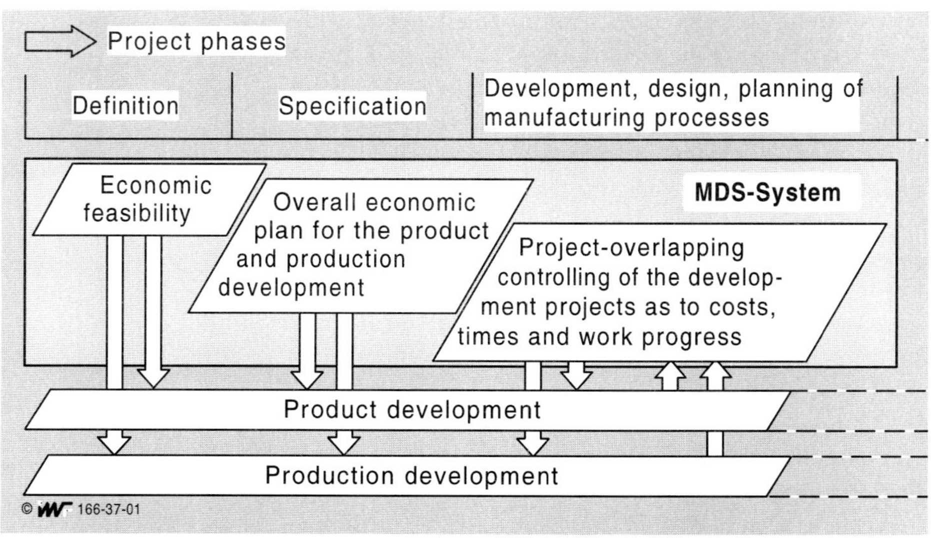

Fig. 5.46: Functions of the MDS System

The project plan is the foundation for subsequent controlling of the product developing process [6]. During this phase, the MDS System supports the manager in updating and representing the project progress. Here, the priorities are the project-accompanying calculation of the product costs defined by the development, design and manufacturing process planning as well as the calculation of the product developing costs that have already been incurred and are still to be considered. The development project ends, once the series production has been started successfully. The MDS System supports the manager in performing the subsequent final analysis. The key parameters of the projects are determined, analysed, visualized and stored as empirical data in the system.

The functions and basic implementation principles of the MDS System are described in the following paragraphs.

Analysis of the Economic Feasibility

During the project definition, the MDS System with the "economic feasibility analysis" provides a comprehensive tool which considers all product costs and all costs of the product developing process for the preparation of a first offer to the customer (see figure 5.48). Here, the product costs include the costs of materials and the manufacturing costs. The investments in manufacturing facilities, testing and measuring equipment form a special group as they are depreciated in respect of the service life. They also increase the manufacturing costs of the product. However, they often contribute considerably to the project-related costs so that they are to be taken into a separate account. The costs of the product developing process include the material costs of the material which is, e.g., necessary for the production of samples and the cost of the personnel. Salaries and wages may arise both internally and externally, if outside services are used. Due to the uncertainties at this early stage of the project development, not only the probable value is to be estimated for each expected cost element, but also a certain range within which these values may vary.

Using experience data while performing the feasibility analysis

The investment costs are fixed by the technical conception and their amount is assessed.

The MDS System supports the calculation of the prospective material cost by searching for similar products or com-

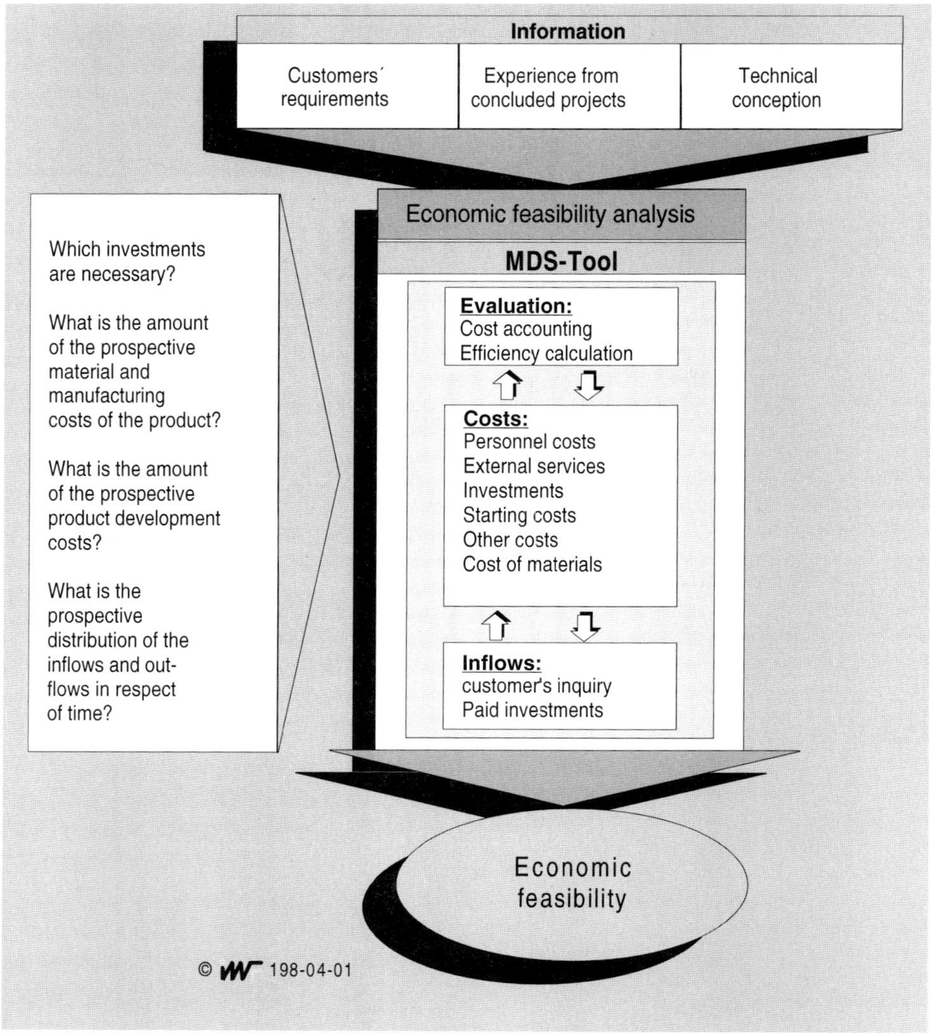

Fig. 5.47: Support of the Economic Feasibility Analysis by the MDS System

ponentries that have already been developed (see figure 5.47). The starting point is the classification of the product to be developed or the appropriate components on the basis of their specification [7]. By means of similarity criteria, such as the functions, performance features, geometry, structure and technology data, the empirical database is consulted so that a similar product or a similar component is determined.

The appropriate parts' list is then adapted to the customers' requirements of the product currently devel-

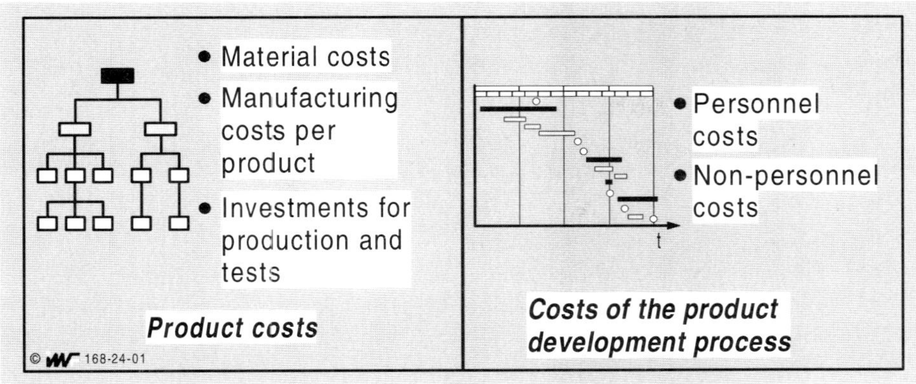

Fig. 5.48: Costs Considered in the Economic Feasibility Analysis

oped in approximation. For the computation of the material costs, the database includes the prices of standard components. The different component prices, depending on time and quantities, are taken in consideration. For assessing the manufacturing costs, the respective methods of production are to be assigned to each component and it is to be stated how much time each operation requires at the respective cost centre. The database includes the wage level and level of the costing rates so that it is possible to make a rough estimate of the costs.

The product developing costs are also estimated with the help of computers. By means of a similarity search, a structural project plan is determined using all the projects stored on the database. The work packages of the determined structural plan are adapted in approximation to the conditions of the project to be calculated regarding expenditure of time, capacity and costs. They also are evaluated by the relevant wage level. Thus, the user and the system successively develop the overall project plan for the product and production development.

The costs of the material required during the product developing process, e.g. samples, are roughly calculated and assigned to the corresponding operations in the project's procedure within the process structure.

Apart from the outflows, the project-related inflows are to be considered in the economic feasibility study. These inflows can be estimated by the quantities required by the customer, the development and investment costs possibly borne by the customer, as well as the intended unit prices.The result of the economic feasibility study represents an estimate of all project-related inflows and outflows. Thus, an efficiency calculation is possible in which both the total cost of the product and the profitability of the project to be executed can be assessed.

Project Planning

Interface to MS-Project for detailed planning of the development processes

By the placing of an order, the customer's inquiry calculated in the economic feasibility analysis becomes a project. The project planning has the task to further particularize the structural project plan developed in the economic feasibility analysis [9]. The work packages and work are fixed in terms of schedules and contents, taking account of the order conditions. This means that existing capacities and time limits which are available in the enterprise are assigned to work packages. For this purpose, the MDS System transfers the structural project plan to the project planning software MS-Project. The MS-Project allows the connection of single operations and the allocation to corresponding resources. The graphic representation of the structural project plan, i.e. the display of the operation connections, capacitiy utilisations and comparative representation of proceedings on a time graph ensure a simple and

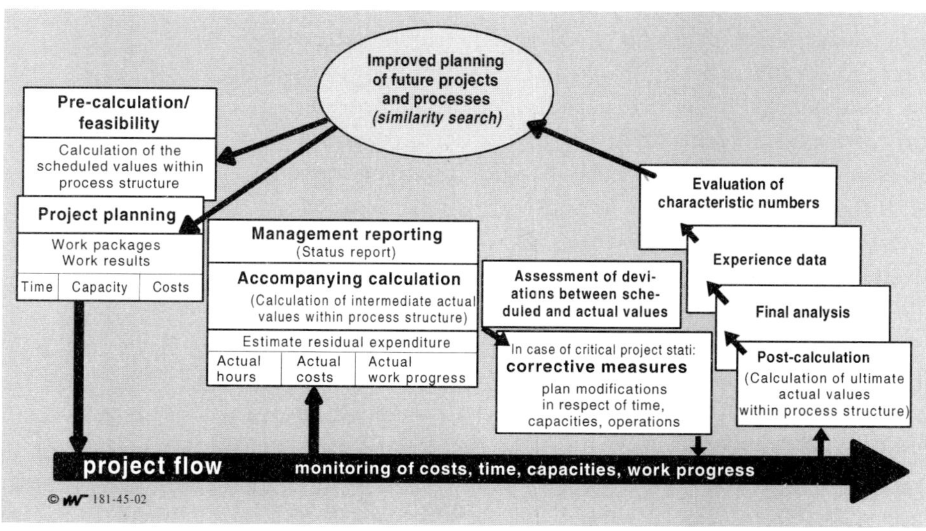

Fig. 5.49: Integration of the Economic Feasibility Analysis, Project Planning and Project Controlling into the MDS System

clear planning of work packages. The indication of time overloads of the capacities, the critical path as well as floating times make it easy for the user to optimize the project plan.

After the planning activities have been finalized, the sophisticated and optimized project plan is transferred back to the MDS System. This is the prerequisite for a project-accompanying controlling regarding costs, timing and work progress in a multiple user operation.

Project Controlling

The MDS System supports the project controlling by its functions "management reporting", "accompanying calculation", "post-calculation" and "final analysis" [6], [8] (see figure 5.49).

The MDS – System supports the development process decider while performing the accompanying calculation

The accompanying calculation determines the actual values for the time needed, costs and progress in the project flow on the basis of activity reports that have been entered so far. It compares them with the scheduled values fixed in the specification phase.

The MDS – System supports the preparation of management reports

The management reporting, on the one hand, serves the control and monitoring of development projects and on the other hand, is an instrument of the management for the long-term planning and control of the development section. It determines operative and strategic characteristic numbers, which are presented to the user in the form of tables and graphics [5].

The operative characteristic numbers enable the individual development projects to be assessed in terms of finances, quality and schedules. The strategic characteristic numbers serve the evaluation of the development section and the rating of the strategic value of new development projects.

Cost – performance index, an example of an operative characteristic number

As an example of an operative characteristic number, the cost-performance index will be presented. The cost-performance index is a characteristic number which allows the project manager to assess the project relating to the budget spent already and the project progress at any time. It is defined as follows (see figure 5.50):

$$\text{Cost performance index (t)} = \frac{\text{performance index (t)}}{\text{cost index (t)}}$$

$$\text{Performance index (t)} = 1 + \frac{\text{work progress (t)} - \text{planned work progress (t)}}{\text{planned work progress (t)}}$$

$$\text{Cost index (t)} = 1 + \frac{\text{cost incurred (t)} - \text{scheduled cost (t)}}{\text{scheduled cost (t)}}$$

Cost-Performance Index

Project: Test

Cons. No.	Date	Phase	Scheduled Costs	Costs Incurred	Planned Work Progress	Work Progress	CI	PI	CPI
1	29.01.1993	1	100.000	120.000	23	17	1,20	0,74	0,62
2	21.02.1993	1	220.000	230.000	45	47	1,05	1,04	1,00
3	07.03.1993	2	370.000	410.000	57	65	1,11	1,14	1,03
4	24.04.1993	2	530.000	505.000	74	82	0,95	1,11	1,16
5	05.07.1993	3	715.000	800.000	124	135	1,12	1,09	0,97
6	02.09.1993	4	930.000	1.000.000	170	182	1,08	1,07	1,00

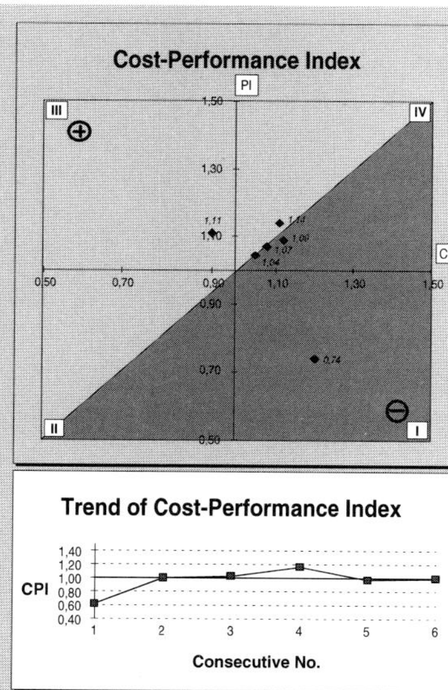

Fig. 5.50: Representation of the Project Status as a Function of the Parameters Work Progress and Cost Incurred

The system acquires the project progress by means of the number of tasks completed at the date considered. This definition is the only useful one, as mostly creative activities are performed in the development section. According to this definition, the cost-performance index is represented in the system as follows:

$$\text{Performance index (t)} = 1 + \frac{\text{number of completed tasks (t)} - \text{scheduled number of completed tasks (t)}}{\text{scheduled number of completed tasks (t)}}$$

The single cost-performance indexes of a project are either represented of the portfolio method or as a function of time.

The portfolio method enables the parameters costs and project progress (performance) to be considered separately. The current cost-performance index, for example, is represented in square IV, if the project has a work progress higher than originally planned and if it has simultaneously caused more costs than originally planned. The diagonal line separating the positive from the negative area, symbolises a cost-performance index of 1. The distance to the single characteristic numbers shows the present status of the project relating to the work progress and the costs incurred.

Once series production has been started successfully and the development project has been concluded, the post-calculation is done. The prerequisite for this is that the complete compilation of all actual data has to be finished [6]. In the subsequent final analysis the data of the post-calculation is compared with the planned data of the pre-calculation at the project start. The reasons for essential deviations are found out and documented. The determined reasons for the deviations are stored together with the first scheduled data and the actual data with reference to the appropriate project tasks. In addition to this, the aggregated values of all accompanying calculations and the post-calculation, such as full cost, contribution margin and efficiency ratios, are filed as empirical data. With the final analysis of the empirical data, the project, process and product are classified by characteristic numbers. When developing new projects and processes, these characteristic numbers allow an effective similarity search.

The MDS – System supports the project manager while performing the post – calculation and the final analysis

Implementation Principles

In this section, fundamental principles are introduced which have been considered in the system implementation.

Consequent Separation of Simulation and Real Data

During the economic feasibility analysis, it is possible that important characteristics of the enterprise are questioned. For the simulation of the prospective product costs, for example, it has to be possible to change the unit prices of bought items and to vary the production costs of components. The latter are supported by the system in offering the possibility of considering production technologies in the analysis that have not been applied to the enterprise yet, but are available on the market.

Therefore, it is necessary that the system provides for a scheduled data base which is semantically totally separated from the actual database and, thus, makes it possible to analyse – taking account of the real data, without changing it – the economic feasibility of the individual product conceptions and to store the results of single analyses. If a product conception is selected to be realized, all the scheduled data assigned to the conception is transferred to the actual database and simultaneously deleted in the scheduled database.

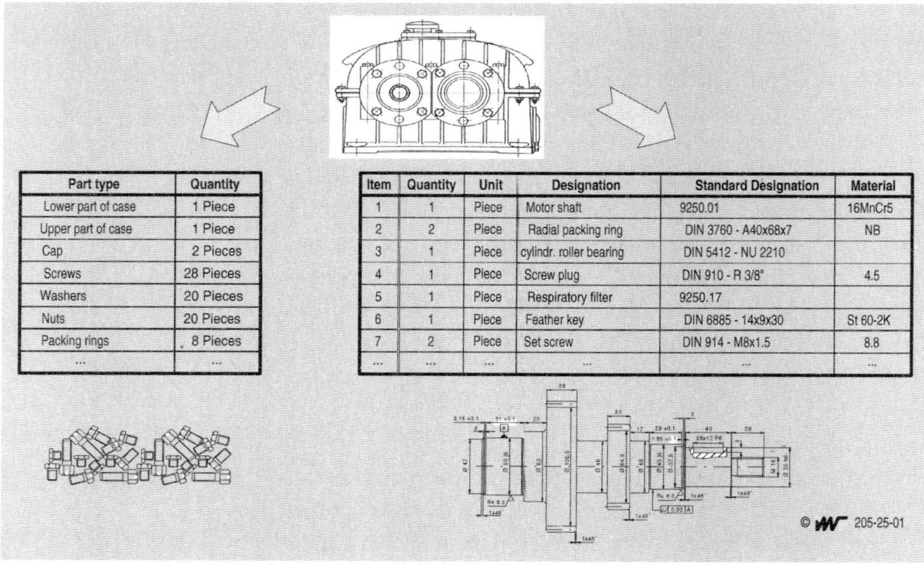

Part type	Quantity
Lower part of case	1 Piece
Upper part of case	1 Piece
Cap	2 Pieces
Screws	28 Pieces
Washers	20 Pieces
Nuts	20 Pieces
Packing rings	8 Pieces
...	...

Item	Quantity	Unit	Designation	Standard Designation	Material
1	1	Piece	Motor shaft	9250.01	16MnCr5
2	2	Piece	Radial packing ring	DIN 3760 - A40x68x7	NB
3	1	Piece	cylindr. roller bearing	DIN 5412 - NU 2210	
4	1	Piece	Screw plug	DIN 910 - R 3/8"	4.5
5	1	Piece	Respiratory filter	9250.17	
6	1	Piece	Feather key	DIN 6885 - 14x9x30	St 60-2K
7	2	Piece	Set screw	DIN 914 - M8x1.5	8.8
...	...	...	...	...	...

Fig. 5.51: Definition of the Part List

Storage of the Product Structure as Preliminary Parts List
As the exact structure of the product to be developed is not defined during the economic feasibility analysis, the system enables the product structure to be defined as a reduced parts' list. This parts' list is successively extended during the development project to complete it (see figure 5.51).

Classification of the Product and Product Developing Process

Classification of products and processes defining qualitative, quantitative and comparative features

The system allows a classification of products and components for one and a classification of phases and tasks of the product developing process for another. The classification is done using qualitative, comparative and quantitative features [3], [1]. Quantitative features have numerical attributes, i.e., characteristic developments. The values can be arranged on a metric scale. A distance can be calculated between two attributes.

Comparative features have to be arranged on a scale. The user fixes the significance of the single features in relation to the other ones and thus draws up an order of ranks.

Qualitative features are not subject to an order. Two attributes are either equal or different. The features are

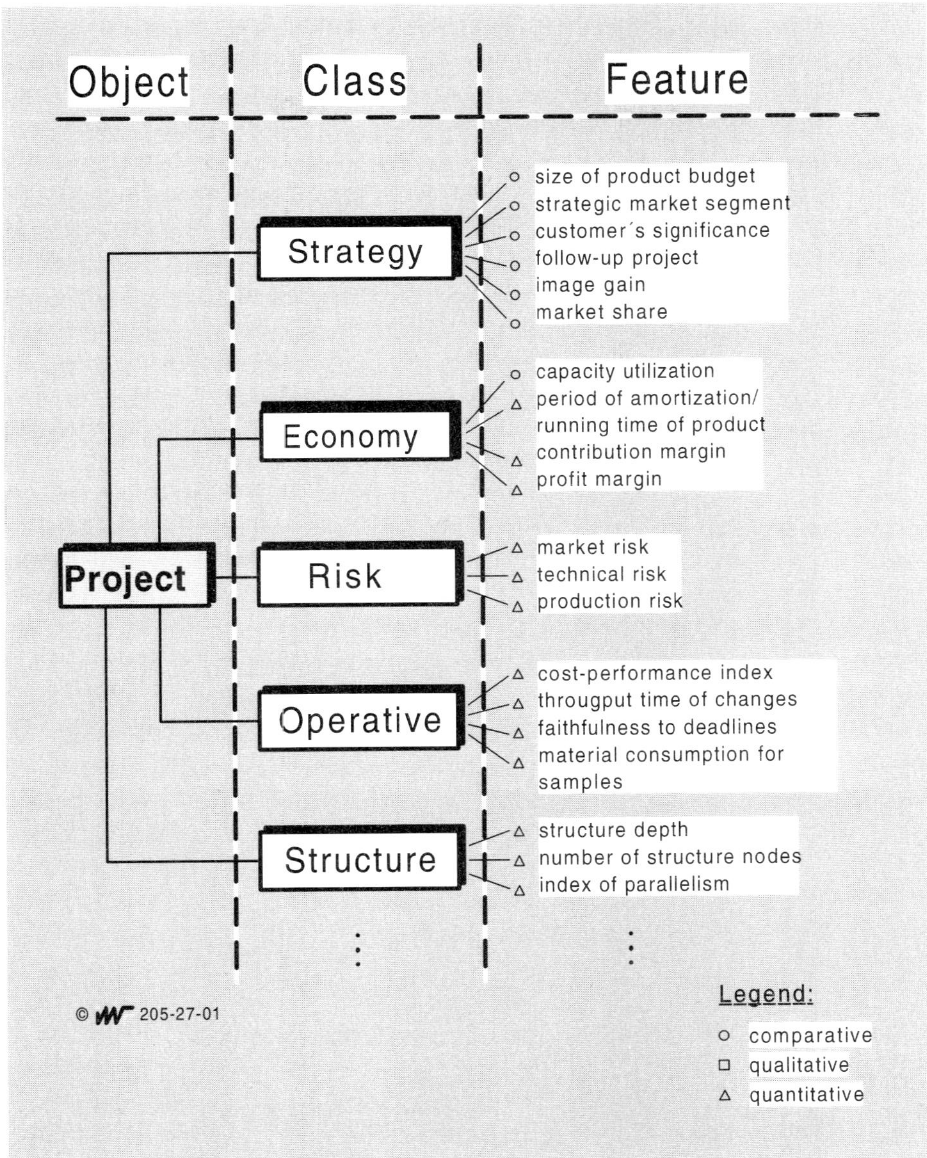

Fig. 5.52: Classification of Projects in the MDS System

assigned to a category. The category colour, for example, can be defined with the attributes blue, red, lilac and yellow.

Figure 5.52 shows the classification of projects in the MDS System.

Similarity search as means to reduce the expenditure of data input and the uncertainty of the calculation results

Support in the Search for Similar Projects and Products

The MDS System supports the user in searching for similar projects and processes as well as similar products and componentries. Thus, the user is able to use empirical data from already concluded projects for the economic feasibility analysis of a new project. The expenditure of the data input into the system can be greatly reduced. In the analysis it is further possible to consider data which, due to experience, could be defined as characteristic numbers of projects and processes and products and componentries, respectively. Thus, the uncertainty of the calculation results can be reduced.

The classification of objects is used for calculating the similarity among objects.

Figure 5.53 shows an example of the similarity calculation of two projects A and B. The calculation starts with the determination of the similarity of the respectively corresponding attributes of the two projects. For quantitative and comparative features, the calculation is done according to the formula shown in figure 5.54 For qualitative features the definition applies that two attributes are similar to each other at 100 per cent, if they are equal, otherwise at 0 per cent. The

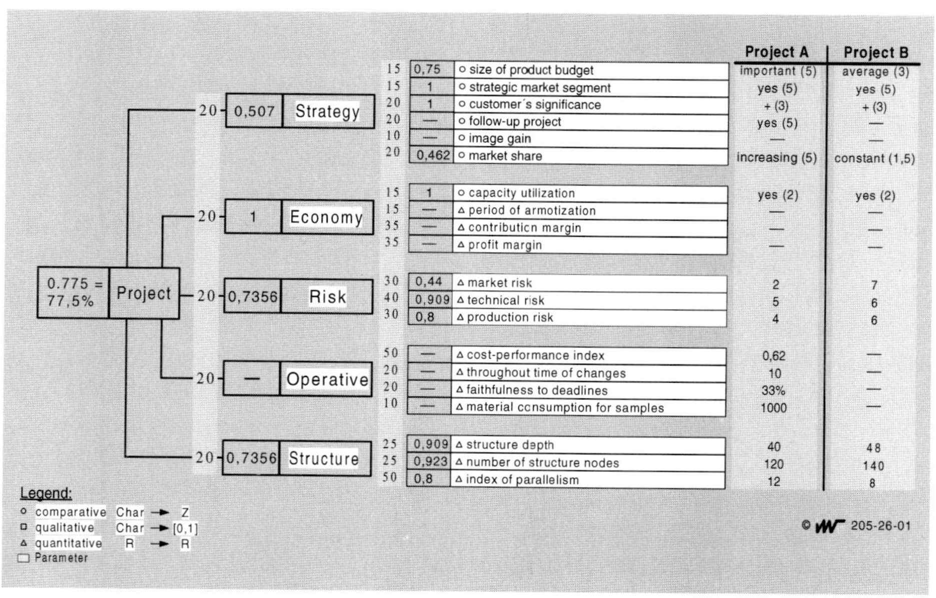

Fig. 5.53: Similarity Calculation of two Projects by the MDS System

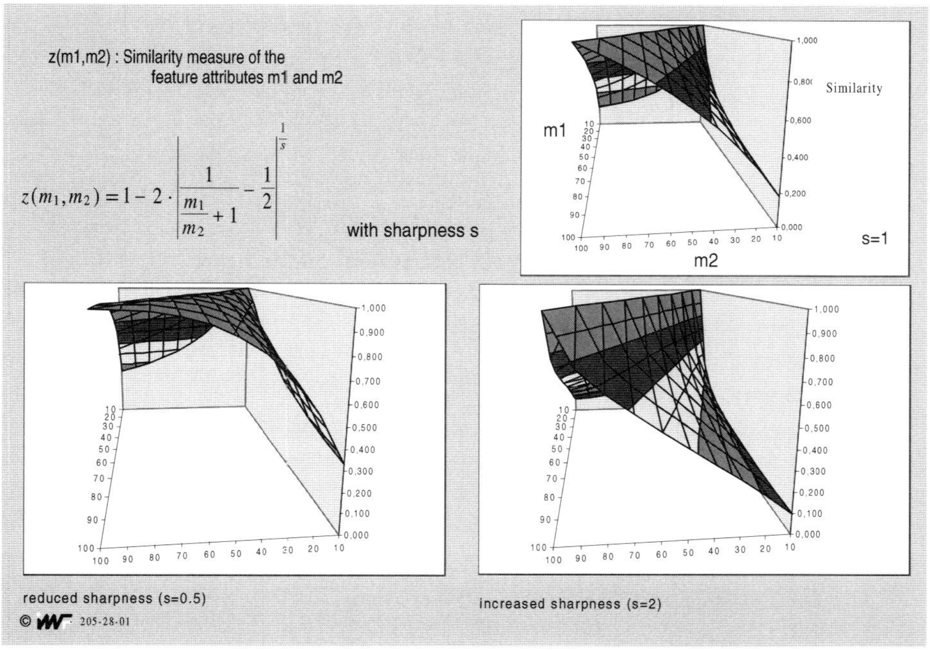

Fig. 5.54: Adjustment of the Similarity Search

value thus calculated at the feature level is subsequently weighted twice. The first weighting indicates how strongly the feature is to be considered within its feature class for the similarity calculation. The second weighting takes account of the varying significance of the individual classes. The user defines both weightings specifically for his enterprise.

Summary and Outlook

MDS is a system which supports the manager during the product developing process, i.e. from the conception phase until the start of the series production. The system enables the user to perform an economic feasibility analysis, the project planning and the controlling of development projects in an integrated form. The system is realized as a client-server architecture and allows the multiple-user operation using PCs under MS-Windows.

The system was developed by IWF within the scope of the EU promoted research project ESPRIT 6896 CONSENS. The industrial realisation of the system is presently worked on in cooperation with a service unit of AEG, the Produktionstechnik und Logistik Consultancy (PLC), at Temic in Ingolstadt.

References

1 Beutler, K: Untersuchung von quantitativen Ähnlichkeitsmaßen bei der Suche nach wiederverwendbaren Ergebnissen aus Automatisierungsprojekten; Institut für Regelungstechnik und Prozeßautomatisierung, Stuttgart 1988
2 Feller, A.H.:Kalkulation in der Angebotsphase mit dem selbsttätig abgeleiteten Erfahrungswissen der Arbeitsplanung, Karlsruhe: Schnelldruck Ernst Grösser, 1992
3 Groll, K.H.: Erfolgssicherung durch Kennzahlensysteme; Freiburg 1986
4 Klein, B.: Simultaneous Engineering: Produkt-Markteinführungszeiten lassen sich verkürzen, Wissenschaft und Technik, Springer Verlag, Februar 1993
5 Mempel, G.: Auswahl von F&E Projekten in: Handbuch Projektmanagement BD. I, Köln 1989
6 Riedl, J.: Projekt-Controlling in Forschung und Entwicklung. Berlin u.a.: Springer Verlag, 1990
7 Roth, K.: Konstruieren mit Konstruktionskatalogen, Springer Verlag, 1982
8 Schmelzer, H.J.: Organisation und Controlling der Entwicklung von Serienprodukten: Praxisnaher Beitrag zum wettbewerbsorientierten Entwicklungsmanagement, Dissertation, Universität Karlsruhe, 1991
9 Schwarze, J.: Netzplantechnik: Eine Einführung in das Projektmanagement, 6. Auflage Herne, Berlin: Verlag Neue Wirtschafts-Briefe, 1990
10 Westkämper, E.; Burgstahler, B.; Korn, G.: Rechnerunterstützte Analyse der technischen und wirtschaftlichen Machbarkeit; VDI-Z 136 (1994), Nr. 7/8
11 Westkämper, E.; Burgstahler, B.; Korn, G.: Integrated Decision Support System for Product Specification and the Introduction of New Technologies; Production Engineering Annals of the German Academic Society for Production Engineering, Carl Hanser Verlag, September 1994

5.6 Conclusion

E. Sleeckx

Now that the contribution of each of the separate tools to the PSI-strategies is clear, we will show how they fit into the overall system

How the CONSENS-tools fit together

5.6.1 The Tools in Relation with the PSI-Strategies

Divida et Impera

SIFRAME, IMS and EPM co-operate to support the overall parallelisation of the product development process. In the normal execution of a product development process, the project is split into parallel tasks by EPM, while SIFRAME supports and controls the parallel execution of the tasks. The IMS keeps track of results of the parallel executed tasks.

PIA provides the possibility to access and use information about the product. This is the precondition for executing parallel processes with the same data. The other modules primarily support parallelisation in the area product. They enable one to work with intermediate results. E.g. KnobieCAD gives information on the overall dimensions of the product in the beginning of the development. From this information, DoPF can start to set up the assembly environment. In doing so the general aspects have been shifted from the end to the beginning of the project. Standardisation has two major aspects: standardisation of procedures and standardisation of data. Standardisation of data is mainly performed by PIA and the design and manufacturing tools. PIA provides a common product model. This enables the reuse of product data and structures from former projects.

The bricks are the same, the houses are different

Tools like DfA and DfM access these structures and data and make use of libraries of standard components. Standardisation of procedures such as, document release procedures, project structures, milestone lists, reporting structures and design processes are done with the help of IMS, SIFRAME and EPM. Building up and using these standards influences, as a side effect, standardisation in the area organisation.

United we stand!

Integration has three levels which should be distinguished:

- Integration on the level of people,
- Integration on the level of applications and
- Integration on the level of data.

The concept of team integrates users with different skills, knowledge and location. This is supported in co-operation with IMS, SIFRAME and EPM. SIFRAME also provides integration on the level of applications. It has the architecture and services to co-ordinate and control design and manufacturing tools. It allows the flexible adaptation and introduction of different tools. Integration of the data level is supported by PIA and individual tools. They consist of a common product model and/or use international or industry standards as PDES/STEP, IGES, SDAI . Finally, the contribution of the different modules to the PSI-strategies can be summarised as follows (see figure 5.55).

	RESOURCES			PRODUCT			PROCESS			ORGANISATION		
	P	S	I	P	S	I	P	S	I	P	S	I
KNOBI ECAD				X	X	(X)						
EPM										X	X	X
MDS	(X)		(X)	(X)	(X)	(X)	(X)	(X)	(X)	X	X	X
RGT				X	X	X						
DfM				X	X	X						
DfA				X	X	X						
DfQ				X	X	X						
DtC				X	X	X						
DoPF				X	X	X						

(X) = optional

Fig.5.55: The contribution of the CONSENS-tools to the PSI-strategy

5.6.2 The CONSENS-Tools and CSE

An Ounce of practice,...

It will be clear to the reader that not all possible tools have been integrated in the CONSENS-prototype eg. Customer-oriented tools like QFD (Quality Function) and production oriented tools like CAPP (Computer Aided Process Planning) have currently and on purpose not been integrated in the system. Also the technological scope, although already large, has not been fully covered. The explanation for this is straightforward if one considers the main strategic objective of the project. This is to prove by practice, that a CSE information system plays a major role in the implementation, the support and the maintenance of Concurrent Engineering in industrial companies. Therefore, it has been a prerequisite to develop a prototype that could be implemented (read integrated) in a company to support its daily activities, and not only work on some theoretical activities in a laboratory environment.

Sesame...

Another major point to justify this limited set of integrated tools is that the system is open, very open. Not only the basic modules have been implemented, but also end-users tools have been successfully integrated ranging from FEA (Finite Element Analysis) to Electronic Lay-out tools. This openness has been a fundamental specification of the system and also one of the important points for its acceptance by industrial companies. The above still doesn't explain why the tools that were integrated were selected. This we will explain right now. From the start, it has not been the intention to develop a fully elaborated CSE system. In fact it would be ridiculous to try, because a CSE system must be adapted to a specific company, so a standard system would never be accepted. Therefore, the aim was to only integrate a limited set of tools and to build a small scale concurrent engineering environment. And here is the point: A Concurrent Engineering Environment. The tools that are selected, cover the field from early design (DtC, DfMA,...) to manufacturing of the product and production facilities. The tools touch the heart of CSE. Eg. Tools like DfM cross the traditional departmental borders and link design with production. Not only does DfM support the CSE strategy, it even forces companies (and hence the supporting information system) towards concurrency. A clear

Beyond the borders

example of the need for concurrency are the different strategies found in DfM and DfA. DfM tends to decrease the manufacturing complexity of a part, whilst DfA tends to increase the functional complexity of a part. Applying both techniques independently would be counter-productive and result in a sub-optimal solution. Therefore, these modules have to be integrated in a concurrent engineering information system. The selection of these tools in the basic system has been a major test for the overall system. The different user scenarios presented further in the text illustrate the complexity and the inter-relation between the integrated and user-provided tools, and show that the integrated tools indeed are representative and fulfil to a large degree the requirements for the pilot system.

6 Case Studies of Successful CONSENS Implementation

6.1 TEMIC TELEFUNKEN microelectronic GmbH

J. Bergner
supported by L. Krickl, A. Voellm

6.1.1 Description of Products and Projects

TEMIC TELEFUNKEN microelectronic GmbH (TEMIC), a subsidiary of AEG Daimler-Benz Industrie, runs manufacturing systems at different locations for the production of electronical parts for a variety of applications. TEMIC offers solutions to automobile manufacturers for passive and active safety elements, driving comfort and operator convenience, as well as systems for reducing fuel consumption and pollutant emission. TEMIC modules also control antilock brake and traction control systems. Numerous automobiles from leading manufacturers come with TEMIC engine and chassis control systems, as well as with comfort features like mirror and seat setting memories.

TEMIC in Ingolstadt is a company with nearly 650 employees that produces:

Products

- electronic control components for antiskidlock braking systems (ABS),
- automotive electronic control units as well as,
- control units for appliances and
- airbag control units.

For the CONSENS project it was decided to focus on the development of an Electronic ABS Controller (EAC). Connected to Ingolstadt are production sites in the Far East and Central America. The products are developed in-house with a high degree of co-operation with the automotive companies. There are four major product lines with a total of nearly eighty different types.

Decreasing cycle times

A typical product consists of approximately 250 parts, 20 to 40 mechanical and 150 to 200 electronical component parts. The lifetime of a product is approximately four to five years, while the mean duration of a development project is about one to two years. These cycle times are decreasing.

AEG PLC

AEG PLC (Production and Logistics Consult) is a consultancy within the AEG Daimler Benz Industry Group. Its scope of activities is for example:
- improvement of logistics processes,
- improvement of production processes,
- implementation of Simultaneous Engineering,
- business process reengineering/redesign.

AEG PLC identifies the reasons for inefficiency and implements solutions to eliminate it.

The CONSENS Project runs at TEMIC with the support of AEG PLC.

6.1.2 Description of the CONSENS Environment

6.1.2.1 General Description of the Current Engineering Process

Product development process

The product development process in use at TEMIC before the start of the CONSENS project was characterised by:
- a mainly serial working-out of the process steps,
- development processes specialised to an individual product,
- separated and partly uncoordinated working methods (imperfect mutual co-ordination),
- many time-consuming reiterating loops within the process steps,
- weak project-management.

Start of CONSENS

One of the first activities at the beginning of the CONSENS project was the redesign of the whole existing product devel-op-ment process starting from the basics. This reengineering was realised under special consideration of the goals:

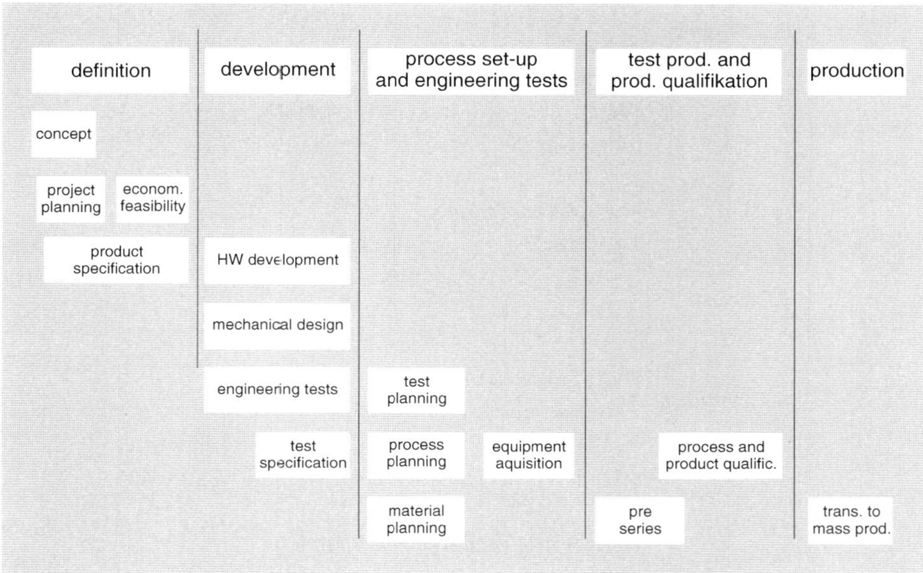

Fig. 6.1: Typical process overview

PSI

- Parallelisation,
- Standardisation,
- Integration.

(PSI) with a coincident strong expansion of the process management authority.

In comparison to figure 6.1, see figure 6.4 which shows the current product development process.

Product development process standard

The current state of the product development process follows a company standard that describes what has to happen during its course. This company standard is however understood as a process which is subjected to permanent quality improvement. After the redesign of the process towards CSE the responsibilities, the inputs and outputs of the different activities and standard cycle times are defined.

The objective of the parallelisation, standardisation and integration of work to do is to ensure the collaboration of different departments to produce products in shorter cycle times with a high quality at lower costs, that fulfil the customer requirements. The following is a description of the organisational runoff and the technical conditions for the release of each phase (phase review).

This company standard breaks down the product development process into several phases. Each phase has a phase review at the end that releases the work performed so far and initiates the start of the work for the next phase. In the following the five phases are described and detailed:

Phases of the engineering process

Phase 1: Conception
The conception includes the technical and economical feasibility study of the new product and results in a quote to the customer.

Phase 2: Specification
The specification phase is used to define the product specification and to predesign synchronised and parallel hardware (electronical and mechanical) and software.

Phase 3: Product Design and Planning
This phase is characterised by the development of hard- and software and the simultaneous planning of production processes and testing. Also equipment acquisition and material planning are calculated in parallel.

Phase 4: Qualification and Test
This phase produces the first batch of pre-series products which are used for process and product qualification.

Phase 5: Start of Series Production
The production facilities are transferred to the factory and the product costs are post calculated. The production commences. The development process is finished.

At the beginning of the specification phase the project manager defines the development process organisation. The generic project plan contains a check list of all important tasks and documents. The project plan related to the actuel project will be used as a guideline.

Project manager

The project manager so far comes from the electronical development department. There is formal co-operation with a project team.

The processes and documents are completed with dates. The milestones as defined in the check list have to be included. The schedule with task, department responsibilities and milestones is distributed to the development, industrial engineering, purchase, sales, sales planning, quality assurance and the business unit management.

Few simultaneousness

As one can see, there was very little simultaneity comparing the old with the new process. The consequence of this was that the whole process, ranging from evaluating product strategies to the start of mass production, took very long and was therefore expensive. Indeed it took longer and was more expensive than necessary. An example is the succession of the check for technical and economical feasibility and the definition of the technical concept in phase 1 (Definition). A better and faster way for these two steps is to carry them out simultaneously as far as possible.

Large number of loops

Another weakness of this product development process was the large number of loops, which appeared during the process phases. This was partly a result of a lack in communication between the separate departments: the "throw-it-over-the-wall" relationship between all departments.

Project-management

The next problem occurring in the "Pre-CONSENS" product development process was the weak project-management. Today the project manager defines at the beginning of the development phase the development process organisation with the support of a form and a standardised project plan. The form contains a check list of all important tasks and documents. Project specific tasks and documents must be added. The project-plans will be distributed to the development, industrial engineering and quality assessment departments for review.

Previously, the project manager was the specialist from the electronics development department with no special project-management skills and no authorisation for the respective project. There was no formal co-operation with a project team. Following this structure most problems arose at the end of the projects and were solved at the last minute by personal resources or money ("fire fighting").

Project Schedule

The development process organisation is used for scheduling the project in a project schedule. The processes and documents are completed with dates. The milestones as defined in the check list have to be included. The schedule including task, department responsibilities, and milestones is distributed to the development, industrial engineering, purchase, sales, sales planning, quality assurance depart-

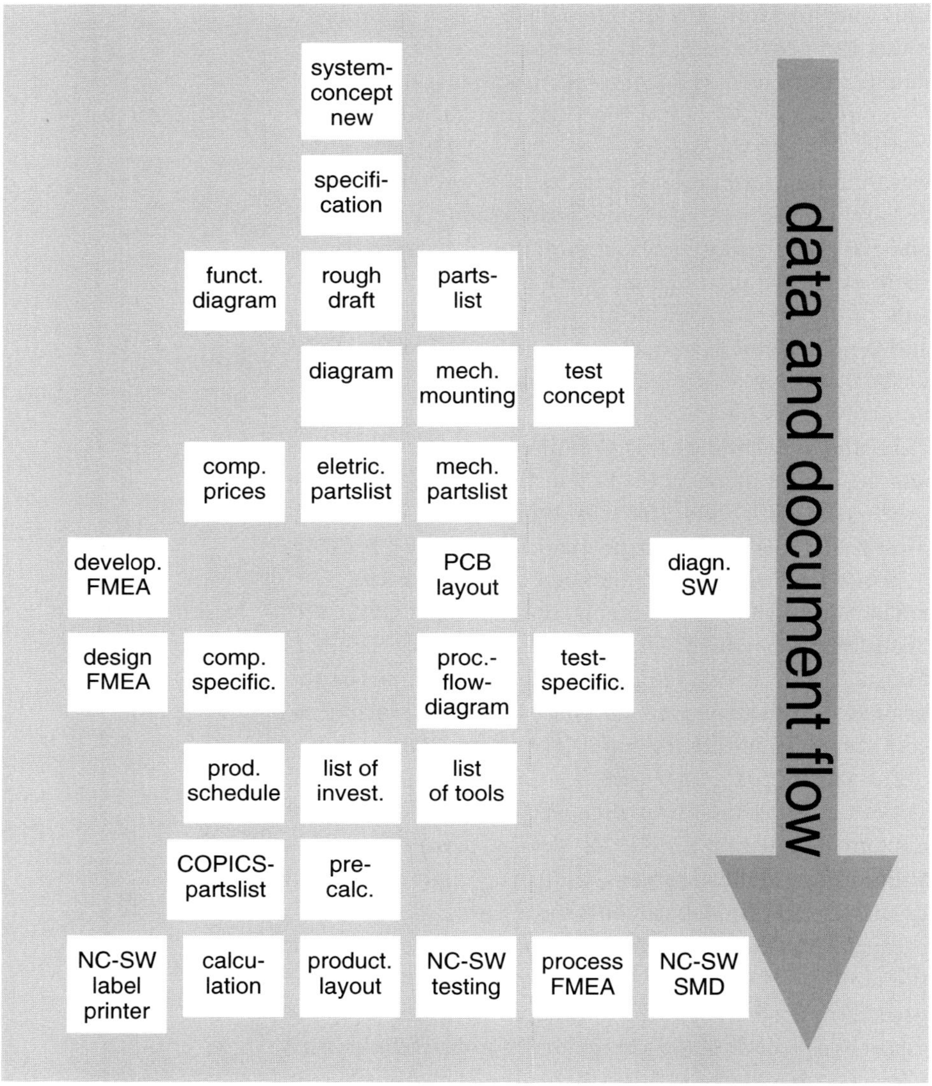

Fig. 6.2: The information flow at TEMIC

ments and the business unit management. During the course of the project the dates and milestones regarding the electronical development department and the mechanical construction department are covered but not those of the following tasks.

6.1.2.2 Improvements and Specific Objectives by Implementing CSE

Areas of CSE

The improvement of the product development process at TEMIC was carried out by taking into consideration the areas of CSE:

- Organisation,
- Processes,
- Product,
- Resources.

Culture

Great importance was attached to the field of Culture which can be seen as an area overlapping the others.

One of the first activities at the beginning of the CONSENS project was the re-engineering of the whole existing product development process. This re-engineering was established under special consideration of the goals

PSI

- Parallelisation,
- Standardisation,
- Integration

with a subsequent strong expansion of the process management authority.

The objective of the parallelisation, standardisation and integration (PSI) of work to be done is to ensure the collaboration of different departments to manufacture products in a shorter cycle time with high quality at lower costs that fulfil the customer requirements.

Key targets for the improved product development process

In other words the key targets for the improved product development process are:

- increased product quality, which meets the needs of the customer,
- reduced product and product development costs to decrease the pay-off period,
- decreased time-to-market to increase the market share.

Removal of the weaknesses of the "Pre-CONSENS" product development process via PSI was the main aim of the CONSENS project.

CSE oriented project-management

One important task was to introduce a CSE-oriented management to the product development process. With the "Pre-CONSENS" project-management, a controlled learning process from former projects did not exist. Management of product development processes was carried out by a specialist from the electronical development department with no special project-management skills and without the support of a project team. Problems with this kind of management of the development tasks are obvious. Following this structure most problems arose at the end of the projects and were solved at the last minute by personal resources or money. It was an important task to introduce a CSE-oriented project-management in the product development process.

A CSE-oriented project-management is a task-oriented form of management. One person is responsible for the success of the task as a whole. This means he must reach the project target. This person is the project manager who is from the electronical development department and has special project training to perform his tasks with regard to leading the project members; he is also responsible for carrying out the project.

There is formal co-operation with a project team. Due to the fact that numerous, complex engineering processes are to be managed in parallel, the task orientation beneath function orientation is necessary in the functional departments such as marketing, development and production etc. Therefore the members of the project team come from all departments involved.

The schedules, efforts, and results of every project are stored in order to be able to compare projects, avoid multiple handling, and to learn from mistakes.

Revelation of irregularities and defects

A special aspect of the product development process at TEMIC is the necessity to reveal every occurring irregularity and defect while testing a new product. This ensures that the security demands of the environment are met for which most products of TEMIC are meant, and the requirements of liability will thus be guaranteed. Meeting these requirements has to be kept in mind during the whole process of developing and implementing the CONSENS tools. Over and above that in the CONSENS area "Product" the documentation of product and production is important.

Flexibility of the product development process

It is necessary to maintain or even to increase the flexibility of the product development process while being supported by Information Technology; this is not easy to fulfil. Often the standardisation and integration of processes with the help of Information Technology trigger a loss of flexibility. However, especially at TEMIC where it is necessary to be flexible to a high degree and to be able to react to customer demands quickly, great flexibility is crucial in the product development process supported by Information Technology.

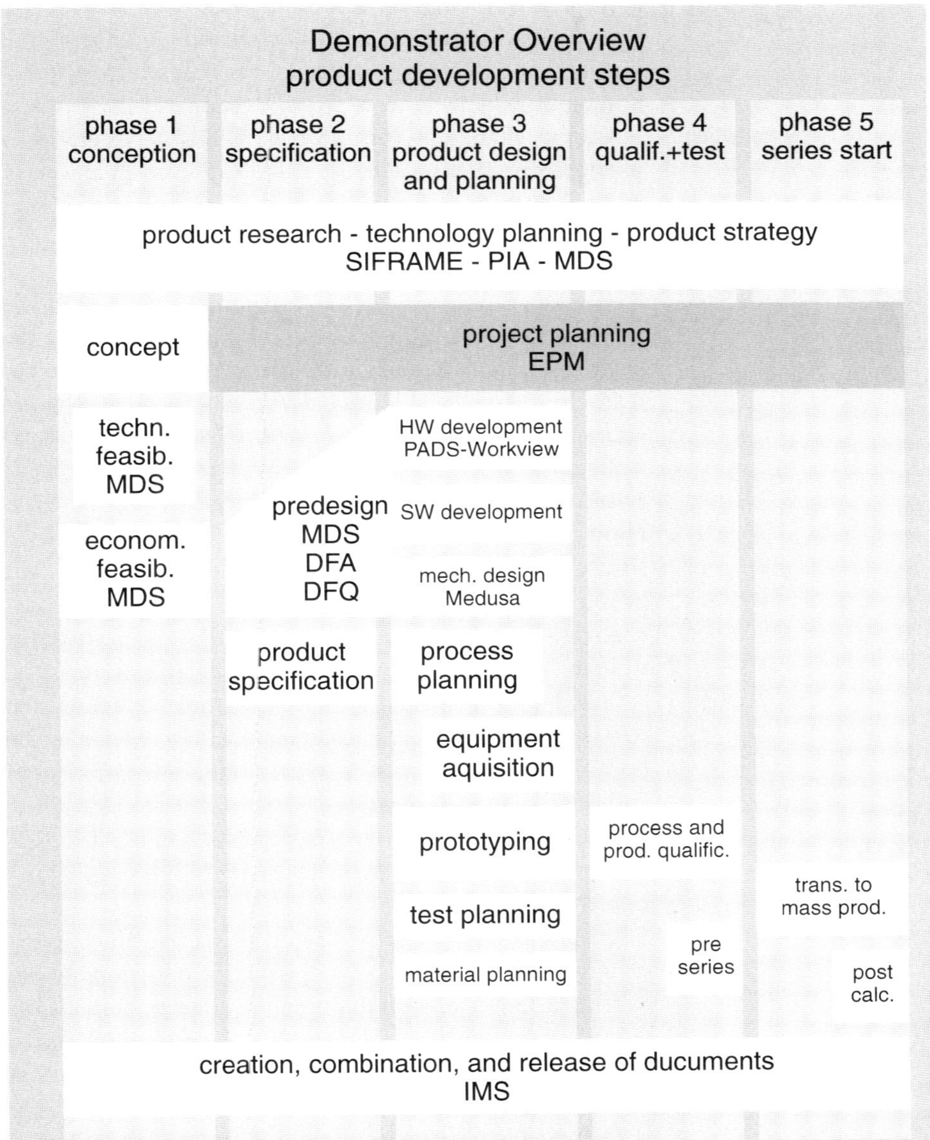

Fig. 6.3: "Post-CONSENS" product development process.

The interaction of employees from different departments within the project team leads to another important aspect of the CONSENS implementation at TEMIC. To integrate departments and employees via matrices or better dedicated cross-functional team work helps to overcome the "throw-it-over-the-wall" relationship between departments and therefore to improve communication in the whole company. Interdisciplinarity and understanding of the point of view of the employees is supported, for example by job rotation of the project leader and project team members.

"Over-the-wall" relationship

Parallelisation

Another main goal within the CONSENS Project was to introduce parallelisation of processes and sub-processes in order to shorten the development time and to reduce time-to-market to raise the market share (figure 6.3 shows the "Post-CONSENS" product development process).

Price and ability to supply

For the employees at TEMIC who are involved in the product development process it is important to retrieve the price of component parts and their availability. Due to the mass production, small price differences may have a great effect on the profitability. Therefore it is important especially for TEMIC – as early as possible in the development phase – to take even minor price differences into consideration and to have the latest price information available. Also, the ability to supply certain component parts might be important for example in the case of the necessity of fast changes. Therefore the developer at TEMIC needs to know whether a specific component part is available or not.

Culture

Introducing a new culture into the whole company step by step was regarded as a special task. No change in the organisational staff would effect a single improvement without changing the employees' views. The re-engineering of the whole product development process must be initiated by the employees themselves if it is to cause any improvement.

Culture, the disposition of people towards behaviour, is an important factor for the successful introduction of CSE. Several different areas of these dispositions are responsible for the success of the various CSE approaches. Two main dimensions of culture are responsible for the success of CSE:

- The management style and
- the working behaviour.

The management style is defined as the disposition of people in a company towards their managers and personnel. The management style has an authoritarian pole and a participative pole. For CSE a participative management style is necessary, which is ruled by matters of fact, experience, and is not determined by the employees` formal positions.

The working behaviour is the horizon of employees. Optimal CSE involves a participational and an inter-disciplinary culture, in which employees are open to arguments from other disciplines (and departments) and are able to evaluate them.

Employees have to learn new ways of solving problems, guiding other people, working in a group. Culture in a company applying CSE is an open-minded way of working, thinking, and living. This target also concerns the relationship with customers and suppliers: Different variants that are desired from the automotive industry cause a difficult position in the market. Therefore TEMIC has to integrate its customers and suppliers in its product development process – on the level of culture, so to speak. Involving customers in the design of specific derivatives as early as possible and involving suppliers in cases of request for changes will ensure the flexibility, especially in short time requests for changes of the customer, which often occur at TEMIC.

6.1.2.3 Description of the CONSENS Product Development Process

CONSENS demonstrator

In the scope of the CONSENS Project the redesigned product development process will be used for the demonstrator in the form of the prototype now in existence. It is adapted to a parallel, standardised and integrated succession (PSI) of the required work steps as far as possible for the demonstrator.

In phase 1, "Conception", the check for technical and economical feasibility already occurs simultaneously with the definition of the technical concept. The "Product Specification" following in phase 2 takes place in parallel with the pre-design which afterwards splits up into

- hardware development (parts list, electrical schematics)

- software development
- design (mechanical parts, housing)

during phase 3 “product design and planning”.

In the same phase

- prototyping,
- process planning,
- test planning,
- material planning,
- equipment acquisition

are performed mainly overlapping.

Pre-series-production and qualification of product and production process will follow.

In the final step “Start of Series Production” (phase 5) the production facilities are transferred to the respective factory and product costs are re-checked by post calculation.

The whole process is escorted by

- product research,
- technology planning,

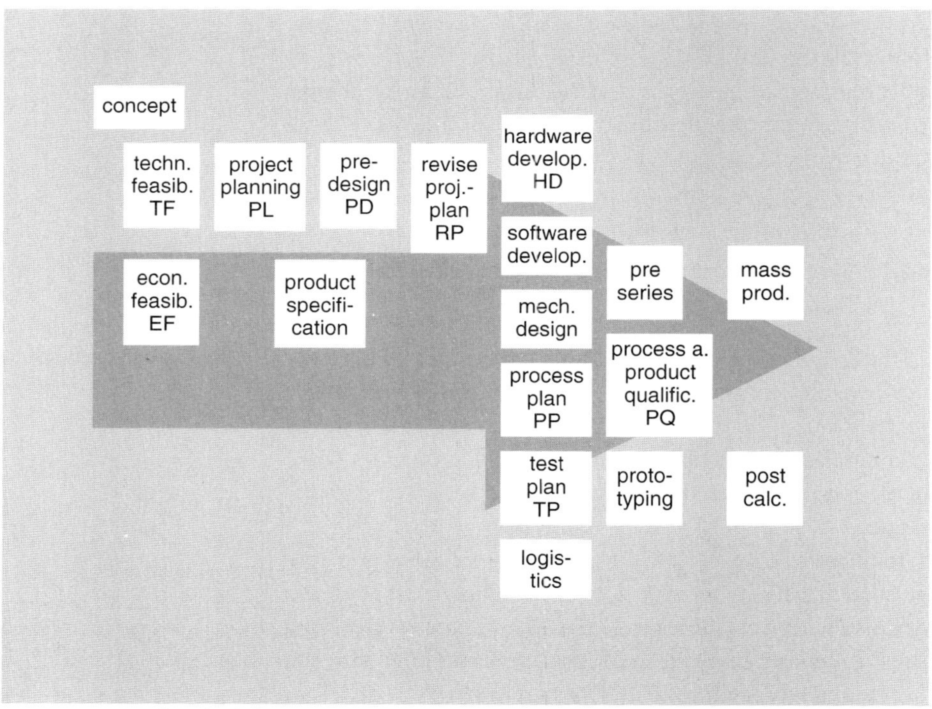

Fig. 6.4: Upper level of the process steps in the product development process

- product strategy,
- the administration of the generated documents.

Project planning also expands over several phases, beginning with specification and ending with series start.

Figure 6.4 shows the upper level of the process steps in the product development process.

6.1.3 Description of Enabling Information Technology

Management

The introduction and testing of the Information Technology at TEMIC was carried out within a demonstrator scenario based on the redesigned product development process to show as much PSI as possible. The demonstrator scenario describes a situation happening during the development process for a new product.

Firstly the management has to check if a completely new Electronic ABS Controller (EAC) has to be developed or if an existing one should be adapted for the use within a newly developed automobile. Economical feasibility through the cost calculation with modules of MDS (Management Decision Support) demonstrates that it will be easier to modify and adapt an existing EAC rather than to develop a new one. Designers and management have to check which modifications to the existing EAC have to be made. They find out that some changes are required, partly coming from special customer requirements, partly arising from technical demands arising during functional tests.

For the execution of these modifications a project plan is generated in CONSENS. Using the available data of the original product (EAC original), at the beginning the data of new components is collected and entered into the system as far as necessary.

6.1.3.1 Management Decision Support

Description of CONSENS modules

An IT-based tool for supporting management decisions supports the following functions:

- documentation of decisions taken during the development process,

- making information available on:
  - production capabilities,
  - technologies available,
  - production capacities,
  - capabilities of engineering and production staff;
- calculation of product and component costs,
- profitability of products,
- supporting the search for decisions taken in similar situations

Improvements by using MDS

and therefore causes the following main improvements:

- methods for estimating:
  - their own technological power,
  - their own situation regarding resources;
- methods for supporting:
  - the design of basic strategies concerning product and production planning,
  - the prediction of risks and chances of product development,
  - the calculation of the main success factors for the engineering process.

6.1.3.2 Engineering Process Manager

Improvements by using EPM

The main improvements by using the Engineering Process Manager are as follows:

- effective project-management by usage of the project handbook,
- optimised project control by definition,
- permanent overview to dates of milestones, risk tasks, resources, costs of development and manufacture,
- uniform information basis for all participants in a project,
- concentrated information for the management,
- clear presentation of arising problems as support for necessary decisions,
- generation of status reports.

6.1.3.3 Design for Quality

Improvment by using DfQ

Using Design for Quality some important main improvements are expected:

- speeding up test planning for new products arising,

- utilisation of the available quality data coming from testing of comparable products,
- optimisation of the test steps in production flow,
- reduction of quality costs with respect to given quality level.

6.1.3.4 Design for Assembly

The Design for Assembly tool is applied during the conceptual phase of electronic products with the target of reducing the effort for assembling products and to minimise component and assembling costs. Suggestions are expected regarding

DfA supports to ...

- possibilities for modularisation of product structure,
- usage of standard components,
- trade-off between component integration and cost decreasing of standard components,
- optimisation of production technology,
- product improvement regarding feeding, gripping, handling, insertion of components.

Furthermore, the minimisation of production costs by means of the simplification and systematisation of manufacture procedures and the increase of the automation degree of manufacturing of electronical subassemblies or parts is supported.

6.1.3.5 Design of Production Facilities

As well as time and cost reduction in the planning of manual assembly work stations, the module Design of Production Facilities gives support in the following main tasks:

DoPF supports to ...

- analysis of work station ergonomics,
- analysis of mounting times,
- chain up of manual assembly work stations,
- combination of manual stations with production automats and robots into production lines,
- integration in production layout,
- determination of work station costs using stored component costs,
- design assembly work station variants based on existing or standard workstations.

6.1.3.6 Change Management and Release Control

Change Management and Release Control was carried out by a special department reporting directly to the top project managers. By using the CONSENS module it is expected that this function will be effectively supported. Therefore the effort with respect to man hours and costs can be reduced considerably, not even taking into account the anticipated improvements in quality and safety.

6.1.3.7 Process Control and Team Co-ordination

The timely delivery of high quality results for the following design or development phase is an extremely important issue in improving time-to-market, cost, and quality. Additionally, there is a strong expectation of a tool to support the product development process efficiently.

6.1.4 Improvements of the Product Development Process with respect to CSE

CSE oriented project-management

Introducing CSE-oriented project-management supports TEMIC in improving time, cost and quality of their products and in doing so, strengthening their market position. As described above, the main improvement through CONSENS is supposed to be caused by the introduction of a new culture in TEMIC. Employees who are unable to deal with this new CSE-oriented culture must learn new ways of solving problems, guiding people and working in teams.

It is obvious that the desired scenario and the expected improvements can only be reached partially, but there is need for a goal for people to be able to change their disposition. There must be no doubt about the direction and the aim of that change.

Improvements that can be reached by introducing the Information Technology developed during the CONSENS project into TEMIC are above all:

- a quick and easy configuration and control of the personal work space (tools, applications, services) for an individual user,

- control of processes with respect to inputs, outputs and process progress,
- experience about the difference between integration and encapsulation of standard tools.

Information sharing and document handling have not been fully supported by IT until now. Only islands of consistent data handling and information sharing presently exist. The contact and special knowledge of data structure and storage of the individual engineers is still a very important means for sharing information and handling documents. TEMIC is expecting further improvements on the level of IT-supported processes, e.g. the implementation of a company-wide information sharing system, that places all information needed for the product development process at the employees´ disposal and means further standardisation of the whole process.

Implementation of a company-wide information sharing system

By using an object-oriented approach for a database (Product Information Archive), TEMIC expects to find solutions in the near future, which are optimised in answering time, flexibility and safety. In addition, the development of a complete or partial STEP model would allow access to data, even if the design tools have changed in the meantime. Furthermore, it is expected that there will be the possibility of shared concurrent access to product information for different engineering processes.

6.2 Dasa Military Aircraft Division

J. Vilsmeier
supported by L. Krickl, A. Voellm

6.2.1 Description of Products and Projects

Projects

The company is engaged in the following projects:
- EF2000 (European Fighter Aircraft)
- TORNADO
- F-4 Phantom
- Airbus
- Ranger 2000 (Joint Primary Aircraft Training System)
- X-31 A (Experimental Aircraft)
- Hypersonic Technology Vehicles.

Involving suppliers and customers

These projects are developed in co-operation with different partners in Europe and the USA. Co-operation means, that a work share is defined at the beginning of a project. This implies, that a lot of interfaces are existing and the procedures to keep track of the versions of the interfaces are a crucial issue in international collaborations. Suppliers and customers are closely involved in the engineering process, too.

For the CONSENS Project it was decided to focus on aircraft structure.

Development takes five to ten years

The lifetime of one product is up to forty years from initial conceptual studies to it going out of service. There is no difference between the product lifetime and project duration because the Dasa Military aircraft division treats a product as a project during the whole product lifetime. However the organisations, taking care of the product are changing. Each phase of the life cycle of an aircraft at Dasa involves other functions and organisations. The typical size of series is several hundreds. During the lifetime of a product the improvement of the aircraft is an almost continuous action. The system´s structure is far less influenced by these improvement-cycles than avionics, electrical systems, and engines. The development of a completely new aircraft usually takes five to ten years, whereas major improvement work may last from several months to a few years.

Typically, the product is produced in one to five different versions for distinct combat roles and is delivered to the Air Forces of three, four or sometimes more countries.

Products developed at Dasa are usually very complex due to the fact that the subsystems like structure, avionics, harness, fuel system, engines etc. are already sophisticated as isolated systems, but the functional dependencies between the subsystems and the many possibilities of interactions between the subsystems can easily produce a complexity that is difficult to handle and to oversee. Structures are developed for high mechanical stresses and a limited life time and are subjected to constraints such as aerodynamic shape, flutter, and efficiency requirements. The interior of an aircraft is filled with electronical systems, engines and fuel tanks, which will influence each other. The combat roles to be performed by the weapon system will normally lead to compromises between the different subsystems. The engineering process consists of a process network including many loops caused by the increasing knowledge and a continuous tuning of the requirements from the customers (Air forces) and the industrial consortium.

Standardised project plan

A standard project structure does not exist due to the large differences between the projects. This is due to the fact that almost every new big project includes break-through technologies. However there is a standardised project plan is used as a rough sketch for the structuring of a new project.

6.2.2 Description of CONSENS Environment

6.2.2.1 General Description of the Current Engineering Process in Consideration of its Weaknesses

The internal process planning uses a scheme consisting of five main phases:

Phase 1: Conceptual Design

This is devoted to conceptual studies. To minimise the technological risk, the key technologies should have been developed to a high degree of maturity prior to the beginning of

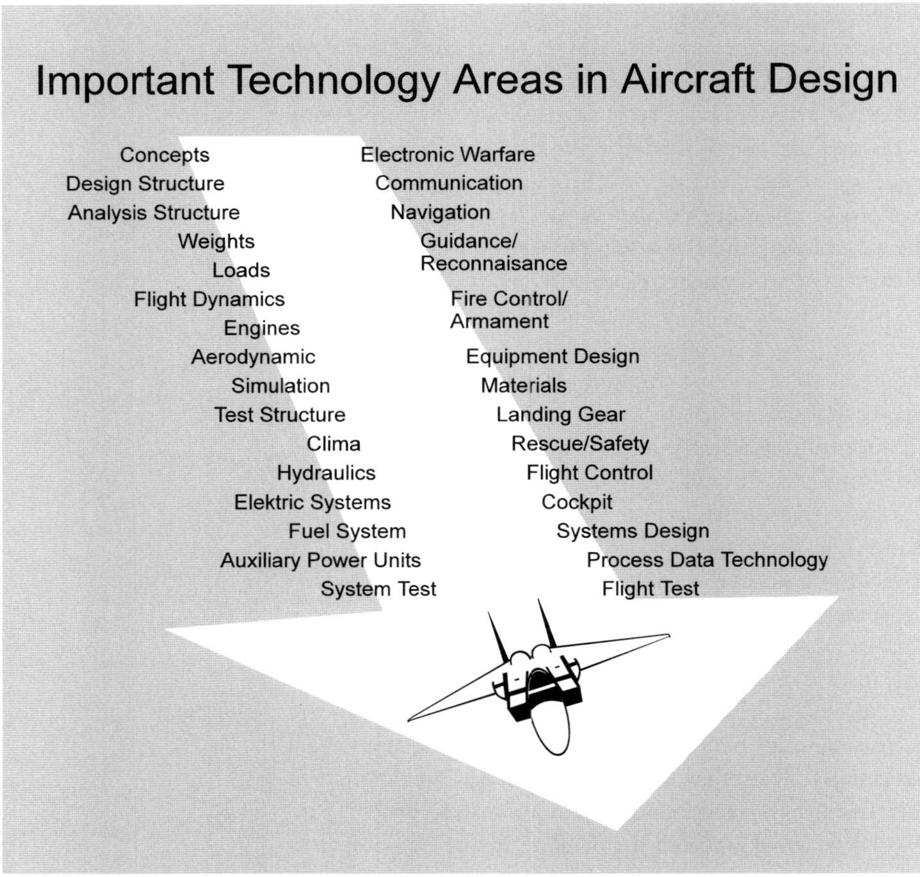

Fig. 6.5: Important Technology Areas in Aircraft Design.

Development in the Conceptual design phase

this phase. Figure 6.5 shows important Technology Areas in Aircraft Design.

In the Conceptual design phase the customer develops together with the management, some planners and some senior designers the concept and project structure for a new product. The major obstacle for the project team in this phase is the integration of functions and subsystems into the aircraft. The result of this phase is a conceptual design of the aircraft and all its components with estimats for all required parameters, a detailed description of interfaces and a project plan.

Phase 2: Preliminary Design

Here, many engineering departments start with their work in a simultaneously. All systems and subsystems are considered in detail and computations are made. The control of interfaces and checking if the product specifications are met, is an important aspect in this development phase. Manufacturing is involved as consultant in order to take the right decisions with respect to manufacturing technologies. Intermediate results have to be exchanged quickly with the other developers. At the end of the preliminary design phase the so-called "stressed layout" will be ready. This means that the aircraft structure is defined in such a way that all critical load paths have been analysed and the internal geometry is defined by many cross sections throughout the complete structure. All interfaces, especially those which are relevant for the work, share are defined.

Involving Manufacturing in the Preliminary design phase

Phase 3: Detailed Design and Manufacturing

Sub-phase: prototypes

The definitions of geometry resulting from the preliminary design phase are the basis for the detailed design. Before the detailed definition of the single parts, the assembly drawings for sub-components, components and major components should be finished. This will help to avoid geometric interference from the very beginning, which may often occur if the detailed design of single parts is done first and the assembly drawings are done afterwards.

At first finishing assembly drawings

Detail design, drafting, stress analysis etc. are working in a very complex simultaneous way. The results of the detailed design phase are manufacturing drawings. As soon as the first drawings are delivered, the manufacturing process will begin. The main activities in the manufacturing locations include manufacturing planning, tool design and manufacturing, production of single parts and a series of assembly steps. Plans for changes have to be distributed very quickly. The information flow between the locations must be well organised. Another important task in this phase is the design of the flight test instrumentation (FTI). All prototype aircrafts which are dedicated to flight testing have to be equipped with flight test instrumentation which usually differs from aircraft to aircraft because each flight test aircraft will have different tasks in the flight test program.

Sub-phase: product investment phase and series production

Realising investments

The product investment phase begins when the customer has placed the purchase order for the aircraft. The investments necessary for series production are made. Due to the results of prototype testing, changes have to be introduced to the aircraft design. In order to reduce the manufacturing costs, the geometry has to be optimised for manufacturability according to the constraints resulting from the different manufacturing methods. At the end of these design-improvement activities, the series production of the aircraft will commence.

Phase 4: Flight Tests

The main task of the engineering departments in this phase is the verification of all parts and assemblies. The support required for the flight test phase is not to be neglected. On one hand the FTI is very likely to be changed due to new requirements for monitoring non-foreseeable parameters. On the other hand the results of flight testing may induce changes to the external and internal geometry.

Phase 5: Customer Support

The importance of customer support

Aeroplanes need special attention throughout their lifetime. The business volume during the customer support phase has become larger than that of the development of a aircraft. Technological and logistic support, maintenance, combat improvements and the extension of the aircraft`s lifetime are almost continuous issues requiring an intensive collaboration between the customer and the aircraft manufacturer. Redesign activities are important during this phase.

These activities have not been included in the CONSENS project.

Start of CONSENS

At the beginning of the CONSENS project the product development process of Dasa was examined for weaknesses which had mainly arisen in the co-operation of different functions and departments.

In the following, the interfaces in the product development process which have been found to be critical are described:

- Surface models and finite element models (a lot of iterations and alternatives have to be handled),
- handling of geometrical interfaces between structural components (e.g. wing box and fuselage, wing box and trailing edges),
- exchange of geometrical data and part-lists between design and manufacture,
- handling of concessions that are a deviation from geometry, material requirements etc. due to unpredictable deviations during manufacturing; a concession has to be signed by drafting, stress, materials and other functions, if important safety aspects are concerned.

Until now, this process network has been conducted by an IT support, which only supports partial product models and takes care only of the final result of the engineering activities in most cases. A typical situation is, that drawings and relevant part lists are registered after their completion into an EDM-system. Nevertheless, the microfiche remains the master document for a drawing, part-lists are already kept in an electronic format. The engineering process itself is not well supported with respect to consistent process and data management.

Support by IT

In the "Pre-CONSENS" product development process problems arise in areas that are important for the success of the whole engineering process. With a growing complexity of the product and, therefore, of the entire project structure and an increasing number of involved employees and departments it becomes more and more important to guarantee that every employee who is involved has access to the latest status of the project. Changes made by the drafting department may cause that different methods have to be used by the stress department for doing computations, for example. Such a need for changes, especially for those arising late in the product development process, are costly and extends the "Time to Market". They can be avoided for they are the result of poor co-operation between functions and departments where the engineering process is open to improvements.

Late changes

6.2.2.2 Improvements and Specific Objectives by Implementing CSE

In the product development process of a modern military aircraft at Dasa the functions “drafting” and “stress” need an especially tight co-operation. Unfortunately, the nature and sequence of the steps inside the internal processes of these two partners are so different that problems regarding the mutual information exchange will arise.

Mutual information exchange

As long as the engineers in different functions knew each other personally, they could give the needed, but incomplete information to their colleagues. With the growing complexity of the aircraft and its structures, the number of people involved has become so large, that the preliminary data is often delivered in an more or less casual and uncontrolled manner to the succeeding process. Furthermore, people often do not know whom to ask for information and because of the tight time-schedule it is tedious and time consuming to go the official way through the organisation.

"Throw-it-over-the-wall" mentality

This problem of preliminary information release and control was considered the worst of all weaknesses in the “Pre-CONSENS” product development procedures and therefore has become an important task for improvement within CONSENS. To overcome the “throw-it-over-the-wall” mentality between the concerned functions it was decided to generate a systematic and standardised method to exchange preliminary information including the probability for later changes during the on going development process. The CONSENS Project will make it possible to test some methods of information exchange between parallel processes. Such a test is not feasible in a productive development environment. The outcome of these tests will help to develop a future systematic method for the exchange of preliminary information. When such a method is implemented and a process handling tool like CONSENS is installed, the processes can be run concurrently and simultaneously to a high degree.

Project-management

Another main aspect of improving the whole product development process at Dasa was the organisation of the project-management. In this area the CONSENS project caused improvements mainly by standardisation of parts of the project structure. The information flow between different functions and the project-management was subject to improvement. Information sharing and document handling at Dasa so far has not fully supported by IT. Only isolated consistent data handling and information sharing processes exist. The engineer`s personal contacts and special knowledge of the data structure and storage is still a very important mean to share information and handle documents.

With respect to structural engineering there are generic documents like material-handbooks including semi-finished products, standard parts-handbooks, test procedures, standards for surface treatments etc. and project specific documents which are shared by different functions of the structural design. These are, e.g., surface models which are the basis for creating a mesh for aerodynamic calculations and first FEM analysis, large FEM data sets which are the basis for stress analysis, structural optimisation codes and dynamics (aeroelasticity, loads evaluation, flutter analysis). Of equal importance with respect to information sharing and document handling are the geometrical data and lists for drafting and manufacturing parts.

Nowadays, scheduling is done with the help of central project-management systems, (like ARTEMIS) which are able to store and compute the amount of information occurring in engineering projects. Costs are estimated from earlier projects and effort values with special factors.The control of time, cost, and quality is done by exchanging results and reporting the project progress of the departments.

Information flow

The information flow in the project is done at the level of the heads of department for technical information together with the project-management for project-management problems.

The project manager in the autonomous project organisation has a position like a manager of a big business unit. He has his own organisation where all functions except personal management are present.

The information flow in the project can be improved by considering the needs of the information receivers more thoroughly. In current practice, detailed specifications existing for reporting and documenting customer requirements. Each project has defined rules for the information flow and information distribution. Basically, all the information is available in the project, but most of this information cannot be queried by electronic means. Of course, all documents in the form input and output of individual development steps are registered by EDM-tools, but very often only META data is kept in these systems which will help the engineer to find the documents required for his tasks. However he has to get a printed copy to look up the information contained in the documents. The problem is, that the engineer very often needs some preliminary information to start his work and the published documents registered in the EDM systems in most cases be available too late. The information management on the project level – which gives more transparency to the employees – could be improved substantially by the CONSENS engineering framework.

PIA – Product Information Archive

To ensure a fast data flow and to guarantee access to all involved functions in consideration of the status of the development project, a common database, called PIA (Product Information Archive) has been installed (see fig-

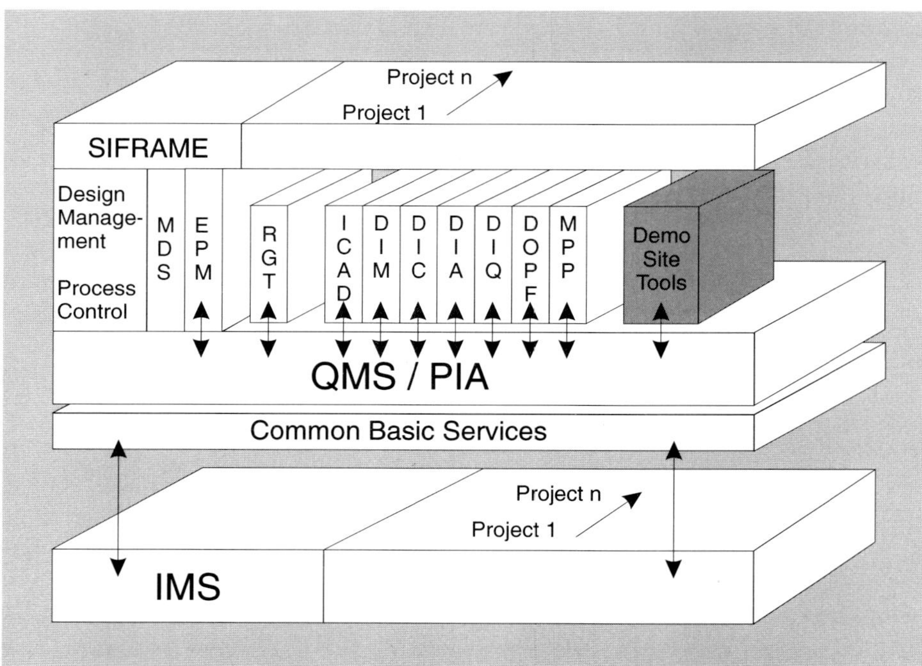

Fig. 6.6: Architecture of IT in CONSENS

ure 6.6), which contains the actual versions of data objects like calculations (e.g. FEM-analysis), drawings (i.e. CAD files), etc. PIA integrates the data flow of different functions and (software-)tools concerning the project. PIA is the answer to the problems outlined before concerning the availability of the latest status of the current to every involved employee resulting from the growing complexity of the projects and the increasing number of involved employees.

Therefore using PIA implies:
- reduction of product and product development costs,
- improvement of overall quality,
- reduction of "Time to Market".

In this context, we refer to chapter 6.2.3, where the enabling information technology is described in more detail.

Co-operation of different departments and functions

The interaction of employees from different departments within the project leads to another important aspect of the CONSENS implementation at Dasa. To integrate departments and employees via matrices or improved cross-functional teamwork means to overcome the "over-the-wall" relationship between the departments and therefore to improve communication in the whole company. Interdisciplinarity and understanding of the point of view of other employees is supported, e.g. by job rotation of the project leader and team members.

6.2.2.3 Description of the CONSENS Product Development Process at Dasa

The Dasa scenario describes the design process during the development and design of a wing as a component of the X31 aircraft. The main components of this wing are a central wing box, which is built from an upper and a lower skin made from carbon fibre reinforced plastics and an internal structure consisting of metal parts, a pair of leading edge flaps, a pair of trailing edge flaps, two flap-actuators with their fairings and a fixed nose skin.

Demonstration of CONSENS

For the demonstration of CONSENS it was decided to select some typical parts and try to run the complete process of (detail-)design. These included some aspects of the preliminary design process and typical parts of the process concerning the consulting function of the manufacturing division in order to assure manufacturing potential in the early stages of the project.

Some parts of the design process are common for all categories of parts. In the following paragraphs, the process-parts needed for stress analysis will be described.

Tasks within the design process

The design of a modern military aircraft is divided into many sub-tasks. The mechanical design is one of these tasks which takes into account all the geometric and functional properties of the structure. Another task is the calculation of stress. There are some other tasks like mass calculation, aerodynamics and flutter etc, but the two issues "design" and "stress" mentioned above need an especially tight co-operation.

FEM

The core of the development of an aircraft structure is represented by the global finite element model (FEM) of the complete aircraft. In this model all relevant data is included, i. e. the aerodynamic surface, the aerodynamic pressure distribution, the geometry of the internal supporting structure, the internal pressures and inertial loads, a range of selected material properties and local thicknesses. The mesh size is made as fine as necessary but as coarse as possible by engineering judgement.

Based on this model, the FE-method determines the internal load distribution, element forces, stresses and strains for further detailed stress evaluations. These detailed stress analyse are the basis for the final definition of geometric details of all parts and certifies the load bearing capabilities of the structure. Another important usage of the global FE-model is the determination of the structural dynamic behaviour with respect to aeroelasticity and flutter. A detailed description of the information technology enabling the FEM-analysis is given in chapter 6.3.3.

The FEM-mesh is given to the loads department, which completes the mesh by introducing the nodal loads for:

- static loads,
- gravity loads,
- thermal loads,
- dynamic loads,

and transfers the completed mesh back to the stress department, where the results of the FEM-analysis are analysed and interpreted.

In case of high load or high stiffness gradients, it might be useful to have a more detailed model to get more accurate answers. In this case, the appropriate part is idealised in more detail. The surrounding structure is either a global one with a transitory region or simulated by interface loads taken from a cut out of the global model. Nevertheless even those models are not stress convergent.

Considering the manufacturing process

When the design of a part has reached a state where the manufacturing process must be considered, the engineering division will have to call on manufacturing advice for information and to initiate a discussion about the best possible part design. To do this, the basic part information (part number etc.) is entered into the planning system for manufacturing called "Netzplan", while the existing geometry and bill of material data are transferred from the working files to the manufacturing mailbox.

"Netzplan" system

Inside the manufacturing division the distribution team will check this data (on-line), and after consideration, add some internal identifications, milestones, priorities and more such data. The "Netzplan" system will send messages to the mailboxes of the appropriate departments and generate intermediate milestones, can open sub-plans if necessary and show the actual status of each step in the process. Furthermore, it can accept "Ready" messages and send answers and results back to the design division.

From the manufacturing mailbox the data will be transferred to a central safety storage, where the working departments can read but not produce any data. The data are in a state of "Calling for advice" and, therefore, ongoing design changes will not be incorporated. From the safety storage, the concerned departments will copy the data to their working files.

The manufacturing planning department will create a preliminary working plan where data that has not yet been defined will be replaced by plausible estimations. If needed, special departments for manufacturing techniques and materials can be called for advice. In parallel, the department for the supply of semi-finished products will check the availability of the material and eventually start the purchase.For many parts some manufacturing jigs will be needed. Therefore, the concerned department will create a jig and tool design concept, using CAD tools and will start with manufacturing, planning and preparation of these tools. For large parts like the wing skin very large and complex jigs and tools will be needed. In comparison to the main manufacturing steps the process of designing and manufacturing these tools will be very time- and money-consuming and has a great influence on the part design.

Based on the inputs of the different departments, the cost calculation department will produce a preliminary cost estimation for the manufacture of the part.

All the results and proposals for alternatives will be returned to the design division. In case changes are required for existing manufacturing machines or the provision of new machines or the change of buildings and similar large investments, the leader of the manufacturing division will be informed.

6.2.3 Description of Enabling Information Technology

EPM

Many of the improvements of the product development process with CONSENS depend on introducing an EPM (Engineering Process Manager) equipment. EPM is the support tool for simultaneous engineering specific project planning, control and communication in the project team with respect to project information. The EPM equipment supports the data flow during the whole process. The object-oriented database SIFRAME constitutes the framework of the CONSENS enabling information technology. The SIFRAME architecture integrates CONSENS tools and existing in-house software in a common framework (see figure 6.6).

SIFRAME

The functioning of SIFRAME is used to install a project specific environment for different user requirements. SIFRAME controls the execution of the activities for defined phases of the product life cycle. It serves as a common environment for teams of designers to perform a co-operative effort and facilitate a parallel execution of their activities. The pieces of information related to the project generally are: teams, process flows of the project and tools to perform the activities in the related process flows. SIFRAME manages the relevant files processed by designers and keeps track of the life and the versions of these files. Information produced during the product development process is stored in the database of SIFRAME. The access and management of that data which are related to the product structure is made possible and controlled by the Product Information Archive (PIA). From the view of information technology the CSE-oriented product development process runs as follows.

6.2.3.1 Processes and Data Flow Related to the Global FE-Model

Input data for FE-model

As input, the data basic geometry and loft (numerically defined geometry of sculptured surfaces) are transferred upon request from the design department in the form of CATIA 3D models. The geometry of structures is transferred from the design department mostly in the form of CADAM 2 1/2D drawings, occasionally as CATIA drawings or 3D models.

All this input data has to be checked and the information not relevant is to be removed (about 80%). Additionally, the material department delivers the appropriate material data: Youngs modulus , stress-allowables etc.; mostly in the form of printed copies of the corresponding standard sheets with hand-written additions. This data is then entered into the stress-internal file management system using ODIN. The calculation of material properties is done using M/VISION, which is a material information system.

ODIN and M/VISION

Based on this data, the global FE-mesh is defined using I-DEAS. This definition includes:

I-DEAS

- **Geometry definition:**
  - grid point co-ordinates (partly derived from the basic geometry),
  - element connections (e.g. web, tetrahedron);
- **Element properties:**
  - element thickness,
  - cross-sectional area,
  - moments of inertia,
  - fibre orientation (if appropriate);

- **material properties:**
  - isotropic/anisotropic material type,
  - temperature dependent properties (if appropriate);
- **boundary conditions:**
  - point constraints,
  - partitioning for sub-structuring.

NASTRAN

In the stress department, the FE-mesh is processed by a solvent (NASTRAN), normally in batch mode.

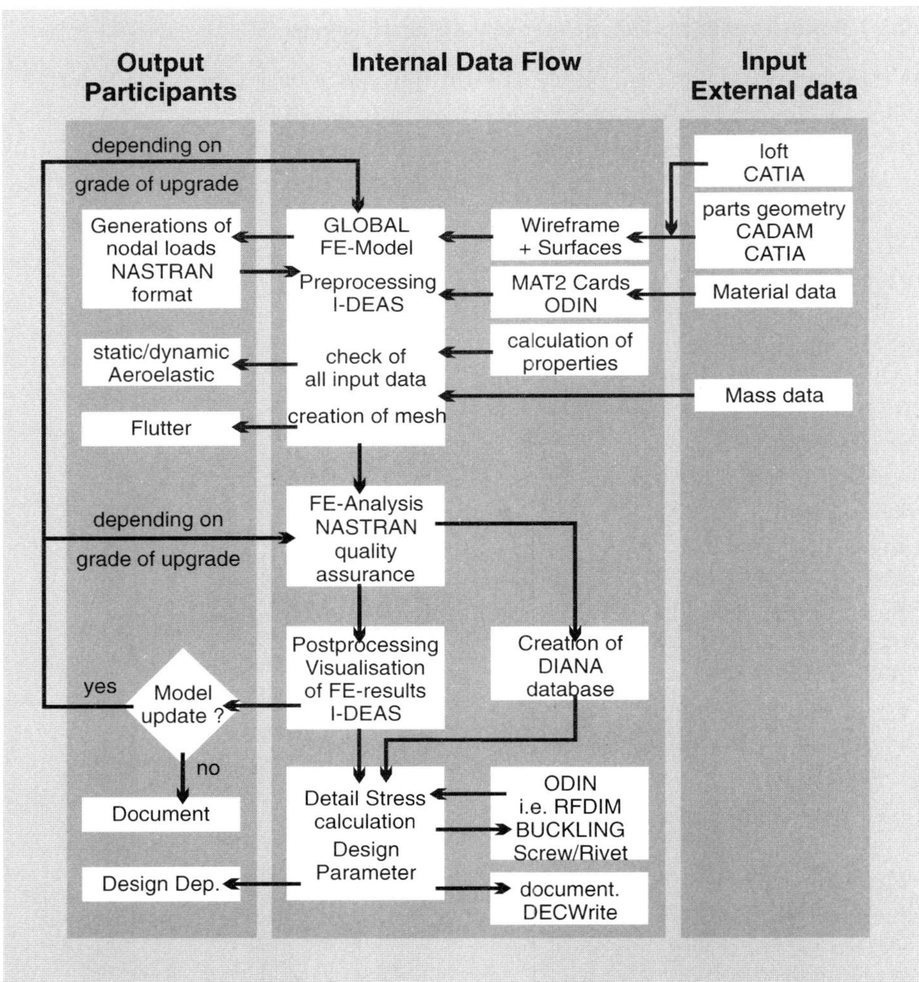

Fig. 6.7: The processes and data flow related to the global FE-Model.

The results of the solver are post-processed to be able to visualise and interpret them.

The results needed most frequently and displayed by graphic means are:

- displacements (enlarged or in colours),
- internal forces or stresses (in colours or arrows),
- critical loads or frequencies.

ODIN is a package of in-house developed software modules with a common graphical user interface and a common database. ODIN can also handle the results of the DIANA program, which works on the NASTRAN results and presents the data in a way that subsequent stress calculations are possible without further data translations.

DIANA

6.2.3.2 Processes Related to Material Selection

Before focusing on the processes related to material selection, it is necessary to describe the structure and function of PIA (Product Information Archive) in more detailed. PIA contains the product data model for the products that are produced by the user companies. The product model is described in terms of complete products with their internal structures and their evolution over time. Customer variants and versions on each level of the product structure will be represented. It will be an object-oriented model describing the product in terms of objects, which are classified according to their qualities. These objects will be: products, assemblies, materials, drawings etc.

PIA – Product Information Archive

The object-oriented structure of PIA simplifies the data management of data, because the complexity of the data can be modelled in very expressive terms. The data can be stored without redundancies and inconsistencies and changes can be quickly made. PIA provides product data for the CONSENS tools so that they can perform their functions. Bills of materials of different types, master-part data for different versions, process plans, and other information can easily be generated from the product data model in PIA. The definition and changes of the database scheme can be made via a graphical user interface.

A part of the CONSENS scenario will be the integration of the material information system M/VISION. A small mate-

M/VISION

rial database will be created and filled with data on the materials used in the X-31 A project. The data flow will be as follows: The design engineer and/or the material specialist will make the choice of material and its form using M/VISION. All of the subsequent processes which require specific information according to their functions about materials like static stress design allowables, material input decks for FEM and the order of material by manufacturing will use the information stored in PIA to identify the material, make a function specific call to M/VISION asking for the material properties needed for the specific process being performed. This information will not be transferred to PIA, because this data is part of M/VISION. This procedure will ensure that the right data is used for the processes and the chance to use the wrong material will minimal compared to the manual input, which is often requested today. If the first choice of material has to be changed for some reason, this information will become visible to all processes involved and force the team involved with the design of a part to repeat the activities affected by the change of material. This is an important contribution to "design to quality". The cases of material test and quality assurance databases will not be considered within CONSENS.

6.2.3.3 Selection of a Surface Protection System for Parts Using ICAD Capabilities and a Dasa Developed KBS Application

Until today the surface protection system has been determined according to the Dasa design practice by the design engineer and the material specialist. The type of surface protection to be used depends on the following characteristics:

- type of materials in contact (e.g. aluminium, steel titanium),
- working environment (fuel tank, dry, erosive),
- type of design (sheet metal, milled, riveted),
- type of structure (fixed, moveable).

WISOS – **W**erkstoff **I**nformations**S**ystem **O**berflächen**S**chutz

Dasa has developed a KBS application called WISOS (material information-system and surface protection) which is now running on a PC under Windows, which is able to determine the surface protection system for about 90% of possible questions, the remaining 10% have to be answered by the specialist. Within CONSENS, the DELOS tool ICAD will analyse the "drawings" to capture the defined parts and to determine the parts which are in contact. For the remaining information required by WISOS, the following possibilities are under investigation:

- interactive input via WISOS,
- definition of the missing parameters in PIA.

The completed information will be used by WISOS to determine the surface protection system and write the result into PIA, if WISOS can provide the answer. If not a message will be sent to the surface protection specialist to solve the problem. This procedure will be a contribution to TQM, because errors caused by the incomplete knowledge of the design engineer will be avoided and the best possible surface protection will be defined for the whole aircraft in order to reduce the risk of corrosion during the service life of the aircraft.

Contribution to TQM

6.2.3.4 Design to Cost: ASCET

ASCET is a design-supporting tool for the evaluation of the costs of a roughly defined assembly group or product. It should be used by the designer in an early design stage, preferably conceiving and scheming, to estimate the cost at a point in time where the potentials of cost reductions are quite high and the effort to improve and change substantial parameters is very low. ASCET specialises in mass producted products and the evaluation of manual assembly operations. A database for all parts, assembly-groups, products and assembly-operations makes ASCET very efficient in the long term.

ASCET allows the estimation of costs in an early state of the project

6.2.3.5 The Specific Design Process Shown in Some Examples

Wing Skin

For the wing skin the specific design process parts are as follows:

Design process parts for the wing skin

- extract given geometry from a 3D-model, define part boundaries, define loads and finite element grid using-CATIA (part of the stress related general process),

- calculate overall layer distribution using LAGRANGE (specific for large fibre parts),
- define sequence of layers using a newly programmed module from IST. This module will be one of the most needed missing links in this process chain,
- re-input all the information into a 3D-model; add definition of main attachments and actuator fittings using CATIA,
- calculate stress and optimise design of layers using ODIN (parts of the stress related general process),
- re-input this information into a 3D-model; deduce boundaries of every single layer and calculate optimised carbon fibre tape-paths using ITLS (specific for fibre parts).

The result of ITLS is an executable data set for a numeric controlled tape-laying machine.

All tools except the one named from IST are already in use at Dasa.

Wing Spar

The wing spar is a large NC-milled part. In general the design process follows in general the scheme described above. Characteristic for this part are several loops in detail design.

Loops in detail design

This is particularly the case for the attachment of the actuators for the leading edge flaps, It is needed because these flaps are designed in parallel and the interaction between the fixed structure and the flaps is complex, depending on the complicated different positions of the flaps relative to the wing.

Additionally, these design loops need to include the stress department to calculate the detailed stresses and hence the part-thickness. The actuator attachments are locally high loaded and the spar must deliver these forces suitable for the surrounding structure. These processes will be modelled by some actions running in parallel:

- part shape design using CATIA,
- stress calculation using ODIN.

Metal Sheet Rib

Parts formed from metal sheets are most common in aircraft design, but differente to classically bent parts most aircraft parts have flanges with curved bending lines which are usually manufactured by a rubber-forming technique. This makes it necessary to build a forming block tool, normally manufactured with NC-mill 3 axis method. The design process must consider these special circumstances and constraints given by this manufacturing method.

The design process for the selected part will be modelled by some parallel running actions:

- part shape design using CATIA,
- calling manufacturing for advice using the specific methods described here.

Actuator Fitting

Design process parts for the actuator fitting

For the actuator fitting, the specific design process parts are as follows:

- extract surrounding geometry from 3D-model, define available space and basic concept,
- calculate optimal material distribution constrained by given forces, the available space and minimal weight using TOPOLOGY from IST the CONSENS partner,
- refine the design to fulfil functional and assembly criteria by changing some features using IST from IST,
- check and optimise the manufacture of the part by changing some features using Detail Design from WTCM.

When the part design is complete, the 3D-models can be merged and the whole design can be checked.

6.2.4 Improvements of the Product Development Process with Respect to CSE

The Dasa demonstrator as described in section 6.3.3 has demonstrated how process and data management could be enhanced with respect to CSE requirements. Of course, the Dasa scenario has included most of the modules related to the engineering framework and some of the tools developed within the CONSENS project to support CSE. On the other hand, Dasa has also integrated some Dasa-specific tools in order to show how those can support CSE by achieving better control of the processes and related data than it has been the case in earlier practice.

Regarding the implementation of SIFRAME the following main improvements have been looked forward to:

Improvements with SIFRAME

- a quick and easy configuration and control of the personal work space (tools, application, services) for an individual user,
- control of processes with respect to inputs, outputs, and process progress,
- experience about different possibilities to integrate standard tools (integrated, encapsulated) and the effort necessary for the different types of tool integration,
- standardisation of typical processes providing improved quality and tractability of data.

With the implementation of SIFRAME, it has become possible to handle complex processes, including different versions and variants more easily and safely. The encapsulation of tools is relatively easy, if an experienced UNIX-user is available.

Improvement of document handling with IT support

At DASA the basis for information sharing and document handling used not to be fully supported by IT. Only isolated cases of consistent data handling and information sharing existed. The contact and special knowledge of the data structure and storage of the individual engineers were still a very important mean to share information and handle documents. Dasa was expecting a major improvement of this situation at the end of the CONSENS project.

By using the object-oriented Product Information Archive (PIA), which is now fully integrated into SIFRAME, the users at Dasa found solutions which were optimised on answering time, flexibility, and safety. In addition the development of a complete or partial STEP model allows the access to data, even if the design tool has changed in the meantime. Furthermore the users at Dasa now have the possibility of shared concurrent access to product information for different engineering processes, which is very important with regard to overcoming the "over-the-wall" relationship between the departments and to avoid multiple handling and, therefore, to shorten the duration of the whole product development process.

6.2.5 Future Improvements

The CONSENS project has paved the way for the introduction of new methods and IT-tool supporting CSE. Consistent process and data management could be demonstrated to be a realistic aim in an integrated engineering design environment. Nevertheless a number of requirements and possibilites for improvements can be identified from a users point of view. These requirements can be categorised as follows:

Requirements and possibilites for improvements

- Integration SIFRMAE-EDM systems,
- SIFRAME functioning,
- Integration of the engineering environment,
- Integration into "general" company IT-applications – integrated and interlinked process models,
- Support of multi-site/multi-company projects.

Integration SIFRMAME_EDM systems

The development of a new product is a dynamic process and at the end of the development the product configuration and change management for such complex product like aircraft should be performed by an EDM-system. A first prototype of check-out/check-in mechanisms between SIFRAME and the IMS could be demonstrated. This topic is very important for Dasa, because Dasa products have a product life of 25 to 40 years. If an aircraft is in production and in service, only visual changes are made compared to the large number of changes that occur during the development process. Aircrafts are largely uniform products and for the aircraft manufacturer and the owner the configuration management of every single aircraft is a continuous and difficult task throughout the aircraft's lifetime. Dasa believes, that this task is a typical domain of an EDM system.

Configuration management as a continuous and difficult task

Nevertheless a significant number of components will be improved and exchanged during the aircraft life time. In order to make improvements, the engineering department will need relevant information on how this improvement can be achieved with respect to existing interfaces. Therefore part of the EDM product model has to be checked-in into the engineering framework has to be checked-out into the EDM-system again. The end standardisation of these mechanisms will require a lot of additional work.

Improvements of SIFRAME functioning

SIFRAME functioning

The following list of demands for improvements of the SIFRAME functioning is reflects the most urgent needs:

- user specific desktop start providing a menu function to restore complete last used display arrangement,
- filter functions for display of objects, e.g. activities in flow, all activites assigned to one team or one design engineer,
- simultaneous display of the attributes of more than one object in parallel,
- if a new object is created, company default standards for all entries to be typed in should be proposed,
- activities in a cell-version-variant should have the attributes "completion date" and "percentage of work done"; a tool should exist to query and display this information in different levels of aggregation thoughout a project,
- context-sensitive on-line help functions,
- the type of graphic display for all structures and activities, etc. should be customisable by the user (e.g. up, down, left, right),
- customisable colours and fonts.

Integration of engineering environment

Three main topics are on the demands list:

- Standardisation of user interfaces, methods of interaction, arrangement and design of menus, buttons, windows, etc.
- Data integration between engineering applications.
- The exchange of information using STEP physical files can only be a first step, in future, application should be able to work directly on step objects.

Integration of cross-company business process environment

In the CONSENS project the main focus has been the engineering and to a smaller extent the manufacturing environment of company. But many other functions are vital for the overall business processes of a company like financial and administrative processes, project-management, cost estimation, and control. The key questions is whether or not the company has the aim to arrive one day at an integrated and interlinked product and process management model.

The key to the success of all hardware-producing industries is the product and its performance/life cycle cost relationship. A methodology and IT-support for the product model management thoughout the life cycle of the product is requested. The product structure therefore, has right in its own and should be the responsibility of the top management. The attributes attached to different items of a product structure should be developed using company-specific process chains. The challenging question is: Should an engineering framework like SIFRAME be used for the dynamic creation and change management of the product the associated processes and data exchange mechanisms for the core of information required across the basic company functions described above.

Support of multi-site, multi-company projects

More and more projects are based on world-wide work sharing agreements. CONSENS has demonstrated that a companywide process and data management is a near term reality. The natural step would be the extension into a multi-company environment. Again, the starting point would be the management of an evolving product model and a multi-company process- and data exchange specification and realisation. As Dasa will be active only in multi-company projects a research project based on this will be proposed in the future.

6.3 HIDROSOREFAME

J. M. Camacho

6.3.1 Description of Products and Projects

The Company

Founded in 1943, near Lisbon, SOREFAME started its activity by manufacturing equipment for hydroelectric and irrigation projects, which were then rapidly expanding in Portugal.

The history

SOREFAME later diversified into others fields, such as railway stock, but the design, manufacturing and installation of hydromechanical equipment like gates, valves, penstocks, screens, trash-racks, etc, remain the main areas of activity, not only for the home market, where about 95 percent of the existing dams are equipped with SOREFAME products, but also for many other important projects throughout the world.

Among them, the Douro river dams and Alto Lindoso, both in Portugal, Cahora Bassa in Mozambique, P.K. Le Roux, Drakensberg and Palmiet in South Africa, Roseires in Sudan, Maria Linda and Xixoy in Guatemala, Song Loulou in Cameroon are only a few examples, worth being pointed out for their importance.

Recently, SOREFAME went through a re-dimensioning process, giving way to two companies. The hydroelectric equipment production is now the main object of HIDROSOREFAME, with 660 employees.

The present and future

Last year, changes in the shareholder distribution brought the company into the international ABB group. Nowadays being re-structured in order to meet Group philosophy, HIDROSOREFAME is the only company of the group with specialised knowledge in this specific area, being considered a Centre of Competence.

Products and Projects

HIDROSOREFAME is specialised in the design, fabrication and erection of:

- Hydromechanical equipment for dams,

    - All types of gates and their operating equipment,
    - Penstocks and steel linings,
    - Valves (spherical, butterfly and hollow-jet),
    - Screens and trash rack rakes,
    - Equipment for irrigation projects,
    - Mobile gantries,
- Electromechanical equipment:
    - Turbines and pump turbines for hydroelectric power stations,
    - Generators and motor-generators.

Sample of product

Several of the gates and penstocks supplied represent special technical challenges, not only due to their dimensions and water heads involved, but also for durability and performance characteristics.

For example, let us look at some data concerning the main equipment supplied for the Alto Lindoso project:

- Steel linings for 2 vertical intake shafts – 5.1 m diameter, 324 m service water head, 2200 ton total weight,
- Two butterfly valves – 5.75 m diameter under 89 m of water head and 279 ton of weight each (which are among the largest in the world),
- Two spherical valves – 3.40 m diameter under 324 m of water head and 335 ton of weight each (the largest valve of this type in the world),
- Two 350 MW power generator units.

The hydromechanical equipment is fully designed and developed "in-house". The electromechanical equipment is manufactured under licence and its project time and object is shared among several partners. All equipment is designed according to the specific customer's requirements.

International engagement

HIDROSOREFAME is presently engaged in projects for Portugal, Iran, Marocco and Brazil. The lifetime of one product is typically more than thirty years from the initial concept to ist full revision or refurbishing, with a project mean lifetime of five years.

6.3.2 Description of the CONSENS Environment

The products developed at the company follow an internal process consisting mainly of five phases:

Phase 1: Conceptual Design
Still performed in the bid stage, it involves senior engineers and designers, outlining the basics principles and the company view of the Technical Specifications issued by the client.

Phase 2: Preleminary Design
This is started as soon as the order is confirmed, it triggers simultaneous action of several departments involved, in order to assist in the design from the beginning.

Phase 3: Detailed Design
Project, dimensioning and stress analysis closely follow the detailed drawing and drafting, in order to release, as soon as possible, the elements needed for the fabrication start-up (when in-house) or subcontractor's contacts. Release control is critical at this phase.

Phase 4: Installation and Testing
Installation, testing and maintenance instrutions should now be ready for the final stage, where very few changes are possible. Furthermore, the verifications and tune-up of several components are performed.

Phase 5: Service and Custumer Support
During the whole life of the product, it is necessary to fulfil the customers' demands concerning updating, expansion of capacities, maintenance and refurbishment.

The next four IDEFo schemes show part of the Design and Engineering Process at HIDROSOREFAME. The first three describe the upper levels, illustrating how the main phases are connected and the last one describes the engineering activities closely related to the demonstration.

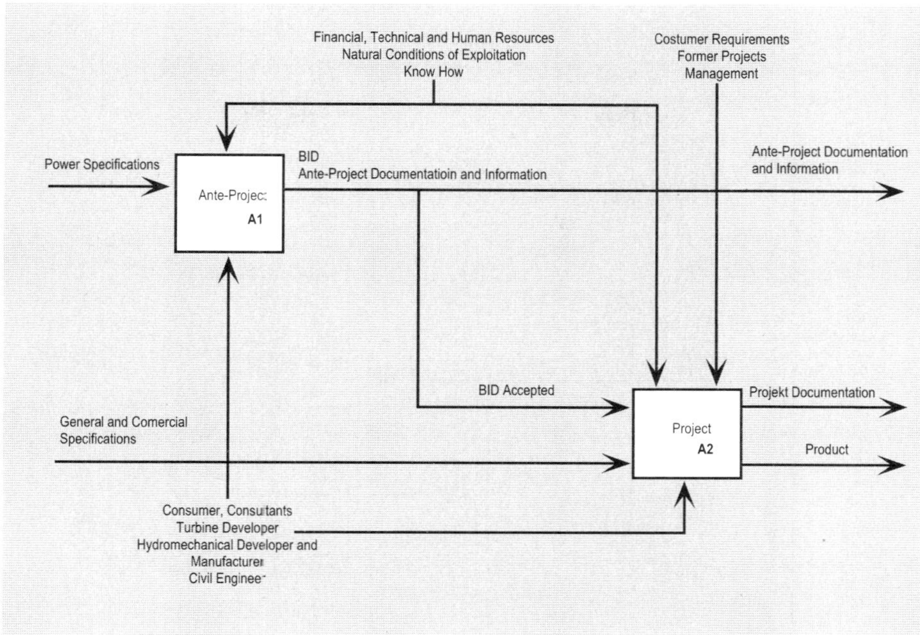

Fig. 6.8: Project of a new product

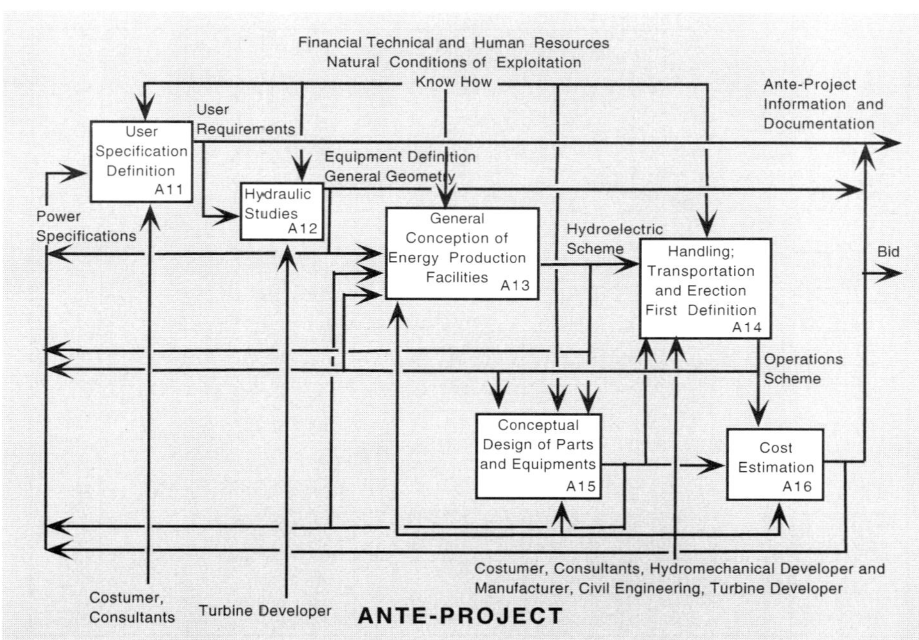

Fig. 6.9: The ante-project A1

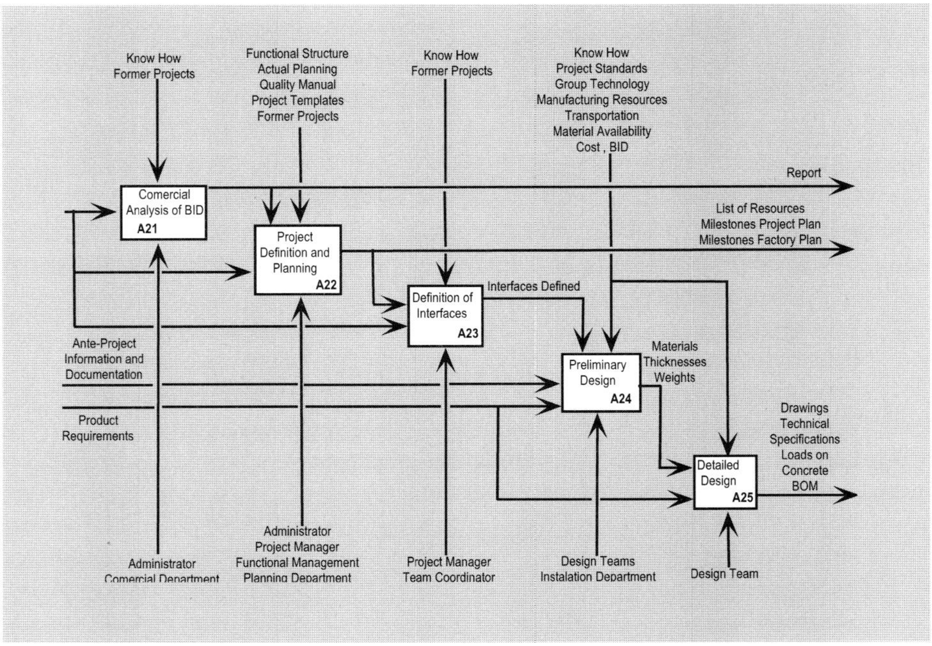

Fig. 6.10: The project A2

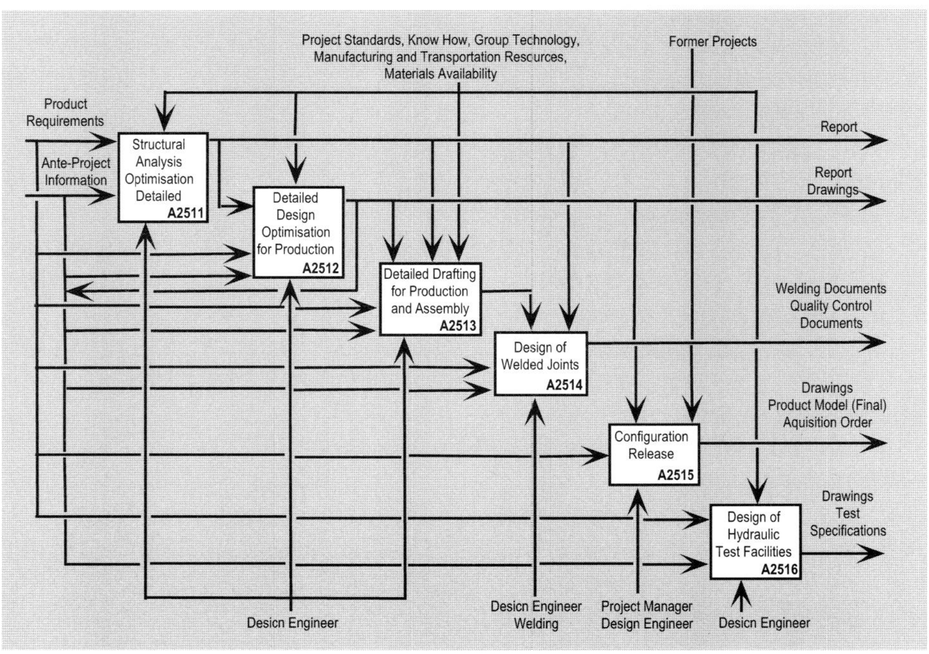

Fig. 6.11: Structural Analysis and design for production

6.3.3 Description of the Test Environment

The HIDROSOREFAME scenario describes the design process during the development and mechanical design of a turbine spiral-casing. Two parts are to be distinguished in the spiral-casing:

- the central part or stationary ring,
- the strakes (volute) which guide the water around the stationary ring.

Short description of the design process

The stay-ring is constructed from cut, shaped and welded steel plates which have been assembled together. The strakes also undergo forming which is carried out on a handling press. In addition, strakes require very accurate marking. For the spiral-casing, the main design process is as follows:

- From the hydraulic studies the part-fixed boundaries and loads are defined.
- The geometrical model is drawn, using the I/EMS from Intergraph.

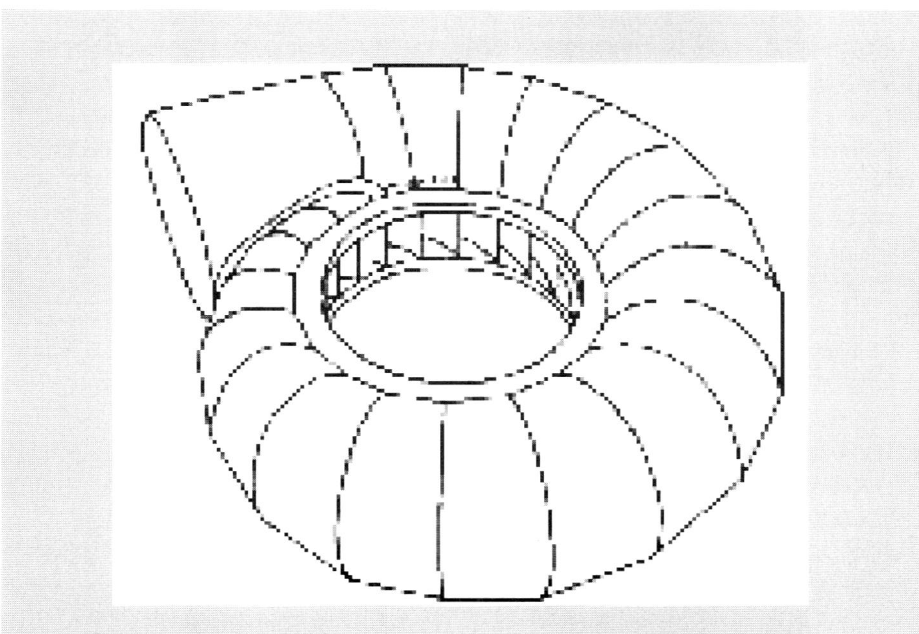

Fig. 6.12: Spiral-casing for a large turbine

- Check design variants for assembly cost using ASCET from IAO.
- Design of the part to fulfil functional and assembly criteria by using IST.
- Detailed Design for Performance.
- Check and optimise the manufacture of the part by using Detailed Design for Production from WTCM.
- Final check merge with other parts of the turbine using I/EMS.

The following schemes will show, in a short descriptive summary, the tasks to be performed in the Demo Scenario.

The detailed scenario include not only the engineering tasks but also the necessary CSE environment. This has been done by using SIFRAME and EPM functions, shaped to use the HIDROSOREFAME culture, organisation, resources and know-how.

The following flowcharts show some of the demonstrator's activities and how they are connected. The tables describe the whole set of activities, the persons in charge, the software modules used and their inputs and outputs.

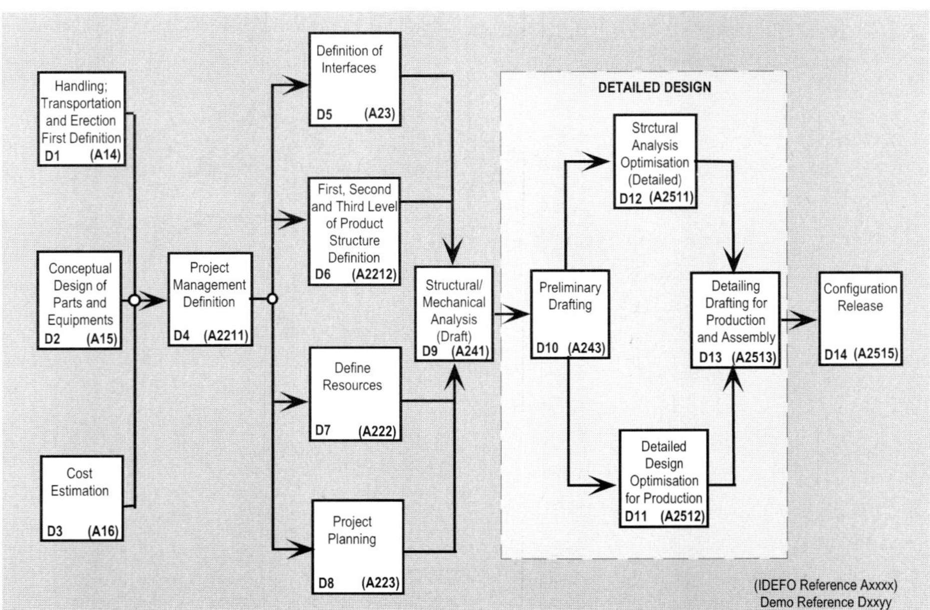

Fig. 6.13: Flowchart of Demonstration

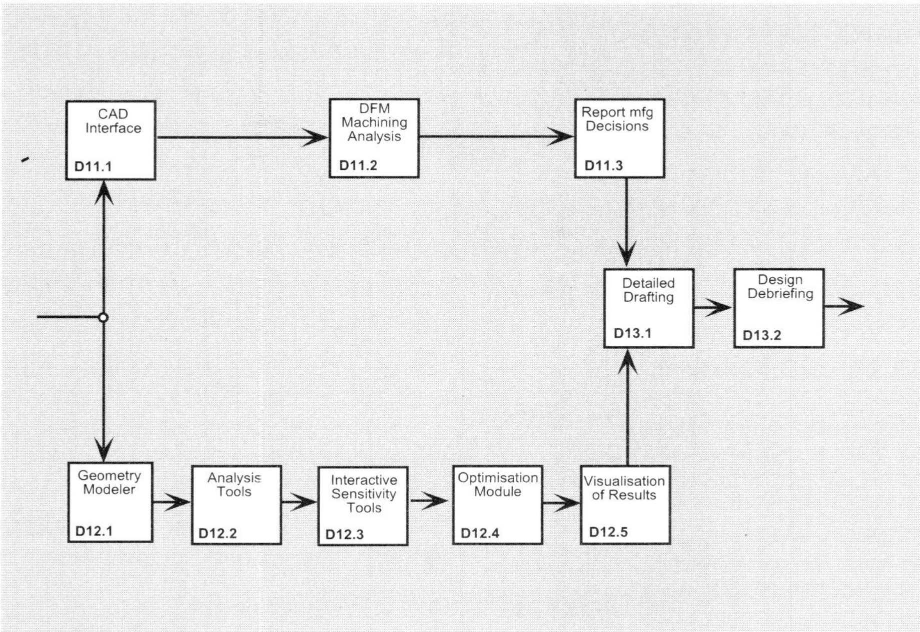

Fig. 6.14: Detailed design

6.3.4 Improvements of the Product Development Process with Respect to CSE

Structural Factors at HIDROSOREFAME

There are many different solutions to reach the aims of CSE. For selecting a particular solution, it is necessary to take into consideration the environment into which the product development process is embedded. This environment can be described by a set of factors, namely ìstructural factorsî.

In order to ensure a structured and consistent procedure, the “The General CSE Matrix” is used as a flexible instrument. The PSI (Parallelisation-Standardisation-Integration) strategy, concerned with the product development process, has a certain interpretation for each area.

Behind each element of the matrix (a CSE element) there is a structured set of approaches, methods and tools that allow to support a specific PSI-strategies in a certain area.

These input factors were collected by a questionnaire filled in by people from different areas. For HIDROSOREFAME this activity involved senior staff directly related to areas such as: commerce, quality, project, engineering and installation.

The results are as following:

Company environment concerning CSE

Product Complexity
Because of the individuality of the product, the design contains key elements that are state-of-the-art or have large numbers of state-of-the-art parts with many sensitive interfaces.

Project Structure
The project size is moderate, requiring hierarchical reporting structures and more formal communications. Subgroups have specialised assignments. Informal communication channels are available.

Project Type
Platform or generational development projects. These are the platform of core development projects. Typically, they have a design life of several years and establish the basic architecture for a set of derivative projects.

Market Structure
HIDROSOREFAME products are useful to each customer as a system, because the company produces and installs all the equipment needed for the production of electricity.

Production Type
Individual production, i.e., every product is adapted to the wishes of the customer. Every dam has its own characteristics, therefore each product has to meet specific requirements and the number of equal products per dam is very small. Consequently, each product is „one-of-a-kind".

Resource Flexibility
Resources are allowed for a limited solution of problems within the process. The current hierarchical structure allows limited resolution capabilities of the staff. Human resources are also limited in number, but are expected to increase soon, especially in the design area. In what con-

cerns to hardware/software resources, there is a network already installed which is not yet working at its full capacity due to the absence of some integration applications.

Technology Structure
These are products that use-state-of-the-art technology to a large extension.

Design Type
The design involves the adaptation of a known design (the solution principle remaining the same) according to the changed task. Here, original designs of parts or assemblies are often needed.

Resource Complexity
Information technologies are used in every area of the company. Managing interdependencies are critical to the infrastructure's performance. With the present process of company re-structuring, these difficulties are brought to light.

Location of Development and Production
This is a multi-departmental organisation in a wide area configuration. The company manages a large number of sub-contractors, some of which are in foreign countries. Therefore, the production sometimes takes place far away from the development. The assembly of very large parts is often done on site, although blank assembly is usually done in the workshop.
Evaluating of these structural factors, the Focused CSE Matrix was generated, showing the company's problem-areas. This matrix is the result of a weighting process, in wich the CONSENS methodology evaluate the importance of each CSE element to the company characterised by the set of structure's factors.

Priority CSE Matrix

Matrix	Project	Process	Product	Organisation	Resources	Culture
Parallelisation	29	**32**	29	26	27	30
Standardisation	24	24	**32**	28	28	23
Integration	26	31	**32**	28	28	**32**

From the analisys of the focused CSE matrix and from the existing knowledgeof the company, it was decided to concentrate the attention on the following problems-areas:

The problem areas extrated from the focused CSE matrix

Culture/Organisation

Considering the company's current situation, part of the ABB Group, the problem-area of culture/integration is pressing and can be analysed at three different levels:

- Firstly, HIDROSOREFAME is involved in a completely new culture (ABB's) and people feel the need for integration.
- Secondly, the company has to integrate its sub-contractors into its own culture.
- Finally, an internal culture integration is needed, through the implementation of a project-management System, and the strengthening of the project manager.

Process/Parallelisation

A part of this problem-area is also a consequence of this situation – processes will take place in time, following a global workplan, in which several processes (as well as projects) will take place at the same time – sub-processes of the same project will be developed simultaneously in several companies of the Group. One of the problems, which is currently being solved, concerns with the non-interchangeability of data between design and product structure management tools and project-management tool.

Product/Standardisation and Product/Integration

Nevertheless, the heavy work already done in the product/standardisation area, using parametric design, group technology and standard part library methods, the company has to re-enforce the work that it has been developing over several years.

These are also problem-areas for the same set of reasons- the need to integrate HIDROSOREFAME's work into a broader project, the need for integration into the work done by others and the common part concept to be introduced to increase performance ratings.

Finally, a word has to be said about the special attention that must be paid to the organisation/integration area with respect to the work done by sub-contractors.

Company goals

Using a structured interview method, the upper levels of the company management were asked, on the basis of TQC-trees (Time,Quality,Cost), to set priorities regarding the main goals concerning time, cost and quality, sub-divided into the aspects of design, production and marketing.

HIDROSOREFAME company goals

- **Time:**
  Reason: each product has a very strict timetable for its completion.
  Value of importance 3
  - Design time: value of importance 5
    Reason: design time is fundamental to the project schedule.
  - Production time: value of importance 3
    Reason: production time is important for the project schedule.
  - Marketing time: value of importance 2
    Reason: marketing is not so important for this activity, but must be considered for the timely delivery of the equipment.

- **Quality:**
  Reason: quality isn't a goal, as the company has achieved the highest quality standards in its products. Therefore, although it is important to maintain this quality level, it is not as important as the other goals.
  Value of importance 2
  - Design Quality: value of importance 4
  - Production Quality: value of importance 4
  - Marketing Quality: value of importance 2

- **Cost:**
  Reason: in a company working in this area, cost control is very important, especially with regard to companies in a very competitive world market.
  Value of importance 5
  - Design cost: value of importance 3
    Reason: adaptative design and product standardisation lead to a lower design cost.
  - Production cost: value of importance 7
    Reason: this is the area in which costs have to be controlled more effectively.
  - Marketing cost: value of importance 0
    Reason: no importance for this kind of market.

Derived from the questionnaire, concerning structural factors, these results were used in the weighting process to define the focused CSE matrix.

Next step on the "CSE Road Map"

According to the next step on the "CSE Road Map", the Priority CSE matrices were defined. These matrices are a result of the reduction process. Their contents are lists of approaches, methods and tools applied to the former selected elements to realise CSE, under consideration of the company specific goals.

These goals were used to prioritise, according to the knowledge of Consortium, the set of tools and methods.

Priority CSE Matrix for Culture/Integration

Elements	Time	Quality	Cost
group feeling	30	48	0
superiors as servants	30	40	0
delegation of responsibilities	**60**	32	0
one-project one-room	51	20	**100**
support group formation	**72**	32	50
visible signs	21	0	0
involving private life	30	24	0
care of brood / parents in company	60	52	**80**

Following the interpretation of the results, it was decided to focus the attention on the CSE elements listed below:

Delegation of responsibilities is to improve

The delegation of responsabilities is somehow a cultural lack within the company. This has come about in the last years due to a strong policy of cost control, which has implied a centralisation of responsibilities. A "support to group formation" is already taking place. Although "Care for the family" presents very high values in TQC factors, the company already has a very supportive environment in this respect.

Priority CSE Matrix for Process/Parallelisation

Elements	Time	Quality	Cost
design for manufacture	54	12	**70**
design to cost	24	12	**105**
design for quality	0	40	**70**
product structure definition	21	12	30

"Design for Manufacture" and "Design for Quality" are methodologies already in use at the company, which are expected to help reducing costs. However, they have to be re-enforced and reviewed due to changes in production and the design environment. The product Structure Definition is a method which has already been in use for many years.

DfM and DfQ is needed

Priority CSE Matrix for Product/Integration

Elements	Time	Quality	Cost
design for manufacture - DFM	54	36	**135**
design for assembly - DFA	54	36	**135**
DFx	54	40	**135**

DFx modules are, in general, expected to help reduce costs. Time can also be reduced with the help of these modules. As the required level of Quality has already been achieved in this area, it is expected that it will be maintained.

Priority CSE Matrix for Product/Standardisation

Elements	Time	Quality	Cost
parametric designs	54	8	30
design handbook	30	48	65
software prototyping	75	20	50
EDI	6	28	65
DFM	33	20	**105**
DFA	33	16	**105**
STEP	15	16	50
standard part libraries	**90**	20	80
group technology	**90**	32	80
product planning	72	24	50

"Standard Part Libraries" have been developed over several years, originally in paper form. They are now available on the CAD System. "Group Technology" is implemented at different levels depending on the equipment type. DfM and DfA are considered to be of great importance for helping the cost reduction in the product standardisation area.

DfM and DfA are important

6.3.5 Future Improvements

For many years, HIDROSOREFAME has had a traditional and supportive environment to Concurrent and Simultaneous Engineering, not in a very formal way but as a cultural trace of the organisation.

People are ready to start with CSE

From this point of view, people are ready to support a more thorough implementation of CSE practices. The new company strategy, for a bigger market share and power, must make important changes in order to accomplish these goals.

In the current company policy, CSE is of great importance in reducing:
- production costs,
- design time and costs,
- and production time.

In previous chapters we have defined the methods to be used:
- design for manufacture and assembly,
- design to cost,
- standard part libraries,
- group technology,
- support to group formation,
- delegation of responsibilities,
- etc.

They aren't isolated from the overall changes in the product development process, but accompany the strengthening of the project managers and the introduction/ implementation of tools for data and process integration.

However, HIDROSOREFAME hopes that the CSE methods will assist the company to face the challenges and competition of the global marketplace.

6.4 ALCATEL BELL

E. Engelborghs

6.4.1 Description of Products and Projects

The Company.
Alcatel Bell is part of the international Alcatel Group. Alcatel is a European company with head offices in Amsterdam and Paris, and Alcatel is the world's biggest telecommunication company with divisions in 23 countries and commercial activities in 110 countries. Alcatel originated from the Alcatel Alsthom (previously CGE (Compagnie Généal d'Electricité)) and the ITT Corporation.

Alcatel Bell, the company

The Alcatel group is active in the domain of switching systems, company communication, transmission, aeronautics, defence, information technology and services and components.

ALCATEL BELL Components Division produces piece parts in a variety of locations throughout Belgium. The division, which works for external customers as well as for companies in the ALCATEL group, has built its reputation on the cornerstones of technical excellence and unrivalled customer service commitment.

The components division

Metal and plastic parts as well as precision tools are developed and produced to the highest standards using the state-of-the-art machinery at the Hoboken plant. Alcatel Bell Telephone is actively conducting research and was already involved in the BRITE projects MODESTI and PROMISES on mould design and manufacturing. Alcatel Bell joined the CONSENS project in 1994 as the Belgian end-user test site.

The Products.
The component division site at Hoboken produces all kinds of metal and plastic components, ranging from mould manufacturing and milled parts, in general over sheet metal to plastic injection parts. This division works for other partners in the Alcatel group as well as for external companies.

The development process

The development process.
The product development process at Alcatel Bell Telephone consists of 3 major phases: the quotation phase, prototyping phase and the development for production phase. The following picture describes this product development process.

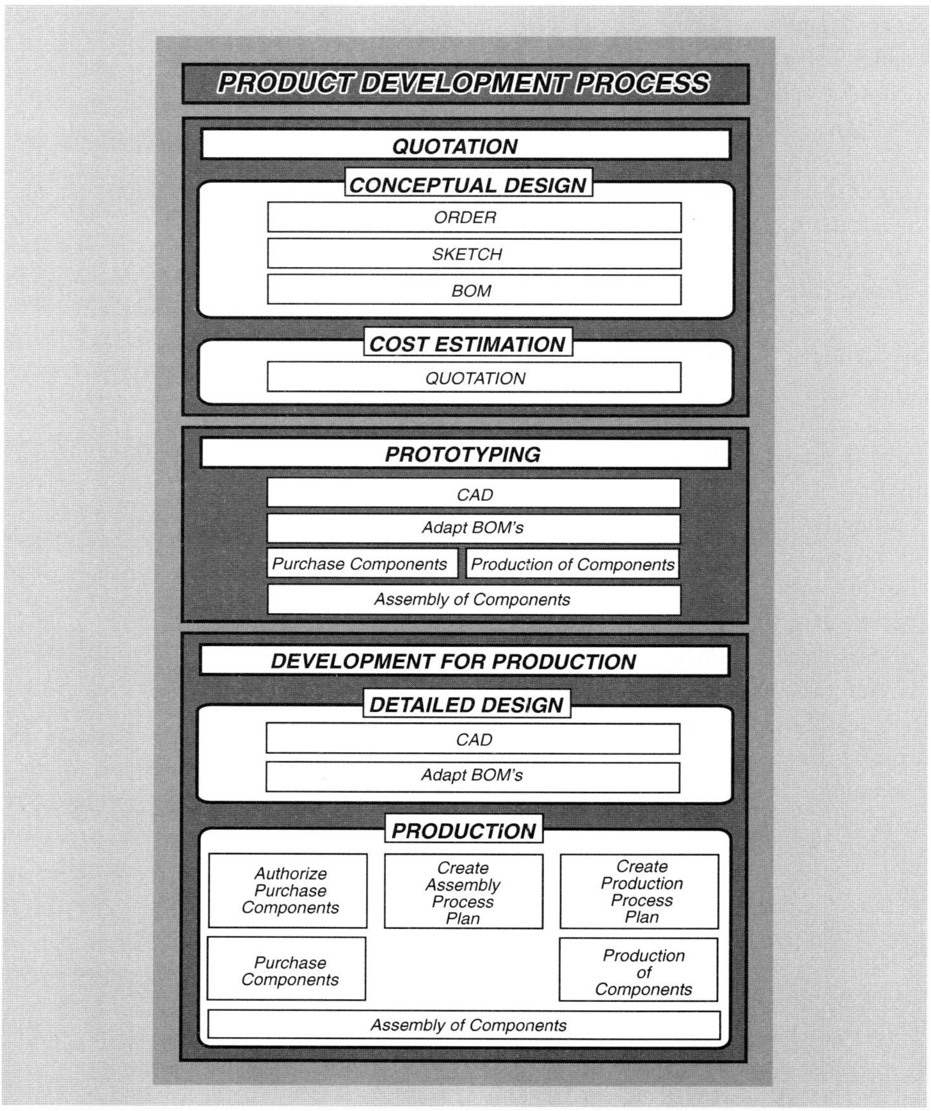

Fig. 6.15: The ALCATEL BELL product development process.

During the bid stage, first a project responsible (Project Leader) is chosen, who will be responsible for this product during the whole product life-cycle. The first task of this project leader is to make a conceptual design. Part of this conceptual design phase is the order acquisition: When Marketing spots a possible new client, the first thing to do is to gather the User Specifications. Next, the project leader makes a sketch of the product in co-operation with the marketing department who has a good view on the user wishes and the industrial engineering department, who has a better view on the technical feasibility (manufacturability) of this project. Based on these initial sketches, a first draft of the production and purchase Bill Of Material is made. These BOMs contain the list of parts to be manufactured inhouse and to be purchased abroad.

The bid stage

A Cost Estimation is made by the project leader, together with the Industrial Engineering Department, the Purchase Department and the Marketing Department. The first three Departments make a cost estimation based on the conceptual design and the BOM. Next, a quotation proposal is completed by the Marketing Department in collaboration with the project leader, taking into account several external influences such as the clients' profile, the competitors etc.

When the complexity and the batch size of the product are important enough (batch size approximately greater than 30), a prototyping stage is inserted between quotation and production.

When marketing has received the confirmation that the customer has chosen Alcatel Bell's bid, the sketches, drawn by hand, are translated in a CAD drawing. Next, the BOM are adapted with some help from the purchase department. The components are purchased by the purchasing department, produced and assembled in the prototyping shop.

Prototyping stage

During the detailed design phase the project leader has to collect the comments on the prototype from the customer (via marketing), the prototyping shop and industrial engineering. On the basis of these comments , the CAD drawings are adapted. A final version of the BOM is developed taking into account the technical & human resources and the possible suppliers. The components to be purchased have to get authorisation from the purchase department, and industrial engineering has to create the production and

Detailed design

assembly process plans before the components can be purchased or manufactured and assembled.

6.4.2 Description of CONSENS Enviroment

The video-server box as demonstration product

The Alcatel Bell demonstrator scenario describes the design process of the video-server box A-1, an assembly of sheet-metal, plastic and milled components.

The demonstration scenario

The development process used in the demonstrator scenario is very like the current development process used at Alcatel Bell for this type of projects. Only the prototyping stage is omitted, since it should be one of the aims of a concurrent and simultaneous system to make a design correct from the first time. The CONSENS modules DfM, DTC (-FCCS), together with the user specific DfA-tool "IDAM", should guide the developer during his design. The CONSENS system introduces the use of group technology in the product development process.

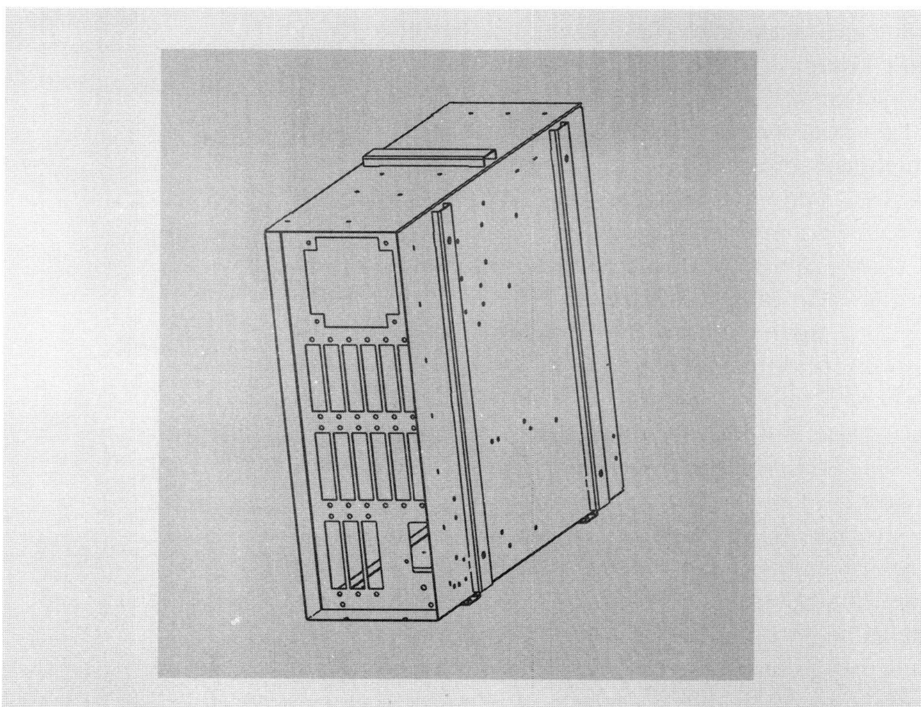

Fig. 6.16: Video-server box A-1.

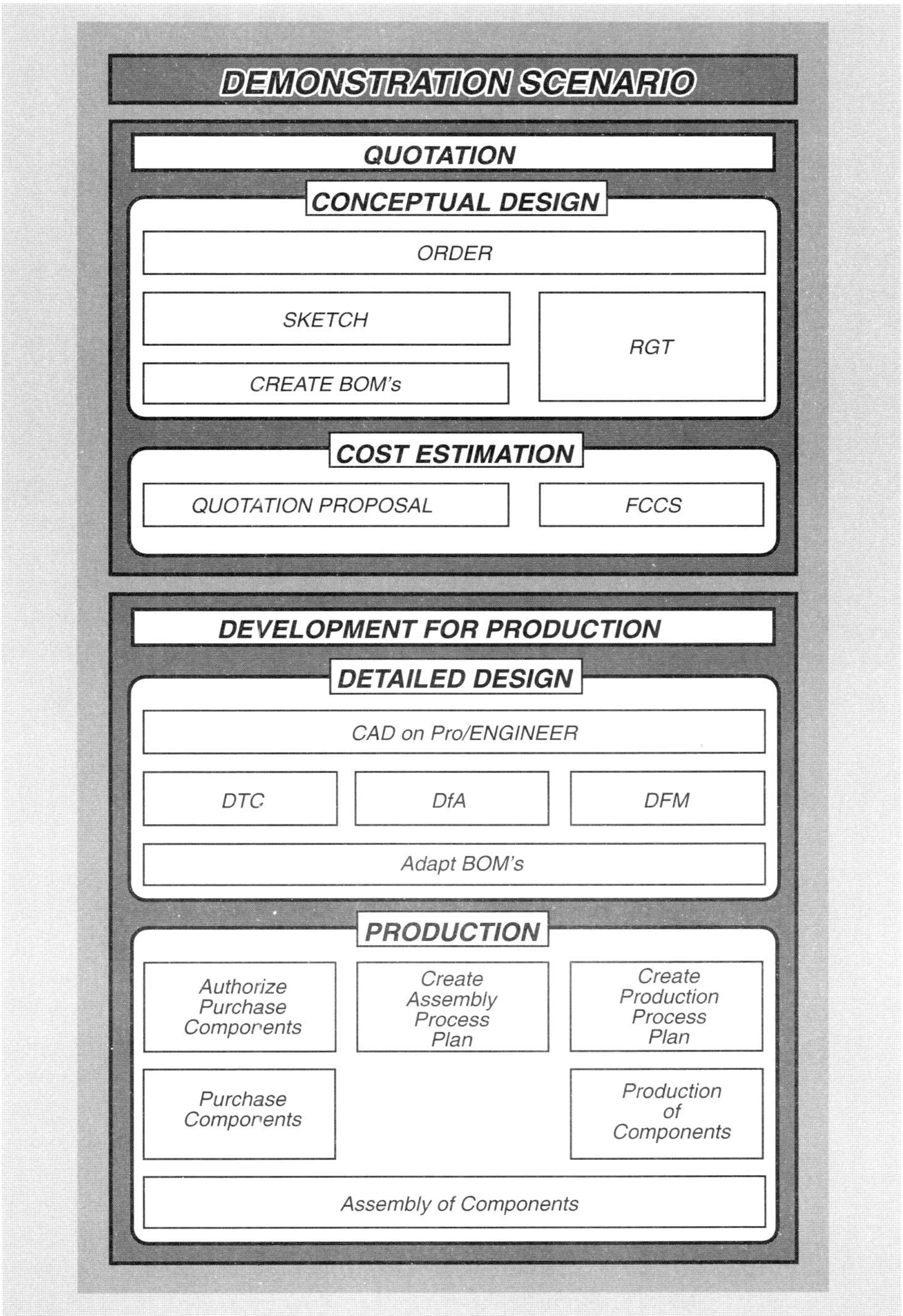

Fig. 6.17: The ALCATEL BELL Demonstration Scenario

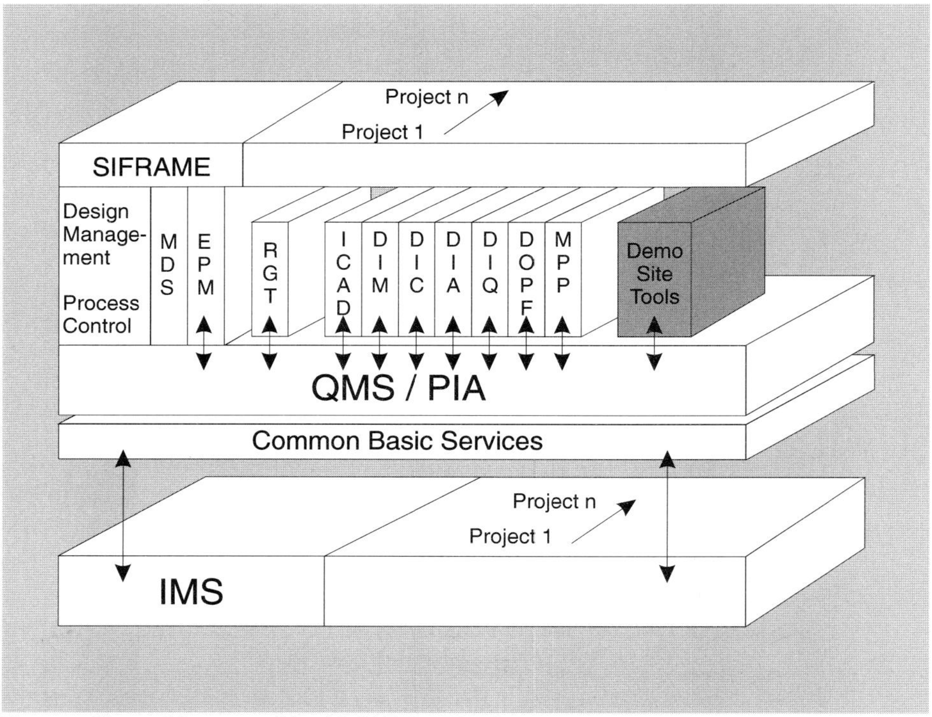

Fig. 6.18: The architecture of the ALCATEL BELL demonstrator

During the bid stage, the project leader searches in the CONSENS Group Technology module RGT for a previous similar design. If he finds a similar project he can use the results of this project in the cost estimation and, when the Alcatel Bell bid is accepted, also in the design phase of the product. The cost estimation will be done using the FCCS module on PC. The BOMs are made on a EXCEL spreadsheet as usual. During the detailed design phase, the product is detailed on Pro/ENGINEER. The DfM, DTC and DfA modules give the project leader some remarks on the manufacturability, cost effectiveness and assemblability of the design. The designer should adapt his design to these comments. This cycle can be repeated as many times as necessary. After the detailed design phase, the normal sequence of creating process plans for production and assembly and the authorisation demand for purchase is continued. In the final phase the components are produced, or purchased and assembled.

The bid stage

Electronic Prototyping & Detailed design

6.4.3 Description of Enabling Information Technology and Improvements of the Product Development

In general, the framework (Siframe together with PIA) should improve the teamwork. But too many integration platforms take too much time to define users, projects, protections, integrate other applications etc. In SIFRAME, the configuration should be much more simplified and easy. The time consuming work of following up a number of projects with respect to input, output and their progress, should be reduced so the project leader has time free to do some more useful work. A very interesting issue is the determination of the break-even points between the two previous remarks: At what number of (similar) projects it becomes profitable to use this integration platform. When will the time you lose when defining the project in the integration platform be gained back during the development of the projects?

Easy project input and update

The use of a STEP-compliant product model gives certain guarantees with respect to the use of this information when the design tools are been changed in the future. The object-oriented database should open possibilities to work in parallel on the same product information without having problems with safety, flexibility and answering time. Change management and release control has been up until now done in a very time consuming and bureaucratic way. The CONSENS system should show the IT-support of change management and release control in an effective way with respect to man hours and costs.

Object-Oriented Database

The order reaction time, the delivery on time and to the right person should be one of the major possible improvements of the CONSENS system. For this purpose, the FCCS system and the RGT systems are introduced in the demonstration scenario. Group Technology (GT) is the concept of identifying and bringing together similar parts in order to obtain advantages from their similarity for designing, manufacturing and process planning purposes. Although Group Technology as a concept is considered to be most valuable, a lot of companies have encountered serious problems trying to implement it.

Order reaction time

The key development in this module is the elaboration of a Computer Automated Group Technology (CAGT) software. By automating the input from CAD, consisting of features and attributes, human interpretation of engineering drawings during the coding process is excluded. A retrieval tool should automate the retrieval of existing similar parts, drawings and manufacturing sequences based on the internal GT representation. The developed internal representation stored in a relational database is a successful start to an engineering database with links to process planning, material requirements planning, manufacturing, purchasing, (...) the two FCCS modules will be used in the Alcatel Bell demonstrator.

Cost Calculation

The FCCS module on PC will be used for a cost estimation during the quotation phase. Making accurate cost calculations is usually a problem for companies when producing in small batches or making piecework. For the moment, cost calculations are usually hand-made on the basis of global rules of thumb and experiences from the past. Accurate results are not always evident in this way. But very often it is possible to formalise these rules of thumb and experiences and put them into a computer model. Such a model can be very simple, offering in many cases more accuracy in cost calculation. By entering a few parameters concerning the global characteristics of a product, the cost can be estimated.

The advantages of a systematic cost calculation are obvious :

- accurate cost calculations can be made very quickly,
- the experience can be recorded.

The "Flexible Cost Calculation System" (FCCS) provides a systematic methodology for cost calculation, implemented in a user-friendly computer program. The program allows the user to enter and change different cost calculation methods (rules of thumb, experiences) for different kinds of products.

The specifications can be summarised in three topics:

Cost Calculation

- Method : group technological -parametric.
- User definable reference pieces or families organised in a tree structure.
- User definable methods linked to reference pieces.

- Screen layout can be adapted by the user (input checks, radio buttons, check boxes, ...).
- As to complex calculations, sub methods can be defined up to 4 levels.
- User definable variables and formulas in a simple way by analogy with spreadsheets.
- Multidimensional lock-up tables for experience values, rates, standard costs, etc.
- Decision tables allow a decision logic to be entered easily
- User definable trees enable support to the user in selecting features, machines, operations,...
- Possibility of retrieving previous cost calculations for comparison.

Quotation Management

- Customer files with user definable attributes.
- Making and printing quotations : user definable format by means of a powerful report generator.
- Classifying and retrieving previous quotations for any combination of criteria.
- Invoicing by means of a report generator.

Design for Production

A third important result of the CONSENS system is the decrease in time to production. The DfM, DfA, FCCS tools are used for that purpose. The DfM module should be used in several phases of the product life-cycle.

Design for X

- In the conceptual phase, the system provides information for specifying the functional requirements of the product. The designer has access to libraries of standard components and product families. Thus, one reduces the number of new designs created. Also, the basic shape of the product is defined during this phase. For Concurrent Engineering, this data is sufficient for starting parallel design sessions.
- During the detailed design of each product, the provided information focuses primarily on the manufacturing information. Based on geometrical and tolerancing information, the system indicates expensive, non-standard operations or impossible specifications. Thus, the design is improved on-line and specific design decisions can be stored.

- During both phases, the operator can start a cost evaluation of the current state of the product design. Using "Group Technology" techniques, he can compare his design with previous, similar products and with the original product specification. Based on the production cost actions can be taken for changing product specification or detailed product characteristics.

Alcatel Bell's DfA tool

The user site DfA tool will illustrate the possible cooperation between DfM and DfA in a CSE environment. The integrated FCCS module will be used together with the CAD system and the DfM module during the detailed design phase. Since product costs are determined approximately for 75% in the early stages of design, all measures for cost estimation and cost reduction should be used in the early stages of the design phase. For the initial cost estimation, a simplified view of products and processes is used. Therefore, the calculated cost is only a rough estimation of the actual costs of the new product. The accuracy of this estimation is directly related to the accuracy of the input data. E.g. if the product description is very detailed, the cost estimation will be accurate, as is the case for the design of single parts. For completely new products or processes, the data are inaccurate and the resulting estimation has a low probability. By integrating this cost calculating (and thus cost indicating) module together with the CAD-system and the DfM module, the ambiguities which arise with human interpretation are avoided and the time consuming work is eliminated.

6.5 AEG Schienenfahrzeuge GmbH

L. Bertling, C. Köhler

6.5.1 Description of Products and Projects

Company Profile

AEG Schienenfahrzeuge GmbH Hennigsdorf (ASF) is one of AEG's traditional locations. As early as 1910, AEG acquired the 770 000 m^2 plot of land near Berlin. In 1913, the first locomotives were manufactured. Today the company employs over 3000 on these premise which have increased to about 1 million m2, and has an output of currently about 350 tractive units per year. Sales in 1994 were DM 430 million. This makes it the largest manufacturer of locomotives in Germany.

ASF offers a comprehensive spectrum of products and services ranging from complete tractive units and components through field service to transport solutions. ASF develops and manufactures complete tractive units with electric, diesel-electric or diesel-hydraulic traction:

Products

- Locomotives, power cars, and multiple-unit trains for long-distance services,
- Locomotives and multiple-unit trains for light and heavy-duty regional and suburban services,
- Industrial and shunting locomotives for industry and open-cast mining,
- Mine Locomotives,
- Special-purpose vehicles.

The range of services comprises:

Services

- Electric and electronic systems and components for main-line, industrial and shunting locomotives as well as for multiple-unit trains and special-purpose vehicles.
- Mechanical equipment, such as car bodies, bogies, traction systems, and interior furnishings for locomotives and multiple-unit trains.
- The performance of maintenance, main overhauls, retrofits, and refurbishing of tractive units.
- Consultancy for transport operators on future-oriented

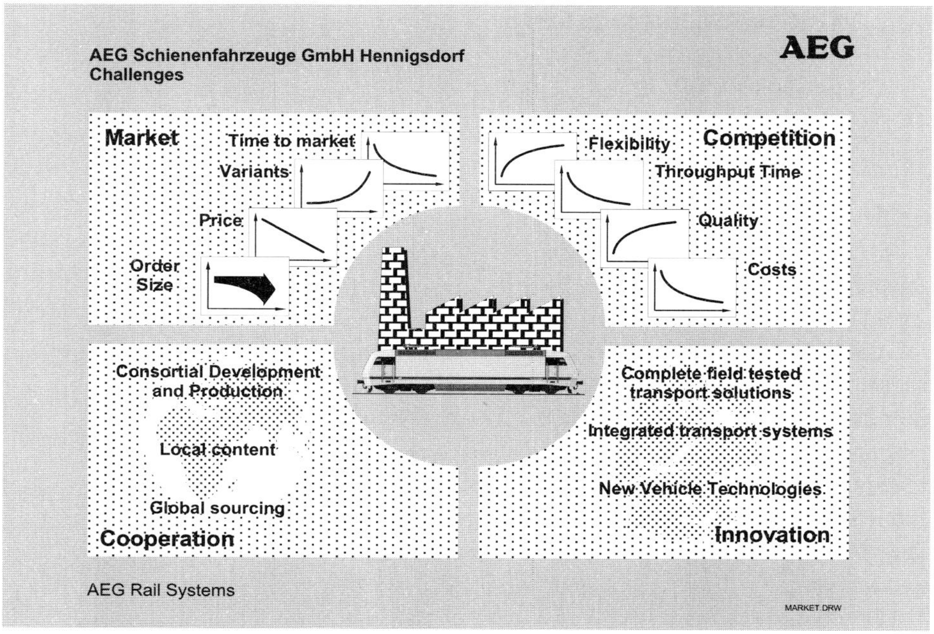

Fig. 6.19: Challenges for ASF

and tailored transport solutions. ASF analyses widely differing field conditions and so provides a sound basis for their client's decision making.

Challenge and Potential:

Market changes

- The market for rolling stock has developed dramatically in the past few years.

Newly emerging markets such as those in the Far East as well as developments in traditional markets, such as the privatisation of the German railways, are major factors influencing this process. These developments have confronted ASF with new challenges. The market demands the production of vehicles in widely varying order volumes (2–500 vehicles) with a large number of variants and with short periods of delivery. The surplus capacity world-wide has resulted in falling prices. Increasingly, the product that customers expect is no longer uniform rail vehicles but rather innovative deployment systems for resolving transport problems. Cooperation with other firms is becoming more and more important against this background. In many markets, products can be sold only if sufficient local content is assured. But even on domestic markets, job orders are often filled by design and/or production consortia. The following factors of ASF performance can be derived from the above factors:

Winning edge

- Provider of complete vehicle systems as a full-range provider (from trams to heavy industrial locomotives),
- Tested application systems for all types of transport (freight, inner city, regional and long-distance passenger),
- Flexible products and production,
- Optimised cost structures (>30%),
- Reducing "Time to Market" (>40%),
- Improving quality (fulfilment of customer demands),
- Capability of technology transfer in cooperating groups.

The development of totally new structures in the production and the design of products and the optimisation of process chains are major means of achieving these goals.

6.5.2 Description of the CONSENS Environment

6.5.2.1 General Description of the Old Engineering Process

Standard process steps

Until 1994, the product development process used at ASF was broken into several steps which are described and detailed in the following:

Phase 1: Draft Specification
The product development process starts with a customer's request for a new product. A draft specification of the main elements (i. e. car bodies, bogies, electrical and pneumatic facilities) is worked out on the basis of documents of similar, already completed projects. This results in a quotation to the customer.

Phase 2: Product Specification
In this phase the product is specified in a rough way. The main components of the product are fixed; this concerns especially the outer design, dimensions and power. Docu-

ments and information are worked out to function as a guideline for the following phases.

Phase 3: Design

The design phase is characterised by the development of the product. First the design concept is worked out. It functions as a basis for the detailed design, FE analysis and cost calculation. Within these processes there is a permanent data exchange in order to get an iterative improvement in the design. The design phase ends with the working out of the manufacturing drawings and bills of material. This results in the documentation of the product. Material with a long requisition time is ordered in this phase.

Phase 4: Operations Planning

The main point of this phase is the planning of the production processes and the beginning of the nc programming. The planning of the production equipment is done in parallel to production process planning.

Phase 5: Material Requirement Planning

Parallel to the operations planning the requirement of material is planned based on the bills of material and the production schedule. This allows early production of prototypes or pre-series vehicles.

Phase 6: Test Planning

In order to ensure the quality of the product and processes the test planning is elaborated parallel to the operations planning.

Phase 7: Prototype or Pre-Series

Prototypes of the product or special components are manufactured and tested. In this phase, experience from pre-series manufacture is iteratively integrated into product design in intensive cooperation between development and production in order to optimise the cost structure and functionality of the vehicle

Phase 8: Series Start

Production starts when the optimisation of the product and production processes has been completed and the cost,

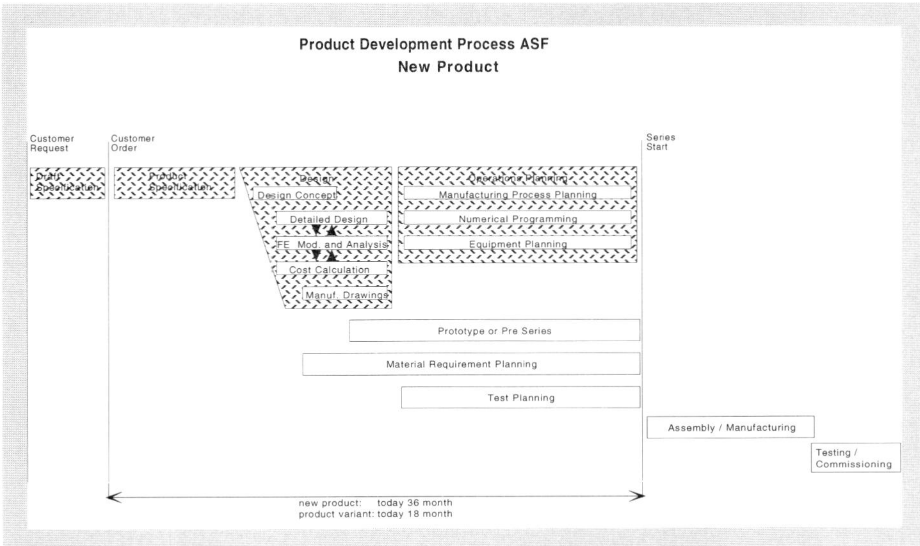

Fig. 6.20: Product development process ASF

deadline and quality targets have been achieved. A phase of continued improvement begins.

Characteristics of the standard engineering process

The current product development process at ASF is characterised by:

- serial working-out of the process steps,
- separated and partly non-standardised working methods between different departments,
- workflows beginning at different times, no coordination of parallel process steps,
- document flow partly without coordination.

6.5.2.2 Improvements and Specific Objectives through Implementing CSE

The redesign of the product development process described in chapter 6.4.2.1 is realized through the implementation of three main subjects. These are:

Main redesign subjects

- a segment-oriented structuring of the production,
- standardised process flows,
- Production and design oriented product structuring.

Segment-Oriented Structuring of the Production

Production centres and segments

The restructuring of production is the first major step towards optimisation of the cost structure, reduction of throughput time and heightening of flexibility. A function-based structure is replaced by Production Centres (PC) and Production Segments PS).

The PCs are:

- Vehicle Structuring,
- Electrical Components,
- Assembly,
- Testing and Commissioning.

The function areas job scheduling, precision control, production of goods and services, quality inspection and internal material flow are integrated in the segments. The segments produce goods and services in self-contained, manageable control units and are responsible for the costs, quality and delivery date of their products. The framework structure selected also includes the "service-providing" spheres of production planning and production logistics. They control the interfaces between production and the firm's other spheres, provide the preconditions for the pro-

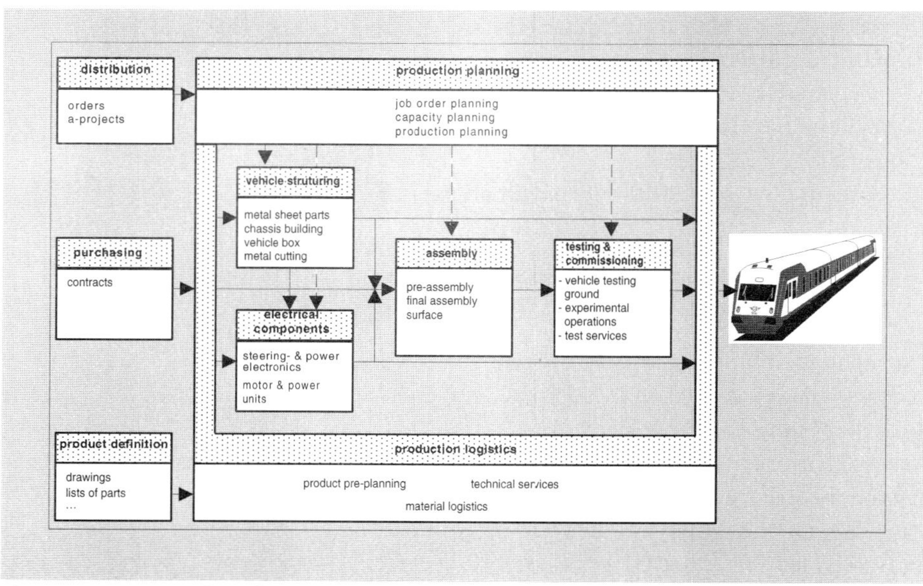

Fig. 6.21: Structure of production centres

duction of goods and services in the production centres and segments, and coordinate their activity from the point of view of the total product. Segmentation takes place in relation to technologies and not product groups. The decisive factors are the widely varying volumes of orders per product group, cost-intensive resources that are repeatedly required at a low relative throughput in the case of product-related segmentation and above all the possibility of segment-related technology transfer within the framework of local content.

Standardised Process Flows

Proceeding from the performance centres, process chains were identified in the firm.

Their optimisation is a further important step towards improvement of the firm's efficiency. At present the process chains are being optimised and integrated uniformly from development to production. They are the basis for the development of standardised process flows and for the definition of a standard product structure.

Process chains

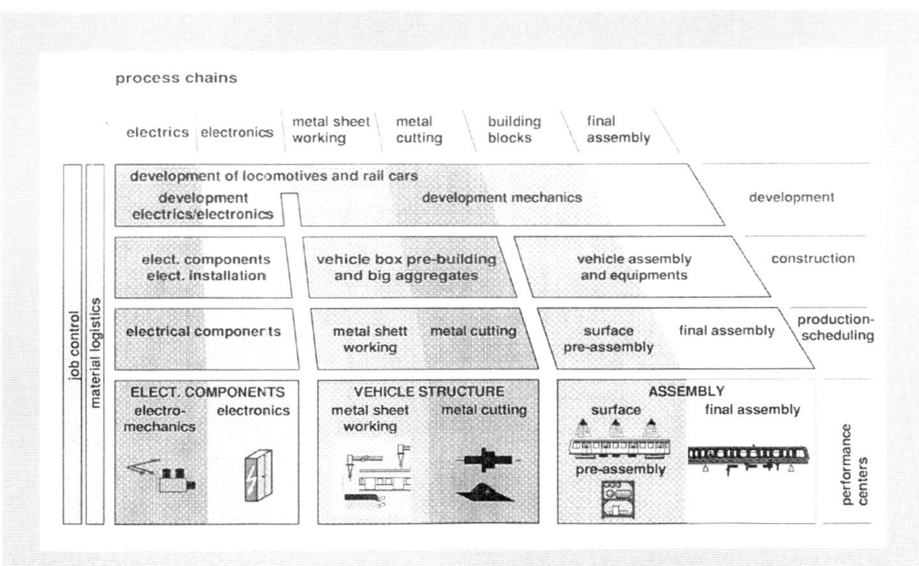

Fig 6.22: The six main process chains

According to the defined process chains and to different types of orders six main process flows can be defined and standardised:

- Electrics,
- Electronics,
- Vehicle Structure,
- Machining,
- Pre Assemblies,
- Final Assembly.

In the process flows the main subassemblies are manufactured. The possibility of simultaneous execution of parallel process steps in different flows is supported by the standardised product structure.

Modern IT application tools are necessary, combined with an integration platform to ensure the integrity and topicality of the handled data and information.

Production and Design Oriented Product Structuring

Product and project structuring

One standard product structure is defined per product group at ASF. This structure is determined jointly by development and production, whereby the function oriented requirements of development and the manufacturing or assembly oriented requirements of production are taken

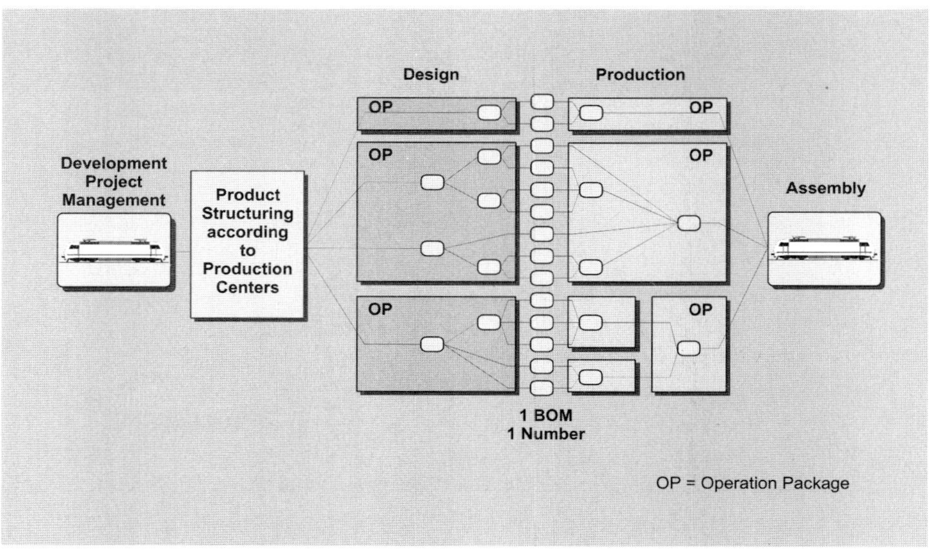

Fig 6.23: Standardisation of product and project

into consideration. The standard product structure obviates the need for the transfer of parts lists, numberings etc. between design and production.

The structure tree is constructed on the basis of production centres and production segments. As a rule, each production assembly group can be assigned unequivocally to a segment of a production centre. Functional relationships are realised through the numbering of the production assembly groups. The standard product structure has been successfully introduced in the case of two new products:

- high-performance locomotive BR 12X and the
- S-Bahn Berlin BR 481.

It is of significance as an organising instrument for all spheres involved in the genesis of the product. Its advantages lie primarily in:

- the avoidance of transmission work and redundant work in development and production,
- consistent information processing on the basis of uniform production assembly groups.

The standardised product structure is also used for:

- definition of "operation packages",
- coordination between subassemblies and external suppliers,
- order scheduling,
- structuring of bills of material,
- standardisation during design and development,
- assembly/manufacturing coordination.

Operation packages

First the operation packages are developed and fixed. An operation package is an internal order to Production Centers and Segments including a determined amount of work inside one customer order. The operation packages are further improved and detailed from the draft product specification till the complete set of production papers is finished.

The product structure is generated and stored in the SAP System. Therefore, bills of material and master part data are used.

6.5.2.3 Redesigning the Product Development Process

The current state of the product development process follows a serial working-out of phases along the process flow of certain subassemblies. In contrast to that the workflow of the development of the whole vehicle is partly in parallel order. For example the development processes of the car body and the bogies are original and in parallel order. Within each of the development processes the workflow is in serial order.

These requirements can only be successfully fulfilled with new methods like Concurrent Simultaneous Engineering (CSE), which have to be supported effectively by information technology. ASF is forced to re-engineer the existing product development process from basic. This re-engineering is done under special consideration of the goals:

Goals

- parallelisation,
- standardisation,
- integration.

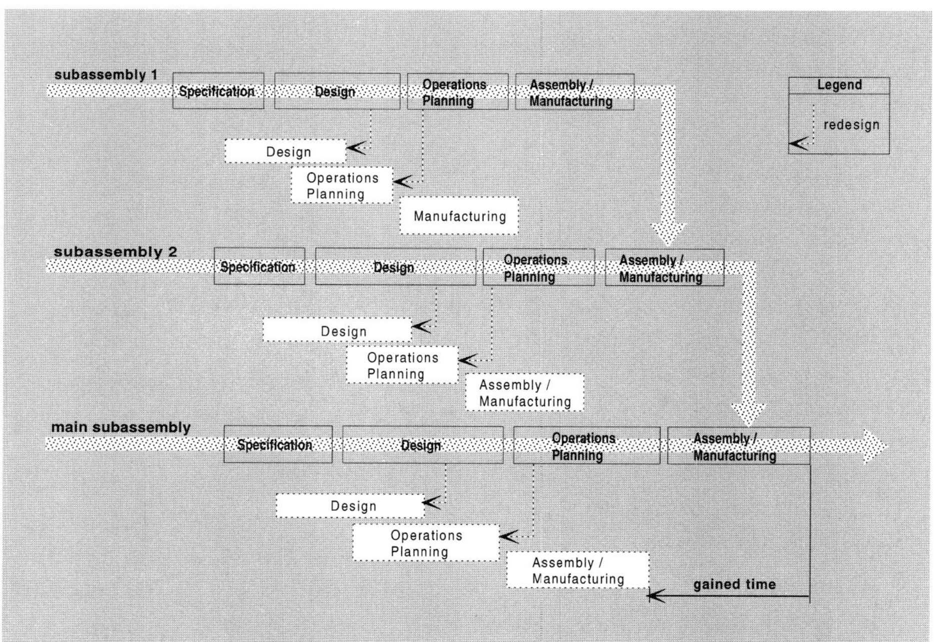

Fig 6.24: Product development on subassembly-level

6.5.2.4 Description of the New Product Development Process

As mentioned above the redesign of the process flows at ASF has to be realised with the parallelisation of process phases in the flows itself to reduce "Time to Market". Hence, the product development process has to be redesigned. This is already being done at ASF in 1995 in relation to the changes in the product and production structure.

In the following, the phases of the new product development process are described:

Phases of the new product development process

The product development process for a new product, such as a new type of locomotive, starts with the analysis of the customer order. A draft specification is made which is part of the quotation. The order is compared to existing quotations. The capability of fulfilling the customer requirements is checked.

With the order entry the project structuring begins. Scheduling and calculations are made. Responsibilities, dates and actions are defined and the order scheduling is finished.

A project structure is developed and implemented for the project. The chosen structure should follow a certain standard concerning phases, actions, teams etc. It stipulates among other things which IT tools are to be used during separate phases of the development process. The standardised project structure enables ASF to enlarge the parallelisation in the development of new products.

A project team is installed which is responsible for the coordination and controlling of the whole project. It consists of members from the engineering, sales, design and operation planning departments.

In the following phase, the technical feasibility is checked. The task structure for the project is elaborated in cooperation between engineering and design. The economic feasibility is checked, too.

Parallelisation in development and design processes

The product is structured according to a standardised product structure which is mentioned before. In terms of product planning the product consists of operation packages. The operation packages are developed in relation to the product structure given by the customer order. The standardisation of the product structure allows further parallelisation in development and design processes for assemblies and subassemblies.

The requirements and the task structure are improved. The draft design concept is elaborated. The next phase includes the detailed design, the FE modelling, analysis and calculation as well as cost calculations. In this phase the design, operations planning, calculation and supply departments are working in direct interaction. Exact cost calculations are very important for ASF because of the changing market situation with its decreasing price level. Cost reduction programmes have to be developed and realised in every phase of the product development process. The results are checked in time during the target costing process.

Manufacturing drawings and first drafts of the manufacturing process plans are made. In the following the process plans are worked out, which means that they are checked, detailed and improved. Furthermore they are customised according to the product structure of the current order.

Bigger production equipment, tools, fixtures and leading devices are developed because of the earlier order points. The development of the smaller ones begins a little while afterwards. NC programming if necessary and test planning are done. Prototypes or a pre-series are produced.

In these phases the examination of manufacturing and assembly methods takes place. Therefore, the manufacturing and assembly methods chosen in first draft are checked for technical and economic feasibility. Corrections are discussed and realised if necessary.

While the series starts the target costing process continues. The manufacturing papers are completed and scheduled, which means in ASF terms that they are finished to order. Meanwhile the batch size optimisation is done. Decisions for outside supply are made, the operation packages are co-ordinated and scheduled between the production centres and segments.

Parallel production

Series start begins with the production of single parts and parts which require longer production times. The parts are assembled and lead into main subassemblies which are themselves assembled to make for instance the car body. The product structure allows parallel production of parts and subassemblies in the production segments and centres.

Series testing begins on subassembly level during the production phase. After the whole vehicle is finished the commissioning starts.

This phase ends with the shipping of the train or vehicle and is itself terminated by the customer acceptance.

The target costing ends with the ex-post costing for product and processes.

Parallel to the operative succession of the product development process are four planning and tracing phases. Midterm scheduling and capacity planning is followed by the capacity levelling between production centres and segments. In these phases for example the operation package scheduling and alignment is improved according to product structure and customer order.

As mentioned above the material requirement planning starts during the design phases depending on early order points for certain materials and purchased parts. It ends during the manufacturing phase.

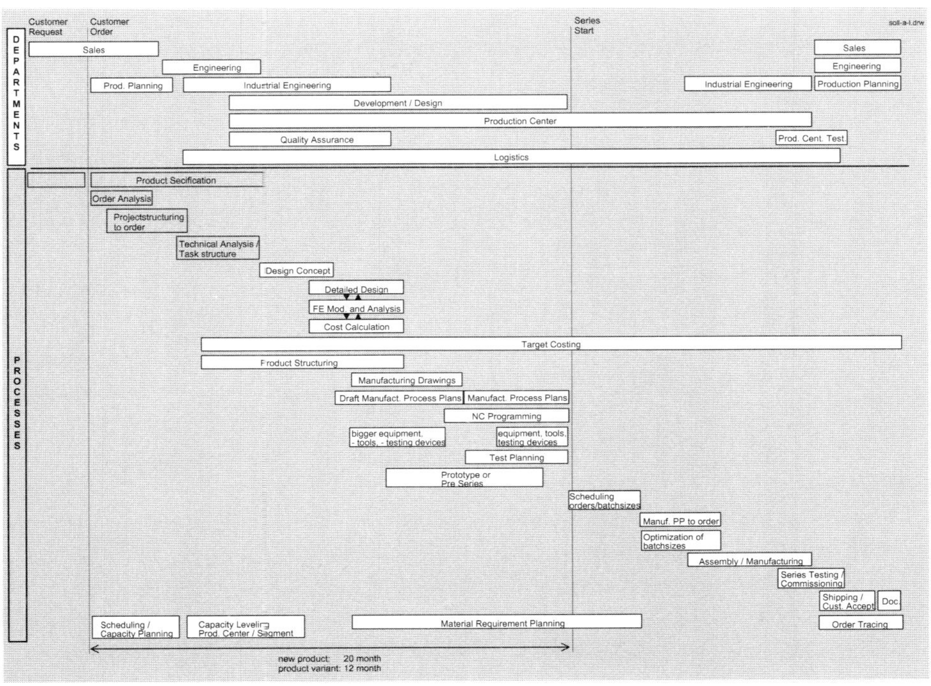

Fig. 6.25: The new product development process

6.5.2.5 The ASF Demonstrator

The ASF demonstrator scenario in the CONSENS project describes a situation happening during the product development process for a new product such as a locomotive. The ASF demonstrator is based upon the new product development process.

The chosen subassembly for the demonstrator is a bogie crossgirder. Normally a locomotive has two bogies carrying the car body. Bogies are welded subassemblies with an O- or H-shape. In the example the crossgirder is the rear part of the O-shape.

The bogie crossgirder consists of the following parts:
- upper girder,
- lower girder,
- ribs and vertical plates,
- brake girder.

In order to manufacture the crossgirder the following production processes have to be performed:

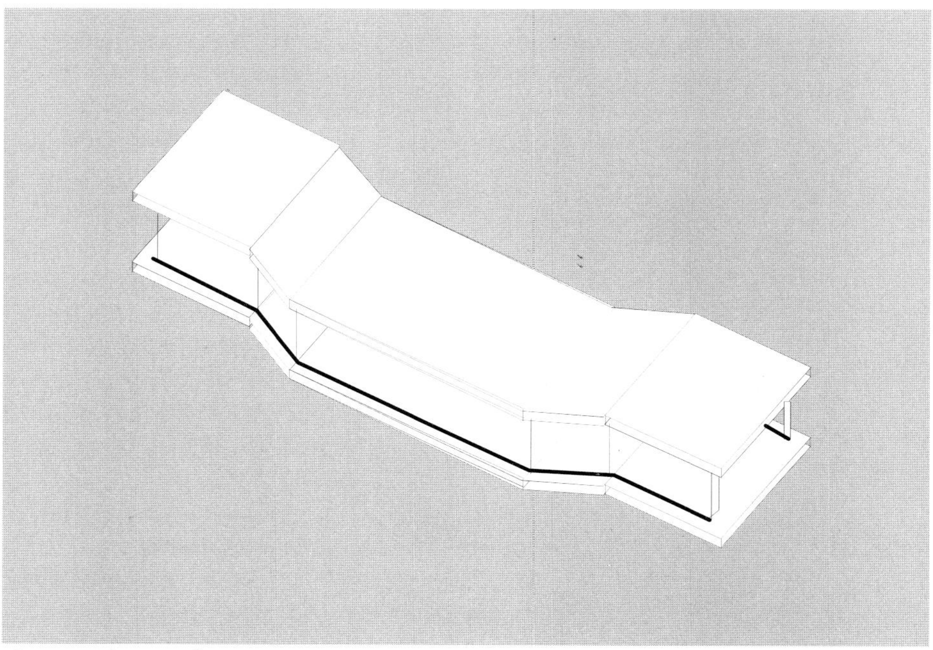

Fig. 6.26: The ASF-demonstrator

- cutting of metal sheets by NC laser cutting machines,
- machining of single parts and subassemblies,
- manualy welding and riveting of subassemblies,
- welding of subassemblies by industrial robots,
- testing of subassemblies and single parts by manualy test equipment,
- quality control after production steps.

Subassemblies such as the crossgirder are produced in smaller batch sizes.

The process of engineering this subassembly includes one complete process flow within the departments:

Phases of the demonstrator flow

- engineering,
- design,
- operations planning,
- production,
- quality assurance.

The CONSENS demonstrator at ASF is consequently concentrated on the tasks and activities in these phases. It refers to the specific requirements of ASF as a job order producer.

6.5.3 Description of Enabling Information Technology

The whole re-engineering process at ASF is supported by information technology. This chapter describes CONSENS IT modules and tools in use at ASF, which will be part of the CONSENS demonstrator.

Integration Platform

The Integration Platform promotes the standardisation of product development processes. At ASF the new development process will be standard for new projects in 1995. To solve some of the problems of the old development process, the introduction of standardised workflows and, for example document handling routines is of crucial importance for ASF.

The implementation of a framework architecture enables the integration of several tools under one common platform. Because of the numerous special and decentralised IT

Decentralised IT Systems

systems in use at ASF this advantage of the integration platform is essential. The framework allows the control of processes with respect to inputs, outputs and process progress as well as the team co-ordination within the product development process. Both aspects are of great importance for ASF because of the redefinition of the responsibilities to be assumed by a project team. The team will continue to monitor the progress of the project with regard to costs, deadlines and quality.

Team work

The introduction of team work profits by the existence of standardised access to IT modules and tools.

Configuration Control

The module "Configuration Control" gives improvements in the organisation of the variant change processing. Because of the multitude of parts and assembly groups that have to be processed, this functionality is of the utmost importance to ASF as a job-order manufacturer if it is to be able to control and/or reduce product costs.

Configuration Control also means the documentation of the valid documents of a product. It prevents multiple handling and supports the data management between modules and tools. This function contributes decisively to allowing parallel processing of various assembly groups.

CADIM (ASF Tool)

The CADIM tool is an Engineering Data Management (EDM) System. At ASF it is used to connect CAD and other tools to the SAP System for the handling and storage of documents and product data.

In the ASF demonstrator CADIM is used instead of IMS. Therefore the CONSENS modules have to transfer a lot of documents between each other and to and from existing IT tools via CADIM. In combination with the PIA CADIM will be the central database in the CONSENS demonstrator.

Process Control and Team Co-ordination

The timely delivery of high quality results for the following process phases is an extremely important issue improving "Time to Market", cost and quality. "Process Control and Team Co-ordination" supports especially the shortening of "Time to Market".

Management Decision Support (MDS)

The MDS tool is used at ASF for supporting project-management and planning. It helps in searching for former projects that show similarities in the product or in the project structure.

At ASF MDS is especially used for calculations in an early project phase. The ASF calculation scheme was implemented for that purpose.

The possibility of making economic feasibility studies is used in the quotation phase.

Key figures of the product development process

With the introduction of key figures for the product development process criteria are available for assessing the performance and economic efficiency of various process phases. The key figures are also of service in the search for similar past projects.

MDS is used for concurrent calculation and post calculation in the target costing process. It thus serves to control the product costs and product generation costs in the course of development, design, production scheduling and manufacture.

Repository Group Technology

Group Technology (GT) is the concept of identifying and bringing together similar parts and subassemblies in order to obtain advantages from their similarity for design, manufacturing and process planning purposes. The RGT tool automates the retrieval of existing similar parts and subassemblies.

Because of the amount of single parts and subassemblies in complex products such as car bodies or complete trains, the RGT tool is of special interest for ASF. The RGT tool helps the designer in decreasing the number of variants of single parts and subassemblies. Because of Computer Automated Group Technology realised in the RGT tool this is done in a very effective way.

Design to Cost

The Design to Cost tool is used for cost calculations during the draft design phase. The tool supports the designer with information about manufacturing and assembly costs. The abilities of the "Design to Cost" tool to calculate costs for single parts, assembly/manufacturing and material are

Reducing product costs

expected to give valuable hints for reducing product costs. Specifically for the reference assembly group selected, scope has been provided for assessing welding processes from the point of view of cost.

Using this tool, the designer can judge whether the design variant being considered remains within the required target costs. "Design to cost" is therefore an effective tool for influencing and reducing product costs even during the design phase.

Design for Assembly

The Design for Assembly tool is used with the goal to reduce effort for assembling parts and subassemblies. This means the minimisation of handling and assembling costs.

Unlike the Design to cost tool, this tool is used in the later phases of the product development process. The possibility of analysing welding processes as special assembly processes has materialised for ASF.

It makes it easier to compare different variants with regard to production costs very quickly and with high exactness. Together with the other CONSENS variant and cost management programmes, this is an effective tool for reducing product costs.

Design for Manufacture

The Design for Manufacture tool from IST supports the designer with sensitivity informations to evaluate the influence of changing design parameters on design performance. This tool is used due to its design parametrisation approach, appropriate for the integration with parametric CAD systems which are in use at ASF, too. The Design for Manufacture tool as well as the Design for Assembly tool provide the designer from the conceptual to the detailed design phase with informations for optimising the product design. For that reason they are used to check and where necessary correct the assembly and production processes initially selected.

6.5.4 Improvements of the Product Development Process with Respect to CSE

The re-engineering process at ASF has been initiated due to changing market conditions like smaller order sizes with many variants, shortening of delivery periods and a decreasing price level.

The foundations for the simultaneous processing of various assembly groups have been laid with the new product structure and the standardised process flows. CONSENS provides a system architecture for the implementation of parallel processing in modern IT systems. With the existing modules and the CONSENS modules, therefore, ASF will be able to parallelise and thus accelerate its product development process in the decisive early phases.

Similar projects

The implementation of a standard project structure at ASF has enabled the company to handle its different projects with a high grade of effectiveness. A standard project structure made it possible to compare a new project in an early phase with former projects in order to recognise similarities. In the CONSENS demonstrator this functionality is implemented and realised with the MDS tool.

The redesigned product development process is concentrated on the functions which are value adding in customers sight. The new process is leaner and more effective than the older one. The implementation of the standardised new process is therefore one main factor making ASF competitive for the future.

Standardising the product development process by means of IT

Within the re-engineering project special effort has been put on supporting and standardising the product development process by IT. The kick-off of processes and sub-processes is given automatically by the IT system as soon as all needed results are available. The document release after finishing an activity is managed by the framework system as well.

Data handling, storage and provision are realised with IT. This includes a logical integrated data base and a standardised product data model.

For single departments the possibility of using decentralised IT Systems which fit especially their requirements is an enormous progress. In the companys view, CONSENS is making the integration possible.

The introduction of the new project structure for customer orders will replace the functional assignment of responsibility practiced so far. This will mean a substantial change in the way orders are handled. Nonetheless, project orientation seems to ASF to be the only way to meet the new demands of the market. The project structure has to be supported through IT solutions such as CONSENS, particularly with reference to the reduction of throughput time and costs.

Increasing motivation

With the introduction of a new project culture responsibility has been put on lower levels of the company hierarchy step by step. New assembly lines have been planned and erected under the responsibility of foremen and workers by introducing their own specific experiences. This has led to an increasing Motivation among the employees.

In several sections teams consisting of different profession groups have been formed.

6.5.5 Future Improvements

In the following some targets concerning future IT systems which are of special interest for ASF are named.

Standardisation of interfaces

One important objective would be the standardisation of interfaces within the whole product development process. This concerns first interfaces of the information technology systems. The integration of specialised decentral IT-systems of several users should be possible without greater problems.

Furthermore the introduction of globally standardised engineering data would be meaningful. Global in this case means first the IT integration of the engineering data of all AEG factories dealing with train production. Furthermore it would be desirable to support the data exchange with local suppliers all over the world. Standardisation in this case concerns questions of data format, carriage and storage media etc.

In addition, the data should be capable of being understood everywhere, that is, the data should be adapted to local standards with regard to language, draughting standards etc. in automated form.

Flexible IT Systems

IT systems that are too large and not adapted to user needs will soon lose their importance for ASF. The future belongs to open, flexible systems that are configured individually.

In addition it would be of great effectivity if there were IT Systems applicable to different data bases. The more open communication interface would mean a great reduction of complexity in the handling with IT.

A structure of data files which would make it possible to open parts of the file for parallel working could expand the process of parallelisation even to the section of information technology in a special way. Simultaneously, other parts of the file that are not open to parallel working have to be protected against unauthorised handling.

7 Future Trends in Concurrent Simultaneous Engineering

D. Koch, J. Vilsmeier, J. Warschat
supported by F. Wagner, A. Petz, A. Bading, J. Wagner

7.1 Conclusion of the CONSENS Project

In CONSENS, a twofold goal has been reached:
- a methodology for implementing CSE principles in a company was developed,
- a software platform and tools for supporting the engineering processes in a CSE fashion were developed.

Levels of co-operation

The case studies in this book have shown the advantages gained already by applying these results to companies. Naturally, however, the end of the CONSENS project is not the end of the developments in the field of Concurrent Engineering. The following sections outline some areas in which research and case studies will have to be done, and in which a concentration of projects is to be expected in the near future. The keyword is Co-operation – to be considered on different levels, from the co-operation in small teams to multi-company cooperation. In general, the trend is moving towards fractal structures in companies, where more responsibility is assigned on teams and core competences. It is accompanied by the necessity of closer co-operation between partners in multi-company projects as well as co-operation with suppliers. This requires effective communication networks more than ever.

7.2 Co-operation in Engineering Based on Core Competences

Reconsideration of concepts

Another upcoming way to tackle the issue of co-operation is based on core competences. This can be seen particularly well in the example of automotive industry. The increasing worldwide competition between the European automotive industry and Japanese and North American opponents forces this industry to reconsider their engineering and manufacturing concepts. Following the current trend, future car companies will mainly be assembly factories especially in the segment of small cars. The car company will define rough concepts, coordinate and harmonise the main development phases while suppliers will be responsible for complete development and manufacture of most of the components, modules and systems.

Concentration on Core Competences

Consequently, a further reduction of development tasks in the car company is required by delegating them to suppliers. All development partners along the vertical supplier chain have to reduce their effort and concentrate on their Core Competences. New development processes have to be specified and rules have to be designed for co-operation and organisation, as well as for time, cost and quality management. This new "Delegated Development", following the "Factory in The Factory" or "Fractal" manufacturing concept, ensures the reduction in complexity, the acceleration and increase in transparency of processes, higher flexibility towards changing requirements and a cost reduction by using the potential and core competences of all partners involved.

Improvement for European industry

The development process of an automobile cockpit will be redesigned in the sense of Business Re-engineering and under the aspects of improvement of information, communication and quality management. Through the application of this re-engineering process, the following quantitative improvements of the involved industrial users and European automotive industry in general are expected:

- Higher innovation rate and reduction of time to market for new cars by 30% to 40%, from nowadays about more than three years down to a minimum of 2 years.

- Avoidance and reduction of engineering changes for complex car modules by 60% to 80% (currently the cockpit is to be modified more than 100 times).
- Increase in flexibility to fulfill 100% of the changing requirements of customers and end users.
- Decrease in costs of complex systems by 20 to 30% due to the focus on core competences of every partner and the cut of overhead costs by 30% to 40% by reduced factor costs of the suppliers.

Redesign methodology

A view to other advanced industrial sectors, like computer and consumer electronics industry, supports these dimensions. A new methodology should contain:

- Benchmarking studies of the current development collaboration in leading industries (and products with comparable complexity), e.g. computers, consumer electronics, etc.,
- Verified re-engineering procedures for the development process of complex automotive systems and its industrial implementation in a complete vertical supplier chain,
- Descriptions of optimised technical and business processes of cockpit development including time, cost and quality as well as the organisation, information management and IT structures – Catalogues of guidelines, measures, metrics and control variables for co-operating companies developing complex automotive parts,
- Human-resource concepts for management and personnel staff.

Additionally, an integration and upgrading of existing and supporting IT systems needs to be done.

State of the art of co-operations

Today, most of the development effort is still provided within the automotive companies themselves or in close co-operation with suppliers. Furthermore, the introduction of platform concepts and the enforced use of existing "carry over parts" (COPs) have reduced the development effort for new cars. However, the market requests a higher rate of innovation and stronger separation of new cars from existing ones.

Initial approaches to meet these market challenges lay in the extensive use of TQM, Concurrent Simultaneous Engineering (CSE) and the integration of suppliers. Following the Japanese example, the Western automotive industry has started to build up a long-term production and partial development partnerships, with only a few system suppliers at the top of a pyramid.

Delegation of development effort

One of the most advanced industrial state of the art in development concepts can be seen at MicroCompact Car AG (MCC) with the Mercedes "Swatch" car and its development and production depth of about 15% to 20% [5]. The car manufacturer keeps only strategic competences and delegates most of the development effort. Automotive benchmarking studies by the consultant company Arthur D. Little [1] and global investigations by Clark/Fujimoto [7] showed that these modern concepts of reduction to core competences and full vertical co-operation [12] is not implemented in the automotive development process yet. First steps in this direction are the delegation of process development to suppliers located at the automotive company's site (e.g. SEAT/Martorell, Mercedes-Benz/ Rastatt and-FIAT/Melfi). On the other hand, the fact that internal development centres of car manufacturers may not exceed the costs caused by the external "Best in Class" suppliers [11]. The North American re-engineering concepts can be seen best in the case study of the successful NEON project [6] and the economic turnaround of Chrysler.

Screening the current literature, investigations by FhG-IAO with 128 German companies [10] and sample interviews with European automotive experts [11] yielded that, nowadays, neither in theory nor in practice does a holistic concept for co-operation and organisational rules, enabling consequent delegation of development tasks to competent partners, exist. On the other hand, automotive industry representatives have stated that the most essential success factor is to combine competences in such a way that they lead to optimal solutions fast [2].

European research projects

The State of the Art of European research can be seen, for example, in the results of the EU projects CONSENS (EP 6896) [8], AMIES (BE 5307) [16], Future Working Structures (BE5276) and the AIT project (EP 7704) [3]. These projects deal mainly with applications of CSE, supplier integration, or the manufacturing phase. Innovative concepts, coming from national research like the concept of the "Fractal Factory" [14], are oriented towards production and, so far, have had little influence on the development of new products [15].

As mentioned above, the most highly evolved industrial application can be found currently in the "Swatch Car" project. Due to the strategic nature of this project, very little information is publicly available yet . But like most of the industrial projects of development restructuring, this approach is strongly oriented towards the car manufacturer, dominating over the interest of the involved suppliers. Opposite development concepts like the strong centralisation of the development in a single centre for companies of the same group have more disadvantages, e.g. inflexibility and bureaucracy, than synergetic effects. The large development center of VW in Wolfsburg can be used as an example.

Continuous Improvement Process

Other reorganisation concepts like Continuous Improvement Process (CIP) are more complementary than competitive. They are necessary for consolidation after periods of radical re-engineering, but CIP alone is not sufficient to induce the necessary improvement of the development performance.

In the field of supporting IT-systems, a wide spectrum of tools is currently available. These CSCW systems (Computer supported collaborative work) can be clustered into four sectors, depending on same/different time and location: meeting tools, conferencing tools, e-mail and workflow systems. Of all the systems not dedicated to engineering applications, but more to a management level, Lotus Notes seems to become a kind of standard. But most of these tools are poorly integrated, the Workflow Coalition is only starting to define general interfaces for these tools.

Innovative approaches particularly will have to address the following issues:

- Business Re-engineering with the focus on a consequent reduction to development core competences in the European car and supplier industry,
- Providing "Best Practice" experience for industry from benchmarking in other, more advanced, industrial sectors,
- Multi-company Business Reengineering of the development process of complex systems along the complete vertical supplier chain,

- New organisational configuration of co-operation between manufacturer and supplier companies described in industrially validated Codes for Good Practice,
- Integrative view on strategy, processes, organisation and human factors, quality, business costs (instead of classical product costs), information management and IT-support.

7.3 Co-operation via a Global Engineering Marketplace

At the summit meeting in Corfu on June 25th, 1994, the Heads of State and Government of the European Union discussed the Bangemann report [4] on the Information Society. The European Council expressed a strong desire to give political impetus to specific application areas set out by the report. Among these were networks for R&D centres, telematic services for SMEs, electronic tendering, administrative networks, and information highways. With regard to manufacturing, such political impetus has to be turned into competitive strengthening of major sectors of European industry in advanced engineering and production.

Information Society

Today, the global marketplace is changing rapidly. Rigid requirements from customers and foreign competitors require a quick response to consumer demands and delivery of high quality products at reasonable prices. The failure to meet these demands in Europe and not achieving shorter time-to-market in comparison with world wide competitors threatens companies with loss of revenue and profit.

In recent years, many European companies invested heavily in production automation and computer integrated manufacturing (CIM). Departments, however, remain responsible only for their own costs. This encouraged everyone in charge to reduce their own costs and to consolidate the consumer-provider relationship between departments while the basic communication bottlenecks of the customer-supplier chain remained. Methods of Concurrent Engineering and Co-operative Work Processes have been employed to tackle this problem, but their implementations have taken place mostly within separate companies. The required advanced engineering may be characterized by the following strategies:

Requirements for engineering

- to increase the level of network co-operation between customers and suppliers as well as co-operating companies by supporting collaborative work,
- to accelerate the engineering process by the extensive use of commonly available pre-designed semi-standards and

sub-assemblies or even standardised parts,
- to concentrate on the early phases of the design process since these mainly determine the product quality and product costs,
- Outsourcing of design and manufacturing capabilities in order to concentrate on those product features where the product differs from those of the competitors and where the market and customers pay for it.

Global Engineering Network

Thus, the vision of a Global Engineering Network (GEN) has been established [9], [13]. GEN is seen as a key factor for the advanced conception and planning of complex products of the European capital and consumer goods industries.

GEN is a scaleable engineering knowledge infrastructure, using trans-European telematic networks and services and information highways to accelerate collaborative work structures.

Although basic components in communication, hypermedia, and network technology are available to exchange information electronically, common services and methods in order to exchange product data over wide-area networks and to facilitate electronic commerce activities to component information, as well as to support collaborative work are lacking. In this context, the vision of GEN has been established as a network connecting industrial suppliers and customers as well as co-operating companies and company departments by services of knowledge and product information brokers. A customer uses intelligent search and retrieval features to have fast access to engineering information, covering products, methods or services. Suppliers can offer their products and the product features by graphics, product data, texts, videos or voice. If required, a customer can even request interactive support in the design process. The broker gives access and offers security, accounting and administration services.

Thus GEN will be the backbone of the solution using public collaborative networks with intelligent nodes and services to provide large scale distribution and control of complex engineering knowledge. Especially engineering knowledge bases accessed by different services from customer and supplier sites provide the "glue" for an integrated way of operation. This defines a working culture of large-scale collaborative engineering capable of fast change, enabling organisations to become active both as SME's from all over Europe as well as large companies located at Europe's industrial centers.

7.4 Multi-company Development Co-operations

Collaboration in aerospace industry

For several kinds of products the development effort is distributed over a consortium of companies on multiple sites. Typically this is the case in the aerospace and automotive industry. Especially in the world-wide aerospace community, a need for collaboration has been established during recent decades, which was driven by the increasing costs for the development of new airplanes. Even Boeing had to look for partners to share the risk, since even the market leader could not alone bear the investments necessary for the development and production of new products. As the national European aerospace companies have become too small to compete alone on the world market, collaborative projects like Airbus and ATR in the civil market and Tornado or Jaguar as representatives of military products have proven to be successful examples for multi-company co-operation in the past.

Inter-company process and data management

During recent years, large efforts have been undertaken by the American aerospace industry to improve competitiveness, and the results of the efforts have been published especially with respect to Boeing's 777 and Lockheed's F22.

International projects are very critical concerning the work share, which means that numerous interfaces within an aircraft have to be defined and kept track of during the evolution of the design. Each partner depends on the results of the other partners, which means that not only the project-management, but also the engineering information has to be shared in a co-ordinated and transparent way. In order to fit these requirements to the overall project schedule, the processes of the partners have to be defined and synchronized concerning the inter-company process and data management.

To achieve this, a distributed, integrated communication and information infrastructure is needed with capabilities that are not available from current commercial systems.

During the CONSENS project, Dasa Military Aircraft could demonstrate that a consistent process and data management through the engineering framework SIFRAME is possible within one company, and that Concurrent Simultaneous Engineering is supported in a comprehensive manner. On the other hand it was found that dynamic process modelling would ease the implementation of complex project scenarios. Therefore, the next step will be to widen the scope of applicability into a multi-site, multi-company environment.

Multi-company Development from design to delivery

The battle in the world aerospace market will become fiercer with new Asian and Third World companies (e.g. NURTANO of Indonesia, Embraer of Brazil) entering the market. As the cost-saving potentials in the manufacturing field have been successfully used in recent years to improve competitivity, such a multi-company extension would dramatically improve the multi-company development work within the early design phases of an aircraft project without losing sight of the downstream processes of the development cycle down to delivery of the product to the customer.

A future system to support such functioning would need:

- a Global Process and Configuration Manager to configure intra-company as well as inter-company work flows,
- a Product Modeller to handle the product structure information as the master copy,
- a link to a Process Modelling and Simulation System, and
- an Information Broker via which the different companies exchange their information. This Broker system could make use of the Global Engineering Marketplace described in the previous section.

7.5 Conclusion

Even with a successful project like CONSENS, many challenging areas of work remain to be tackled, some of which have been outlined here. The main emphasis will presumably lie in the fields of increased co-operation between distributed partners, within a single company, between a company and its suppliers, as well as in large multi-company consortia. In order to support the co-operation, new technologies and concepts like CSCW, new network technology, and distributed information systems, will be used.

References:

1 Arthur D. Little, Improving the Product Creation Process in the Automotive Industry, Internal presentation, Wiesbaden 1991

2 F. Wressnigg, Kooperationen effizient gestalten, In: Automobil-Industrie Sondernummer, Mai 1994

3 AIT Consortium: Management Overview of the First Project Phase, ESPRIT Project 7704 (November 1993 - April 1994), unpublished, Stuttgart 1994

4 Martin Bangemann, EU Commissioner: "High Level Task Force Report on the Information Society in Europe"

5 R. Böhmler, Swatch Auto - Diskrete Hinweise, In: Wirtschaftswoche Nr. 52, 1994

6 M. Brackon, The Neon Story, In: Proceedings of CMS 3rd Quarter Meeting Dec. 1994, Consortium for Advanced Manufacturing International, Scottsdale, Arizona 1994

7 K. B. Clark / T. Fujimoto, Product Development Performance, Harvard Business School Press , Boston 1991

8 CONSENS Consortium: CSE Methodology and Implementation Guideline. ESPRIT III Project CONSENS EP 6896 Deliverable H 0.2., unpublished, Munich 1993

9 European IT Industry Roundtable: "Europe and the Blobal Information Society - Proposals for the Implementation of Priority Applications by the European IT Industry", H. Heckl (Siemens Nixdorf) (Ed.), November 1994, p.7

10 G. Wasserloos, Zuliefer-Integration heute, Datenerhebung FhG-IAO, Stuttgart 1994

11 Moderne Industrie, Fortschritte in der Automobilentwicklung, Entwicklungsperspektiven für Automobilhersteller und Zulieferer, Stuttgart 10. und 11. Oktober 1994

12 C. K. Prahalad / G. Hamel, The Core Competence of the Corporation, Havard Business Review 1990

13 Ulrich Rethfeld: "Manufacturing and the Information Highway". Keynote address at ESPRIT CIME, 10th Annual Conference, 5.–7. October 1994, Copenhagen

14 H.-J. Warnecke, Die Fraktale Fabrik: Revolution derUnternehmenskultur, Berlin 1992

15 H.-J. Warnecke, Aufbruch zum Fraktalen Unternehmen, Praxisbeispiele f. Denken und Handeln, Berlin 1995

16 F. Wagner / A. Petz, AMIES- Automotive Multi-company Integrated Engineering System, M. Klittich (Ed.), Proceedings of the 2nd Open Workshop on Business Process Management Based on CIMOSA Offenburg. EEISS, Eschborn 1995

Editors and Authors

Editors

Hans-Jörg Bullinger

Dr. Hans-Jörg Bullinger started his career at Daimler-Benz in Stuttgart in the field of manufacturing. He studied at the university of Stuttgart, where he received a Master's degree and a Ph.D. in Manufacturing.

As a member of the Fraunhofer-Institute of Production Technology and Automation in Stuttgart, Dr. Bullinger was responsible for numerous applied research projects in the fields of design and manufacturing with the aim of increasing the productivity of PC-assisted systems. After two years of lecturing at the University of Hagen, Dr. Bullinger became a full professor at the University of Stuttgart. Besides his chair at the University, he is the head of the Institute for Human Factors and Technology Management (IAT) at the Fraunhofer-Institute for Industrial Engineering (IAO).

In 1991, Dr. Bullinger received an Honorary Doctor of the University of Novi Sad and an Honorary Professor from the University of Science and Technology of China (Hefei). His publications in industrial engineering, work science and applied informatics range over 600 articles and books.

Joachim Warschat

Dr.-Ing J. Warschat (1949) studied Mechanical Engineering at the University of Stuttgart, followed by a Ph.D. Presently, he is the Head of the Department for R&D-Management at the Fraunhofer-Institute (IAO). His research is focused on: Simultaneous Engineering, Rapid Product Development, Computer Supported Co-operative Engineering, Project Management, and CAD/CAM. His publications in these fields amount to more than 100. He is a lecturer at the University of Stuttgart for Simulation, Methods of Simultaneous Engineering and Project Management, CAD/CAM.

Authors

Jens Bergner

Jens Bergner (1955) studied Mechanical Engineering at the Technical University of Braunschweig, Germany, with special interest in industrial and business organisation. After finishing his master's degree in 1984, he joined a world leading CAD/CAM company as a sales supporter and sales representative. In 1993, with several years of experience in consulting and project management, he has changed to the AEG Production and Logistics Consult, a consultancy within the AEG Daimler-Benz Industrie. As a project manager for a business re-engineering project within the TEMIC Telefunken microelectronic, a subsidary of AEG Daimler-Benz Industrie, he is responsible for the implementation of Simultaneous Engineering at the TEMIC Automotive. Besides, he is the Technical Manager and Workpackage Leader within the ESPRIT project CONSENS.

Stefan Berndes

Stefan Berndes (1966) studied Aerospace Engineering, Philosophy and Political Sciences at the University of Stuttgart. After completing a degree in engineering he worked at the Fraunhofer-Institute (IAO) in Stuttgart as a consultant in the fields of Lean Management and Concurrent/Simultaneous Engineering. He published several articles in co-operation with Prof. Bullinger, Dr. Warschat and Mr. Stanke about Concurrent Simultaneous Engineering and held lectures at conferences.

Lutz Bertling

Dr. Lutz Bertling, ASF/TPH2-HE is the head of the car bodies' and structures' profit center (Production of car bodies, subassemblies, single parts incl. processes, production controlling and quality assurance). Working fields: Business re-egineering, optimisation of process chains.

José Manuel P. F. Camacho

José Manuel P. F. Camacho studied Mechanical Engineering followed by post-graduated studies in Computer-aided Design and Production Aided by Computer. He also completed a master's degree in "Science on Management and Industrial Strategy". Presently, he is the project Manager at LEMAC – Instituto Superior Técnico – in Lisboa. He is responsible for the Design Area of penstocks, valves and irrigation equipment at HIDROSOREFAME.

Uwe Eigenmann
Uwe Eigenmann studied Cybernetics at the University of Stuttgart. Since 1992, he has been working as a consultant in the field of Industrial Engineering and Ergonomics at DELTA Industrie for Informatics GmbH.

Erik Engelborghs
Erik Engelborghs finished his degree in Mechanical Engineering at the Katholieke University of Leuven (Belgium) in 1991. Since February 1993, he has been a research engineer at the Mechanical Engineering Department of WTCM-CRIF, the collective research centre of the Belgian metalworking industry. Since then, he has been working on the CONSENS project, especially on the integration aspects of Group Technology with Design for Manufacture and Design to Cost modules. Currently, he is working on a Design for Recycling project TOPROCO calculating the recycling costs of electric and electronic products.

Joachim Frech
Joachim Frech studied Mechanical Engineering at the University of Stuttgart and Industrial Engineering at Northwestern University, Evanston, USA. He has more than five years of experience in the area of cost-information-systems for mechanical engineering. His current research interests are in the fields of simultaneous recycling and cost conscious design.

Klaus Jeschke
Dipl.-Ing. Klaus Jeschke, geb. 1964, studied Mechanical Engineering at the Technical University of Braunschweig. Since 1990 he is working as a research engineer at the Institut für Werkzeugmaschinen und Fertigungstechnik (IWF), Braunschweig. His special subjects are quality management and computer aided systems for quality assurance. He

coordinates the research group "Null-Fehler-Produktion in der Prozeßkette" and he is responsible for research and consulting projects belonging to quality management at the IWF.

Sabine Kessler

Sabine Kessler, born August 12, 1965 in Augsburg, Germany. After her study oe electrical engineering at the Fachhochschule München she joint Deckal AG, a milling machine producer in Munich, where she worked in the R&D department. 1992 she started working for Siemens Nixdorf Informationssysteme AG, München. Since then she deals with SIFRAME in several aspects, especially in the project and presales enviroment.

Dorothee Koch

Dipl. Math. Dorothee Koch, M.Sc. in Computer Science, graduated from the University of Trier (Germany) in 1987 and from Stanford University (USA) in 1989. Since 1989, she has been working as a researcher and Project Manager at the IAO, Stuttgart, in the fields of object-oriented and relational database design and applications. Other than that, she has given presentations at various international conferences. Currently, she is the Technical Manager for the IAO in the EU-funded ESPRIT project CONSENS, which is concerned with implementing Concurrent Simultaneous Engineering in industrial companies.

Christian Köhler

Dipl.-Wirtsch.-Ing. Christian Koehler, AEG/PLC-F is working as a Project Engineer in the inhouse consulting department of AEG named Production and Logistics Consult or PLC. Working fields: Business re-engineering, Value engineering, Cost reduction programmes.

Goy Korn

Dipl. Inform. Goy Korn (1964) studied Computer Sciences at the Technical University of Braunschweig (Germany) from 1984–1989. Since 1990, he has been a research engineer at the Institute for "Machine Tool Manufacture and Manufacturing Engineering" (Technical University of Braunschweig) in the department of Production Systems. His main activities are focused on Quality Management and Simultaneous Engineering, as well as product and production development planning.

Raimund Menges
Dr. Raimund Menges studied Aeronautical Engineering at the University of Stuttgart. From 1986 to 1990, he worked at the Fraunhofer Institute for Industrial Engineering (IAO), Stuttgart, as the head of the Department of Computer Aided Engineering (CAD) Systems. In 1992, he received his Ph.D. in Manufacturing from the University of Stuttgart.

Since 1990, he has been the Managing Director of DELTA Industrie for Informatics in Fellbach.

Pietro Raiteri
After completing his master's degree in Artificial Intelligence, Pietro Raiteri worked in software development and knowledge engineering. Presently, he is the project Manager of some ESPRIT and BRITE projects and the product Manager in Italy for inference and remedy products.

José L. Trigo Santos
José L. Trigo Santos obtained his Ph.D. in Mechanical Engineering in 1987 from the University of Iowa, USA. He is currently Associate Professor at Instituto Superior Tecnico, in Portugal. Main areas of interest are: design sensitivity analysis and optimisation, computational methods in engineering and concurrent engineering.

Eric Sleeckx
Eric Sleeckx finished his degree in Mechanical Engineering at the Katholieke University of Leuven (Belgium) in 1982. Presently, he is the Supervisor of national and European research projects at WTCM (Research Center for the Belgian Metalworking Industry) in the fields of Product Development.

Alexander Stanke
Alexander Stanke (1965) studied Cybernetics at the University of Stuttgart. Since his master's degree in 1991, he has been working at the Fraunhofer Institute (IAO) as a consultant in the fields of Lean Management, Concurrent Simul-

taneous Engineering and Customer Management. Presently, he is the head of the market focus unit of business development and of the R&D-Management Center at the IAO.

Steven Tilley

Steven Tilley finished his degree in Mechanical Engineering at the Katholieke University of Leuven (Belgium) in 1982. His professional experience includes: From 1988 he has been a Research Engineer at WTCM., Dept. of Mechanical Engineering, Leuven (Belgium). From 1988–1990 : He was involved in several National and European projects in the area of Process Planning for sheet metal products : PRESS, BRITE 2406. From 1990 he has been involved in the Design For X projects: Design to Cost for sheet metal, BRITE DEFMAT (4669), Esprit CONSENS (6896).

Josef W. Vilsmeier

Josef W. Vilsmeier (1950) finished his degree in Mechanical Engineering at the University of Munich in 1977. Since 1978, he has been working as a Development Engineer for Messerschmidt-Bölkow-Blohm, now Daimler-Benz Aerospace AG. He is the Group Leader of the Composite Materials Department, Manager of the Structural Technologies Department, Project Manager of European Projects and the Senior Advisor of IT-based Business Processes.